RETHINKING POLYESTER POLYURETHANES

Emerging Issues in Analytical Chemistry

RETHINKING POLYESTER POLYURETHANES

Algae-Based Renewable, Sustainable, Biodegradable, and Recyclable Materials

Series Editor
DAVID HESK

Editor
ROBERT S POMEROY

ELSEVIER RTI INTERNATIONAL

Elsevier
Radarweg 29, PO Box 211, 1000 AE Amsterdam, Netherlands
The Boulevard, Langford Lane, Kidlington, Oxford OX5 1GB, United Kingdom
50 Hampshire Street, 5th Floor, Cambridge, MA 02139, United States

ISBN: 978-0-323-99982-3

For information on all Elsevier publications
visit our website at https://www.elsevier.com/books-and-journals

Publisher: Susan Dennis
Acquisitions Editor: Charlotte Rowley
Editorial Project Manager: Aera Gariguez
Production Project Manager: Sruthi Satheesh
Cover Designer: Matthew Limbert

Typeset by STRAIVE, India

Dedication

I would like to dedicate this book to everyone who seeks to make a change in an effort to improve how we as humans interact with the planet. I am frequently reminded of the following quotation from anthropologist Margaret Mead: "Never doubt that a small group of thoughtful, committed citizens can change the world; indeed, it's the only thing that ever has." Another quotation, a favorite of Steve Mayfield's, comes from Will Rogers: "If you find yourself in a hole, stop digging." Combined, these two ideas sum up what ridding the world of plastic pollution will require.

Contents

3. Renewable, sustainable sources and bio-based monomers

Bhausaheb S. Rajput, Anton A. Samoylov, and Thien An Phung Hai

III

Redefining the analytics

4. Biodegradation: The biology

Natasha R. Gunawan, Michael T. Read, and Woodrow R. Brown

5. Polyurethane processing and degradation: The analytical chemistry

Marissa Tessman, Berk Kuntasal, and Miheer Modi

6. TEA and LCA of bio-based polyurethanes

Matthew Wiatrowski, Eric C.D. Tan, and Ryan Davis

IV

Reformulating polyester polyurethanes

V

Reimagining polyester polyurethanes

11. The bioplastics market: History, commercialization trends, and the new eco-consumer

Thomas Cooke and Robert S Pomeroy

12. The future of biobased polymers from algae

Stephen P. Mayfield and Michael D. Burkart

Contributors

Woodrow R. Brown University of California San Diego, La Jolla, CA, United States

Michael D. Burkart University of California San Diego, La Jolla, CA, United States

Frances Carcellar University of California San Diego, La Jolla, CA, United States

Thomas Cooke Algenesis Materials, Cardiff, CA, United States

Ryan Davis National Renewable Energy Laboratory, Golden, CO, United States

Natasha R. Gunawan Algenesis Materials, Cardiff; University of California San Diego, La Jolla, CA, United States

Thien An Phung Hai University of California San Diego, La Jolla, CA, United States

Berk Kuntasal University of California San Diego, La Jolla, CA, United States

Stephen P. Mayfield University of California San Diego, La Jolla, CA, United States

Miheer Modi University of California San Diego, La Jolla, CA, United States

Nitin Neelakantan Algenesis Materials, Cardiff, CA, United States

Robert S Pomeroy University of California San Diego, La Jolla, CA, United States

Naser Pourahmady University of California San Diego, La Jolla, CA, United States

Bhausaheb S. Rajput University of California San Diego, La Jolla, CA, United States

Michael T. Read University of California San Diego, La Jolla, CA, United States

Anton A. Samoylov University of California San Diego, La Jolla, CA, United States

Ryan Simkovsky University of California San Diego, La Jolla, CA, United States

Eric C.D. Tan National Renewable Energy Laboratory, Golden, CO, United States

Marissa Tessman Algenesis Materials, Cardiff, CA, United States

Matthew Wiatrowski National Renewable Energy Laboratory, Golden, CO, United States

Author's biographies

Woodrow R. "Woody" Brown is originally from Ventura County, CA. He graduated from the University of California San Diego in 2021 with a BS in ecology, behavior, and evolution.

Michael D. Burkart is a professor in the Department of Chemistry and Biochemistry at the University of California San Diego. He started his career there in 2002 and is the Director of the Center for Renewable Materials; the Associate Director, California Center for Algae Biotechnology; and the Teddy Traylor Faculty Scholar. Dr. Burkart obtained his BS in chemistry from Rice University, Houston, TX, and his PhD in organic chemistry from The Scripps Research Institute, La Jolla, CA. That was followed by a National Institutes of Health Postdoctoral Fellowship at Harvard Medical School, Boston, MA.

Frances Carcellar was a transfer student from Daly City, CA. She completed her BS degree in environmental systems: ecology, behavior, and evolution in 2021 at the University of California San Diego. She is a research associate at Genomatica in San Diego.

Thomas Cooke recently joined Algenesis Materials as President, after 7 years as Vice-President of Product at Reef. He was responsible for Reef's global product line strategy and execution, including product management, design, and development. Prior to Reef, he spent 16 years at Vans, where he held numerous roles in product management and development, most recently as Director of Footwear for the Core Channel, which included the Syndicate, Pro Skate, Surf, and OTW footwear lines. Tom holds a BS in mechanical engineering from the University of Virginia.

Ryan Davis is a group manager and process research engineer on the Economic, Sustainability, and Market Analysis team of the National Renewable Energy Laboratory (NREL). His main focus is on techno-economic analysis and life cycle assessment for biomass conversion technology pathways, primarily to hydrocarbon biofuel products.

Natasha R. Gunawan received her BS in environmental chemistry and MS in chemistry from the University of California San Diego. Her master's thesis was done in collaboration with Algenesis Materials, studying the biodegradation processes of renewable polyurethane foams in the natural environment. Her work focuses on the processes of polymer biodegradation.

Thien An Phung Hai completed his bachelor's and master's degrees in material science and physical chemistry at the University of Science—Vietnam National University and his DEng in polymer chemistry at the Kochi University of Technology, Japan. His primary research interests are the synthesis and application of conducting polymers and copolymers for ionic detection, and modification of the interface of biomaterials by grafting conducting polymers to enhance their properties. He is a postdoctoral researcher at the University of California San Diego focusing on designing and developing polymer material from renewable bio-sources to address new applications.

Berk Kuntasal is originally from Turkey. He attended Santa Monica College before transferring to the University of California San Diego, where he completed his BS in chemistry in 2021. He is continuing at UC San Diego and expects to complete his MS in chemistry in 2023.

Stephen P. Mayfield is the Director, California Center for Algae Biotechnology; Co-director, Food & Fuel for the 21st Century; and Distinguished Professor of Molecular Biology at the University of California San Diego. He obtained BS degrees in biochemistry and in plant biology from California Polytechnic University in San Luis Obispo in 1979, and a PhD in molecular genetics from the University of California Berkeley in 1984. From 1984 to 1987 he was a National Institutes of Health Postdoctoral Fellow at the University of Geneva, Switzerland. In 1987 he joined The Scripps Research Institute and became a professor and Associate Dean of the Graduate School before leaving to join UC San Diego in 2009.

Miheer Modi was born in India and raised in Singapore and Malaysia before coming to the University of California San Diego in 2016 to purse his BS in biochemistry. He is expected to complete his MS in chemistry by December 2022, focusing on analytical chemistry and plastic biodegradation. He hopes to work in industry postgraduation, focusing on renewable materials.

Nitin Neelakantan is Director of Product Development at Algenesis Materials. He received his BS in chemistry from the University of California Irvine and PhD in materials chemistry from the University of Illinois Urbana-Champaign. His work is primarily on the formulation and process development of bio-based, biodegradable polyurethanes for use in commercial products.

Robert S "Skip" Pomeroy is a teaching professor at the University of California San Diego. He is an analytical chemist and works in several research centers within the university (CAICE, CalCAB, FF21, and the

Center for Renewable Materials), serving as an educational lead and chemical analyst. He obtained his BA in chemistry from the University of California San Diego, MS in analytical chemistry from the California State Polytechnic University Pomona, and PhD in analytical chemistry from the University of Arizona and was a postdoctoral student in the Marine Physical Laboratory at Scripps Institution of Oceanography. He also served as the R&D lead for Southern Grouts and Mortars for 10 years.

Naser Pourahmady received his PhD in organic and polymer chemistry from Oklahoma State University. After 2 years of postdoctoral work in polymer chemistry and 3 years as an assistant professor at Missouri State University, he moved to the chemical industry. He spent 32 years with BFGoodrich/Noveon/Lubrizol in a variety of research and management roles and recently joined the University of California San Diego as a visiting scholar.

Bhausaheb S. Rajput joined Professor Michael Burkart's group at the University of California San Diego in August 2020. His work is on the synthesis of new monomers from renewable resources to make polyurethanes for various applications. He obtained his MTech and MSc from KBC-North Maharashtra University, Jalgaon, India, and his PhD from the CSIR-National Chemical Laboratory, Pune, India, under the supervision of Senior Scientist Dr. Samir H. Chikkali. He works on the synthesis of sugar- and plant oil-based renewable and degradable polymers and copolymers for various applications.

Michael T. Read is an undergraduate at the University of California San Diego. He expects to complete his BS degree in biology in 2023.

Anton A. Samoylov immigrated to the San Francisco Bay Area after being born in Russia. He is currently pursuing a PhD in chemical engineering at the University of Arizona after completing his BS in chemical engineering from the University of California San Diego in 2021. The focus of his research is to prioritize renewable and sustainable technology, focusing first on biodegradable polymers and currently on novel photovoltaics.

Ryan Simkovsky is an assistant project scientist in the organized research unit on Food and Fuel for the 21st Century at the University of California San Diego. He obtained his BS in biology from the California Institute of Technology and PhD in biology from the Massachusetts Institute of Technology and was a postdoctoral fellow at The Scripps Research Institute, La Jolla, CA, prior to his work as a postdoctoral researcher and project scientist at UC San Diego. His work focuses on all aspects of developing and producing algae and algae-based products.

Eric C.D. Tan is a senior research engineer at the National Renewable Energy Laboratory (NREL). He provides leadership spanning many research areas, including sustainable process design, techno-economic analysis, and life cycle assessment. He is an American Center for Life Cycle Assessment-certified professional and holds a PhD in chemical engineering from the University of Akron and an MA in sustainability from Harvard University.

Marissa Tessman received her BS and MS in chemistry from the University of California San Diego, where she developed algae-based polyurethane rigid foam for surfboards and gained experience with a wide variety of analytical techniques. She is currently a research scientist at Algenesis Materials, where she is developing robust methods of measuring polyurethane biodegradation for Algenesis' polyurethane flip-flop foam, films, and adhesives.

Matthew Wiatrowski is a process engineer at the National Renewable Energy Laboratory (NREL) in Golden, CO. He is a member of the biorefinery analysis team, doing techno-economic analysis (TEA) of various biofuel pathways, and is responsible for all aspects of TEA, including conceptual process design, process modeling, and optimization in Aspen Plus, and refinery economic analysis.

Foreword

UC San Diego was founded in 1960 with a vision of its founding leaders to create an experimentally driven campus that would shape education and research at public universities. In 2015, we developed an eight-word strategy for our campus: student-centered, research-focused, service-oriented public university. Students, faculty, and staff were given the freedom to interpret what these words meant to them. There was no top-down mandate. This empowers the university and our community members to take responsibility for making the world a better place. The benefit is developing students who can solve problems, lead, and innovate in a diverse and interconnected world. The work described in this book is one of the results of this strategy, and a strategy of leveraging our academic teaching labs in the Department of Chemistry and Biochemistry to promote the development of focused teams of undergraduate researchers. This provided a nucleus for dedicated students, faculty, and public and private sector groups from multiple disciplines to come together to propose a potential solution to the sea of plastic that now plagues our oceans. *Rethinking Polyester Polyurethanes* recognizes the tremendous positive impact plastics have had on our lives, and it offers a perspective on the challenge of replacing petroleum-based plastics with algae-based renewable, sustainable, and biodegradable materials.

Drs. Mayfield, Burkart, and Pomeroy have embraced the eight-word description for UC San Diego and have engaged students, industry, and the community in their efforts to protect our planet through innovation and entrepreneurism. These eight words fueled their mission to take concrete, bold, and creative action to solve one of the world's most challenging problems. Lastly, we love their shoes!

Pradeep Khosla
Chancellor, UC San Diego

Robert Continetti
Distinguished Professor of Chemistry, UC San Diego

Acknowledgments

Thank you to all students, faculty, staff, collaborators, family, and friends shaping my daily life. To UC San Diego, my alma mater, and the Department of Chemistry and Biochemistry, thank you for all of the support, or none of this would have happened.

PART I

Overview

CHAPTER

1

Rethinking plastics

Robert S Pomeroy

University of California San Diego, La Jolla, CA, United States

Introduction

Plastics are ubiquitous. Their extensive development and use have enriched our lives. They are also the principal material in the Great Pacific Garbage Patch, the most significant accumulation of ocean plastic, located in the central North Pacific.[1,2] Viewing an image of the Garbage Patch often elicits a visceral response, a sense that what has happened is unacceptable and something should be done about it. Lau et al.[3] modeled the flow of plastics and the impact of interventions such as recycling through global systems. They projected a massive buildup in the environment and estimated that mitigation will require coordinated global action of the 4 Rs: *reduce* plastic consumption; increase *reuse, recycle,* and *recover* by waste collection.[4] They point out the need for accelerated innovation in the plastic value chain. We never anticipated the environmental impact of these remarkable new materials. If we had, what should we have done differently?

The use and demand for plastics took off in the 1950s, and that is when the question should have been addressed.[5] Polyester plastics had a minor flaw in that they were subject to hydrolysis,[6] which means chemical breakdown due to reaction with water. Ester linkages are easily hydrolyzed by reaction with dilute acids or alkalis. To reduce hydrolysis, the chemistry of the polymer chains was converted from ester linkage [–C–O–(C=O)–] to ether linkage [–C–O–C–]. This small change rendered the plastic made with polyether resins much more stable to hydrolysis. It also reduced the viscosity[6] of polyether resins, which made them easier to use industrially. Though logical at the time, this one decision did not address a different and more critical question: Should all plastics be less amenable to breakdown? Should we not have thought to match the lifetime of the plastic to the lifetime of the object? The cost of petroleum was, and is, so low

Rethinking Polyester Polyurethanes
https://doi.org/10.1016/B978-0-323-99982-3.00007-9

that it effectively renders all plastics disposable even though the environmental lifetimes of polyether plastics are in the hundreds of years. Is it sensible to make a drinking straw intended for a single use to last 500 years? No one foresaw the Great Pacific Garbage Patch!

Geyer[7] estimated that "8,300 million tons of virgin plastics have been produced. Half of this material was made in just the past 13 years. About 30% of the historic production remains in use today. Of the discarded plastic, roughly 9% has been recycled. Some 12% has been incinerated, but 79% has gone to landfill. Shortest-use items are packaging, typically less than a year. Longest-use products are found in construction and machinery. Current trends point to 12 billion tons of waste by 2050." Very few plastics are biodegradable, and none of the commonly used ones are.[8] They need to be reused or recycled, or the energy value recovered by incineration. Incineration is complicated by the impact of emissions on health.

This chapter will introduce the terms, concepts, and economics of how oceanic garbage comes into being and how its growth might be mitigated. Plastics and the materials from which they are made are introduced. Past practice, future alternatives, and a basis for comparison are presented. The chapter concludes with a brief description of what can be done and what stands in the way. The case will be made that there is value in rethinking one sector of the plastics industry by producing algae-based polyester polyurethanes.

Plastics: What are they?

Plastics are approximately 100 years old and have made an indelible mark on the world.[9,10] It might be argued that we are living in the plastics age. Another way to view the situation is that every plastic water bottle ever used is still somewhere on this planet in one form or another. Plastics have been celebrated at world fairs and expositions[11] and reviled for the negative impact on the environment.[1] They are everywhere. They enable us to live cleaner, healthier, safer lives. They are found in clothes, food and drink containers, toys, the screens we watch, and modes of transportation, and are ubiquitous in the household. They are remarkably inexpensive, and some are virtually indestructible. It is these last two characteristics that led to the problem. The good news is that they are inexpensive and last a long time; the bad news is that they are inexpensive and last a long time. Modern plastics will easily survive 500 to 1000 years in the environment.[12] In the late 1950s and early 1960s, the cost of making them decreased dramatically. The low cost of petroleum as a source material enabled cheap mass production. Low cost makes them disposable at worst and economically difficult to recycle at best.

"Plastic" comes from the Greek *plastikos*, meaning fit for molding.[13] Their malleability allows them to be cast, pressed, or extruded into a variety of shapes, such as films, blocks, slabs, fibers, tubes, and bottles, with a wide range of physical properties, such as foams, adhesives, sealants, and hard and soft solid objects. Plastics are derived from a wide range of synthetic or semisynthetic materials. They have myriad uses and are incorporated into a broad spectrum of consumer and industrial applications. Low density, toughness, and chemical stability make some plastic products lighter, stronger, and more durable than those constructed of wood, ceramic, or metal. Plastic materials can range from colorless to any color imaginable. Unlike metals, they are excellent thermal and electrical insulators. They have an enormous range of functions, as manufacturers can mold them into virtually any shape. Physical characteristics can be enhanced by incorporating other materials to make composites that expand the range of applications even further.[14–16]

Plastics are polymers (many units) which consist of chains of monomers (single units). Plasticity can be adapted by altering the chemistry and arrangement of the monomers and the method employed during manufacture. This adaptability is the hallmark of plastics and accounts for their widespread use. They are organic materials, like wood (a polymer of β-glucose) or wool (a polymer of amino acids from the protein keratin).[17] They are made from chemicals derived from the refining of crude oil.[18] The refining will be detailed later. There are two broad classes of plastics: thermosets and thermoplastics.[19] Thermosets, upon solidification, adopt a permanent shape. They are amorphous materials with a near-infinite molecular weight, which is one giant molecule. Thermoplastics range structurally from amorphous to partially crystalline, with molecular weight between 20,000 and 500,000 amu (atomic mass unit). The partially crystalline structure makes plastic pliant, capable of being heated and remolded repeatedly.

There is a distinction between plastics and polymers. Plastic is a specific type of polymer comprised of a long chain of polymers. Polymers are composed of smaller, uniform molecules. Examples of polymers that are not plastics are silicones (which are based on silicon rather than carbon), proteins, and DNA. The polymers used in plastics are made by linking monomers together, creating large molecules composed of multiple repeating units. Joining identical monomers establishes a class of polymers called homopolymers.[19] Different monomers linked together are called copolymers.[19] The structures of homopolymers and copolymers can be straight chains or branched chains. There are about 45 unique plastics in common manufacture, with properties varied by the manufacturing process.[20] Characteristics can be tuned by making changes in the physical processes and chemical makeup to accommodate the end application. When manufacturers change or modify things such as the molecular weight distribution, the density, or the melt indices, they alter the effectiveness and create plastics with many specific properties and therefore many different uses. Plastics are classified by the acronyms[21] describing the monomer used to produce them. For example, polyethylene terephthalate (PET) is the **p**olymer of **e**thylene glycol and **t**erephthalic acid joined together via a condensation reaction that links the monomers by an ester bond. PET is an example of a polyester. Other examples are high-density polyethylene (HDPE), polyvinylchloride (PVC), polypropylene (PP), polystyrene (PS), and low-density polyethylene (LDPE). The monomers to produce these polymers come almost exclusively from crude oil.

Petroleum: What is it, how is it formed, and how is it transformed?

"Petroleum" is derived from the Latin "*petra* (rock) and *oleum* (oil)."[13] It is a broad term for the organic liquids and gases that form after kerogen is heated and compressed over long periods.[22,23] This mix can be somewhat complex. The main gaseous component is natural gas, primarily methane; the main liquid component is crude oil. Crude is a mixture of hydrocarbons, the exact composition depending on the algae and the amount of heat and pressure applied over millions of years. The conditions to which the biomass was exposed also contribute to variations that are found in hydrocarbons and crude. Crude is about 13% hydrogen and 85% carbon by weight; other elements are oxygen (1%); nitrogen (0.5%); sulfur (0.5%); and metals such as iron, nickel, and copper (each less than 0.1%).

Because of this variation, the petroleum industry classifies crude by the location of its origin, its sulfur content, and its relative weight or viscosity (light, intermediate, or heavy). Light oils have up to 87% hydrocarbons, while heavier oils typically have less than 50% and contain larger quantities of other elements. The petroleum industry uses a metric established by the American Petroleum Institute, the API gravity, to measure the density of crude. Crude ranges in color—yellowish, reddish, tan, or even greenish—but is typically black or dark brown. The color is due to the variations discussed before and makes crudes from different sources uniquely identifiable. Those with lower sulfur and metal concentrations are lighter in color, sometimes nearly transparent. The relative sulfur content is denoted by "sweet" (low) or "sour" (high). For example, Brent and West Texas crudes are sweet. The density of Brent is lower than that of other crudes, so it is light and sweet. The density of West Texas is intermediate, hence the acronym WTI. Oil from Saudi Arabia has high density and high sulfur, so it is heavy and sour.[24]

Petroleum formation

Petroleum was formed in the Carboniferous period. Starting 450 million years ago, algae and plants lived in shallow seas, which eventually dried up, leaving the biomass behind in basins, where it was decomposed by bacteria. Over time, sediments gathered in these basins, compressing the biomass and subjecting it to high pressure and temperature. Petroleum is found in the rock formations associated with the sedimentary basins where the seabeds used to be.[24] The deserts of the Middle East, for example, were once shallow seas and today have some of the greatest proven oil fields in the world.

Over millions of years, the decomposition and the high pressure and temperature transformed the molecules associated with algae into a waxy material called kerogen. Continued heat and pressure over time caused the kerogen to alter its chemistry, compensate for the changing conditions, and convert into hydrocarbons through catagenesis (Fig. 1.1).[24] Different combinations of heat and pressure change the chemical composition of the hydrocarbons, hence the diversity of crude. The resulting materials exist in three physical phases: gaseous, liquid, and solid. Because the sedimentary source rock is porous, the low-density gas and liquid fractions migrate through it. Natural gas and crude rise until they encounter an impermeable layer such as shale or salt. Petroleum engineers use seismic

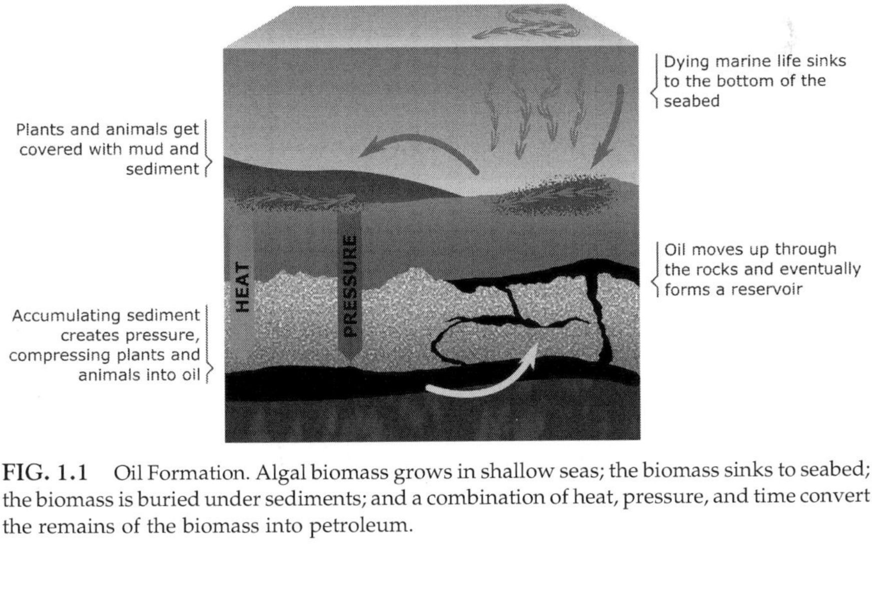

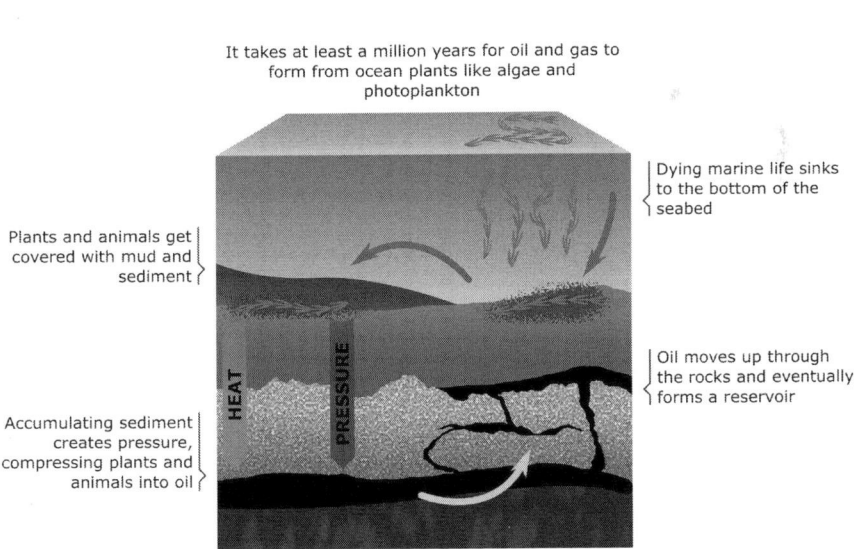

FIG. 1.1 Oil Formation. Algal biomass grows in shallow seas; the biomass sinks to seabed; the biomass is buried under sediments; and a combination of heat, pressure, and time convert the remains of the biomass into petroleum.

reflection to locate rock structures that might hold trapped crude. A small explosion is set off, and the sound waves travel underground, bounce off the different formations, and return to the surface for detection. The time and strength of the returning waves allow identification of geological formations likely to create a reservoir.[25,26]

Although petroleum seeps exist, such as the La Brea Tar Pits[27], they are rare, and, having lost the more volatile components, are usually heavy. Most petroleum is collected by drilling boreholes and pumping it out, some by processing tar sands. The amount of crude in a reservoir is measured in barrels or tons. A barrel is 42 US gallons. Producers in Europe and Asia use the metric ton, which is 1000 kg or 2205 pounds or 7.33 barrels.[28]

Petroleum is not a sustainable resource, and supplies will be exhausted. Some time in the near future the world will reach "peak oil" (highest production level). Many experts predict that that could happen as soon as 2050.[29–32] In our energy-intensive world, finding alternatives is crucial to the future global economy.

Petroleum refining

Refining [33,34] separates crude into different products for use, such as fuel and asphalt. The primary method is fractional distillation, in which components are isolated based on differences in the phase transition temperature of each fraction. This is a physical process, not a chemical one. The fractions vary in composition on a gradient and are collected based on differences in vaporization (boiling point). The crude is heated in a distillation tower that has trays. There is a temperature gradient, from hot at the bottom to cooler at the top. The fractions that reach the top are the most volatile, those at the bottom the least volatile. As the crude is heated, vapors of the volatile components rise to different levels and condense back into liquids, which are collected on the trays. The refined materials can be sold directly as propane or kerosene or subject to chemical transformations to yield more desirable products. For example, long-chain hydrocarbons may undergo catalytic cracking and reformation. Cracking breaks the chains into small hydrocarbon molecules. Catalytic reforming then connects the molecules into higher-value reformate products such as 2,2,4-trimethyl pentane, the critical component of antiknock gasoline. For polyesters, ethylene is converted to ethylene glycol, and p-xylene is used to make diacids. These conversions are very energy intensive and work against atom efficiency.[35] This will be considered in the section on green chemistry and LCA/TEA (life cycle assessment and techno-economic analysis).

The carbon cycle

Carbon is the fourth most abundant element in the Universe. On Earth, it is the essential element associated with living things. It is so important that an undergraduate chemistry major spends two semesters of study dedicated to the chemistry of carbon, known as organic chemistry. Earth's largest reservoir of carbon, approximately 65,500 billion metric tons, is stored in carbonate rock. The rest is cycled among ocean, atmosphere, plants, soil, and starting in the 1850s, fossil fuels. Its constant flow in and out of these reservoirs is the carbon cycle.[36–38]

Carbon is released through volcanoes, soil erosion, and evaporation. It is absorbed by plants, passing through the ecosystem via the food web and released back to the soil, water, or atmosphere. Atmospheric carbon—carbon dioxide—plays a key role in regulating the Earth's temperature and makes the planet habitable. Most of the carbon is sequestered in the form of carbonate rock and petroleum. However, since the Industrial Revolution, fossil fuels have been intensively extracted and burned for energy. The combustion of oil releases the carbon sequestered underground and upsets the carbon budget of the atmosphere. It unbalances the carbon cycle because of the differential rates of release and sequestration. The intensive use of petroleum has in 150 years removed a substantial fraction of all of the carbon ever sequestered. Sequestration took place over geologic time: 10% in the Paleozoic age (541–252 million years ago), 70% in the Mesozoic (252–66 million), 20% in the Cenozoic (65 million to the present). Clearly there is a severe imbalance between the rates of release return.

What is green chemistry?

The American Chemical Society states that [39] "the concept of greening chemistry developed in the business and regulatory communities as a natural evolution of pollution prevention initiatives. In our efforts to improve crop protection, commercial products, and medicines, we also caused unintended harm to our planet and humans." The goal of green chemistry is to create products that are environmentally sustainable and economically viable. It is cleaner, cheaper, smarter chemistry. It is pollution prevention at the molecular level. The design of environmentally preferable products can reduce waste, prevent costly end-of-the-pipe treatment, lead to safer products, and save resources such as energy and water. These practices should favor renewable materials over petroleum and incorporate chemicals to break down products into innocuous substances after use. There is also a simple fact that cannot be avoided: the petroleum will run out!

Green chemistry can be quantified. Though industry has yet to produce a unified set of metrics, ways to quantify greener processes and products have been proposed. These are applied to mass, energy, and hazardous waste reduction or elimination; the use of more sustainable feedstocks; and minimization of end-of-life impact.

The United States Environmental Protection Agency (EPA) has established 12 principles of green chemistry.[40]

(1) **Prevent waste:** Design chemical syntheses to prevent waste. Leave no waste to treat or clean up.
(2) **Maximize atom economy:** Design syntheses so that the final product contains the maximum proportion of the starting materials: waste few or no atoms.
(3) **Design less hazardous chemical syntheses:** Design syntheses to use and generate substances with little or no toxicity to humans or the environment.
(4) **Design safer chemicals and products:** Design chemical products that are fully effective yet have little or no toxicity.
(5) **Use safer solvents and reaction conditions:** Avoid using solvents, separation agents, or other auxiliary chemicals. If you must use these chemicals, use safer ones.
(6) **Increase energy efficiency:** Run chemical reactions at room temperature and pressure whenever possible.
(7) **Use renewable feedstocks:** Use starting materials that are renewable rather than depletable. The source of renewable feedstocks is often agricultural products or the wastes of other processes; the source of depletable feedstocks is often fossil fuels (petroleum, natural gas, or coal) or mining operations.
(8) **Avoid chemical derivatives:** Avoid using blocking or protecting groups or any temporary modifications if possible. Products use additional reagents and generate waste.
(9) **Use catalysts**, not stoichiometric reagents: Minimize waste by using catalytic reactions. Catalysts are effective in small amounts and can carry out a single reaction many times. They are preferable to stoichiometric reagents, which are used in excess and carry out a reaction only once.
(10) **Design chemicals and products to degrade after use:** Design chemical products to break down to innocuous substances after use to not accumulate in the environment.
(11) **Analyze in real time to prevent pollution:** Include in-process, real-time monitoring, and control during syntheses to minimize or eliminate the formation of by-products.

(12) Minimize the potential for accidents: Design chemicals and their physical forms (solid, liquid, or gas) to minimize the potential for chemical accidents, including explosions, fires, and releases to the environment.

Green chemistry is a proven systems approach that reduces adverse human health and environmental impacts and builds a strategic pathway to a sustainable future. The rest of this book will focus on chemistry using renewable, sustainable, biodegradable materials which do not persist in the environment. Algae are the principal source of the biomass that produced petroleum. Is it possible to replace the fossil fuel-based chemical world with one based on algae?

Why algae?

Algae have emerged as one of the most promising long-term, sustainable biomass sources for fuel, food, feed, pharmaceuticals, and other products.[41–43] Like other green plants, algae, when grown using sunlight, take up carbon dioxide (CO_2) and release oxygen (O_2). Algae and cyanobacteria are attractive because of their large diversity and proven ability to thrive in a wide variety of ecosystems, many of which do not require arable land or potable water. These organisms are both robust and efficient, having evolved over billions of years. Compared to land plants, algae do not form structures such as leaves and roots. They have developed disease resistance and a rapid growth rate. Light energy captured in their chloroplasts is stored in the form of oil and does so more efficiently than any other known natural or engineered energy harvesting process.

Here are ten reasons why algae are a promising new source of fuel and other products.

(1) They grow fast.
(2) They remove CO_2 from the atmosphere.
(3) Certain strains are oil rich.
(4) Cultivation does not require arable land.
(5) Microalgal biomass can be used for fuel, feed, and food.
(6) Macroalgae can be grown in seawater.
(7) They absorb pollutants and can purify wastewaters.
(8) Their biomass can be valorized by making other useful products.
(9) Farming it is open to all: decentralized.
(10) The industry creates jobs.

Making polyester polyurethanes from algae has the most significant impact on green principles 2, 4, 7, 9, and 10. Concerning atom economy, converting algae to petroleum involves the loss of oxygen and nitrogen from the molecules. The hydrocarbons obtained through refining now require putting the oxygen and nitrogen back onto the molecules to make them into monomers suitable to produce polymers. It would be much more efficient to gather the molecules from algae, with oxygen and nitrogen already in place, and to develop mechanisms to acquire the monomers from these. Green principles 4, 7, and 10 are incorporated naturally through algae as the source material. Green principle 9 is an active area of research. The National Science Foundation Centers for Chemical Innovation[44] are focused on significant, long-term fundamental chemical research challenges. Two of the US National Science Foundation Centers for Chemical Innovation involve such work: the Center for Selective C-H Functionalization (http://nsf-cchf.com/) and the Center for Sustainable Polymers (https://csp.umn.edu/).

Algal biomass as a source of global energy has enormous potential. Algae oil, known as green crude, can be converted into a biofuel. High growth rate and density mean it occupies far less space than other feedstocks. About 40,000 square kilometers (15,000 square miles) of algae—less than half the size of the US state of Maine—would provide enough biofuel to meet all of the country's petroleum needs. Additional benefits include absorption of pollutants, release of oxygen, and the fact that it does not require potable water or arable land.

Life cycle assessment (LCA) and techno-economic analysis (TEA): How is sustainability measured?

An LCA is a systematic analysis of environmental impact from source materials, manufacturing, delivery, and disposal.[45–48] It models the environmental impact of all facets of industrial production and provides insight for businesses, policymakers, and other organizations in making informed decisions to advance toward sustainability. Every product is born and lives a life, and when it is no longer useful, its life ends. The life cycle that describes most manufacturing follows a linear model: material extraction, production, packaging, distribution, use, end-of-use, and waste treatment or recovery. There is an alternative model known as cradle-to-cradle. It seeks environmental neutrality in a circular economy, where the end-of-life phase feeds directly into a new life cycle, often through a value retention process like remanufacturing. Whatever the economic model, the LCA is a tool for quantitatively identifying the environmental impacts of a product by including all material and energy flows and gauging the impact of those activities on the environment. It provides metrics to quantify the inflows (e.g., energy, water, resources, land) and outflows (e.g., emissions, wastes, products) that occur throughout the manufacturing process. It quantitates the flow of energy and materials in and out of the system. These are tangible, objective measurements, keeping track of metrics such as mass, volume, and energy utilization. These values represent the life cycle inventory (LCI). An LCA helps determine to what extent these material exchanges with the environment are detrimental to natural ecosystems and human health. For example, a specific volume of petroleum may be used to produce one plastic fork. This value is what is tabulated in an LCI. The life cycle impact assessment (LCIA) quantitates how much the fork contributes to, for example, atmospheric CO_2 concentration.

A TEA (the "A" can also stand for assessment)[49–52] is a systematic analysis of economic impact. It uses process modeling and engineering design, factoring in capital, operating expenses, and revenue. The feedback directs project development and delineates when and where there is a need for investment and research and development expenditures. The goal is process and financial optimization. In practical terms, it speaks to the commercial viability of an endeavor. The development of a sound LCA aids in a correct TEA. Together these techniques answer the questions "Should we do this, and at what cost?" and "Under which market conditions will the technology or product be viable?" A TEA report should identify the advantages and disadvantages and the critical factors—technological, commercial, regulatory, environmental, and social—that inform investment decisions.

The Corfam story: An object lesson

Patent leather is made by coating the flesh side of the leather with linseed oil until a high gloss, waterproof, lacquer-like finish is obtained.[53] Many steps and much time are involved, making it expensive. In the 1950s, DuPont started working on a polyester resin to coat fabrics to create a synthetic substitute for patent leather.[54,55] It was made by placing a polyurethane coating on a fibrous polyester base. By 1962 DuPont employees had wear-tested 15,000 shoes. The company introduced this substitute to the shoe industry at the Chicago Shoe Show in 1963 under the trade name Corfam. DuPont also coined the term poromeric, a combination of "porous" and "polymeric." Corfam was featured at the 1964 World's Fair in New York. Approximately 7.5 million pairs of the shoes were sold by 1969.

Synthetic patent leather does not breath or stretch like genuine patent leather, but the military took to the material because of its relatively low cost and ability to maintain a high gloss. In 1969 competitors began making synthetic patent leather from polyvinyl chloride (PVC). It had many of the same characteristics as Corfam and was less expensive because of the broader demand for PVC in other applications. Despite selling between 75 and 100 million pairs of Corfam shoes, in 1971 DuPont, along with Tenneco and eventually B.F. Goodrich, got out of the poromeric-coated fabric business because of competition from the cheaper PVC, of which over 160 million pairs were sold. Many viewed Corfam as an example of a synthetic material that did not measure up to its initial hype.[56,57] But did it? PVC, like all commonly used plastics, is nonbiodegradable. The only way to dispose of it permanently is incineration. Environmental scientists have raised concerns that burning PVC forms dioxins, creating environmental and health concerns. Polyester polyurethanes can be made biodegradable, and incineration is less suspect in the formation of dioxin. What if the polyesters used were biodegradable and the cost analysis (LCA and TEA) in 1971 had considered the end-of-life expenses in terms of proper disposal and environmental remediation? The Corfam story might have a different ending. Instead of a financial flop, it suffered from not using LCA and TEA to direct research and development and create a biodegradable version to push back against PVC. Despite the low cost of PVC shoes at the point of sale, the overall cost–benefit analysis is negative.

Microplastics: A global problem

These are small particles of plastic debris in the environment with diameters ranging from microns to several millimeters. They also vary in density and shape, generally either spherical or elongated fiber.[58–62] The nonpolar surface and large surface area allow them to absorb and transport toxic contaminants to sediments and organisms within the ecosystem. They do not biodegrade, and therefore persist. Their presence has direct and indirect impacts. The direct impact is by ingestion, and it affects organisms from worms to whales. The indirect impacts include habitat alteration by changing the movement of water through soils, bioaccumulation and biomagnification of toxins that partition on the plastic surfaces, and alteration of an ecosystem by negatively affecting species that play a role in the balance of that ecosystem.

Without LCA/TEA, the end-of-product-life expense is ignored. This lack of financial and environmental accountability has led to worldwide ocean pollution, and the waste continues to mount (Fig. 1.2).

Polyester versus polyether polyurethanes and biodegradability

Polyester polyols are mixtures of diols, triols, and dibasic acids.[63,64] A stoichiometric excess of the polyol over carboxylic acid results in a hydroxy-terminated polymer. Current polyester polyols include ethylene glycol, 1,6-hexanediol, 1,4-cyclohexanedimethanol, and neopentyl glycol

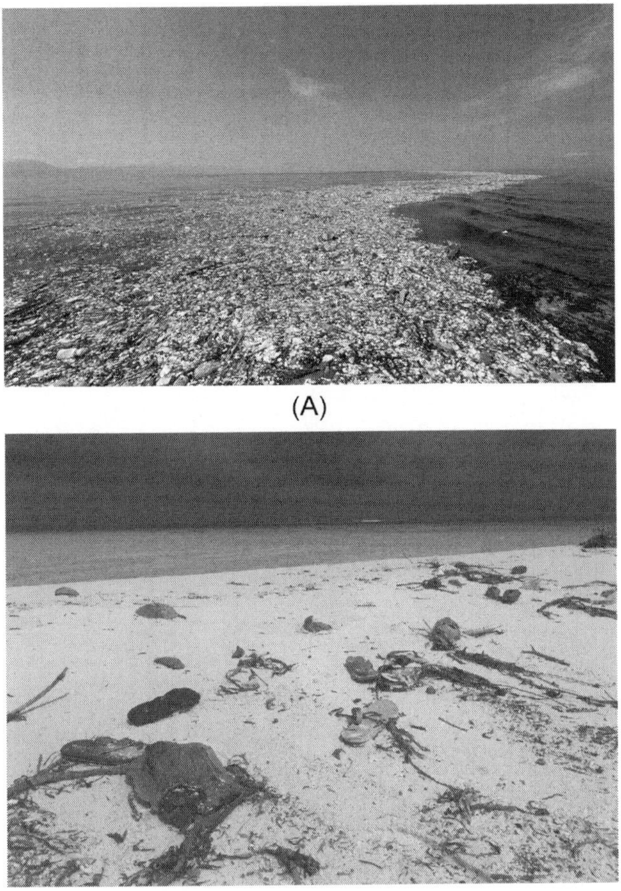

(A)

(B)

FIG. 1.2 Visible plastic pollution. (A) In the ocean (photography credit: Caroline Power). (B) Water bottles and flip flops washed up on the beach of one of the uninhabited Maldives Islands 700 km away from mainland Asia. *Photography credit: Stephen Mayfield.*

(2,2-dimethylpropanediol). The usual diacids include isophthalic, terephthalic, and adipic. Crosslinking of change is accomplished by using branched, trifunctional monomers such as trimethylolpropane (TMP, 2,2-di(hydroxymethyl)-1-butanol). The monomers are selected for the desired characteristics of the finished product. For example, the use of unbranched, straight-chain diacids (aliphatics such as adipic acid) leads to flexible coatings with low damage resistance. The use of aromatic diacids results in materials more resistant to damage (hydrolysis chiefly) but at the cost of rigidity. Blending these two types or mixing different diols can create a spectrum of physical properties between the soft, vulnerable materials and the rigid, more chemically resistant materials.

Polyols fall into two chemical families: those with repeat ether linkages (polyethers) and those with repeat ester linkages (polyesters).[63] The most common class of polyol in current polymer formulations is polyethers. They are synthesized on a very large scale from materials obtained through petroleum refining. Common polyester polyols are also synthesized from petroleum. The main ingredients from the refining of petroleum dictate the types of base polyols available. Other types of polyols can be synthesized, but the cost increases because the manufacturer does not use the cheap commodity-based monomers from petroleum refining.

From a manufacturing perspective, polyether polyols have lower viscosity for equivalent average molecular weights, making storage, dispensing, and mixing easier than with polyesters. The higher viscosity of polyester can be overcome by increasing the temperature or adding a viscosity modifier. Comparatively, either of these solutions leads to higher production cost for the polyester over the polyether. Consequently, the polymer industry chose the cheaper, more durable (less reactive), easier-to-use polyethers. The problem is that these materials do not biodegrade (Fig. 1.3).

The best biodegradable polymers are either polyamides or polyesters.[65,66] The reason for this is the millions of years of evolution where

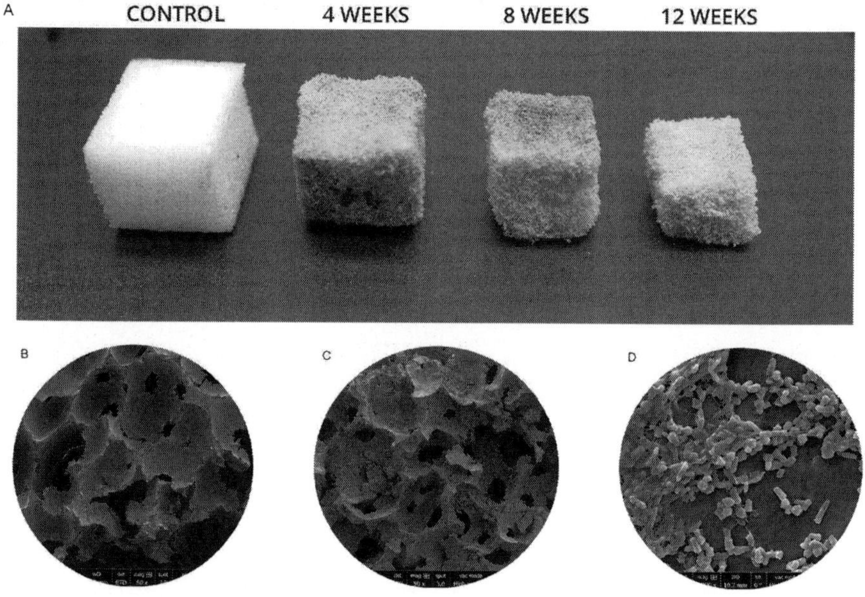

FIG. 1.3 Biodegradation of foam samples and scanning electron microscope images of polyester polyurethane foams in soil. (A) The degradation process in soil at 0 (control), 4, 8, and 12 weeks. (B) Image of the control. (C) Image of the same foam as the control placed into soil for 4 weeks. (D) A zoomed-in image of (B) showing the bacteria on the surface.

biological systems selected amide and ester linkages because of the stability under one set of circumstances and reactivity under another. Catalysts called esterases and lipases are excellent at breaking down bio-polyesters (energy storage molecules), and proteases are excellent at breaking down bio-polyamides (proteins). Bacteria and fungi also evolved to create these catalysts, enabling them to use these molecules made by others as energy sources. By judicious selection of the diacids and diols, synthetic polyesters can be so constructed that they are biodegraded by opportunistic microbes.

Polyurethanes are an important sector of the plastic economy.[67] They have a broad spectrum of characteristics and are used to make furniture and bedding foams, thermal insulation, elastomers (elastic polymers), footwear, straps (watch bands to bungee cords), coatings and sealants, and adhesives. They have excellent biocompatibility, durability, and resilience, but it is their biodegradability that is important for our purposes. They are made by reacting three components: a diisocyanate, a polyester diol, and a polymer chain extender. They can be derived from petroleum and still biodegrade. However, there is a rich set of chemical processes available to derive them from bio-based materials and make products that are renewable, sustainable, and biodegradable without compromise in performance.

The mechanical properties of biodegradable polyurethanes can be enhanced by adding materials to make a composite. These enhancements improve strength, chemical resistance, resilience, and processability.

Biodegradation means that polyurethanes disposed of carelessly will not persist in the environment for centuries. However, let us consider genuinely recyclable plastics. Plastic recycling requires harvesting, cleaning, and reprocessing into new products.[68,69] Primarily because of the crystalline structure of plastics, the energy required to recycle, and the low cost of virgin plastic, less than 10% is recycled. In some cases, it is simply ground up and placed into new plastics as inert filler. The result is often inferior to virgin plastic from petroleum. The holy grail of plastic recycling is depolymerization (Fig. 1.4). This can be done chemically or enzymatically. Now the opportunity exists to create a circular economy. Gather waste plastic, depolymerize, and remake any polyester polyurethane desired with all the quality of the virgin material (Fig. 1.5).

Why is the change so slow? The economics of petroleum

Two key features of the American system of manufacturing are interchangeability of parts (the separation of concerns) and mechanization.[70,71] Separation of concerns is a design principle for dividing a task into distinct sections such that each addresses a separate concern. It uses well-separated

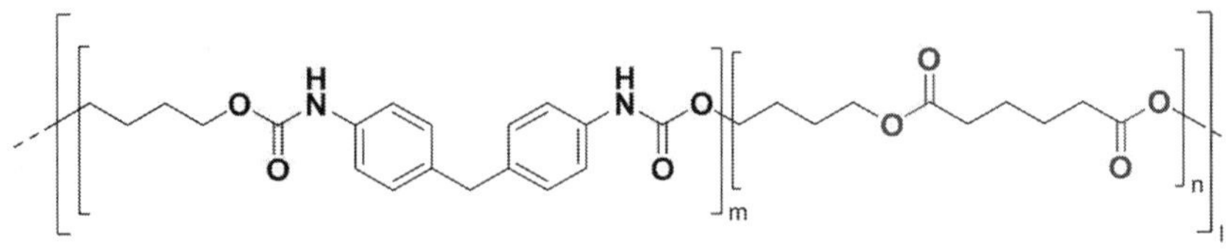

FIG. 1.4 Segment m represents the reaction of the diisocyanate with a diol; the urethane bonds are highlighted in red. Segment n represents the polyester polyol formed from the reaction of a diacid with a diol; the ester linages are indicated in blue. The average molecular weights of the m and n segments typically range from 50 to 10,000 amu and are varied to produce different properties. The bracket with the l subscript represents the number of combined m and n subunits linked together.

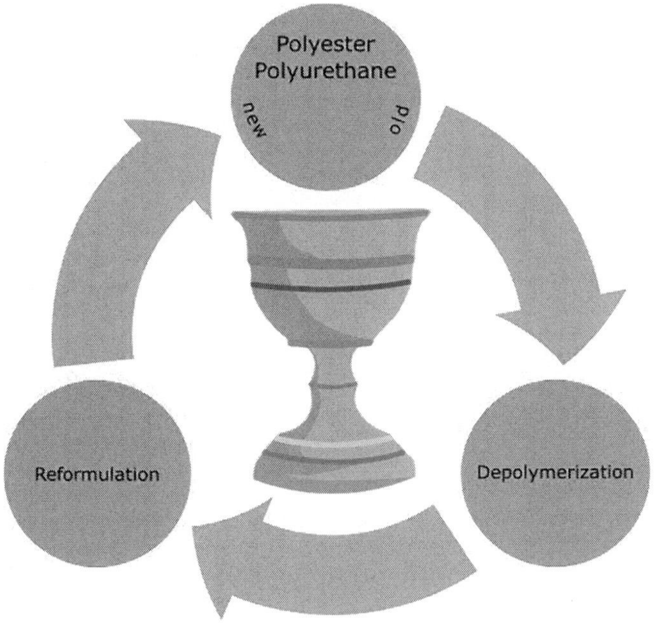

FIG. 1.5 The holy grail of plastics: a circular carbon economy that maintains quality and performance and is economically rational.

individual sections that can be developed and updated independently. This modularity gives rise to simplicity, allows reuse of developed technology, and enables modification of one section without knowing the details of other sections and making corresponding changes to those sections. The petroleum-based polymer industry is an example of the separation of concerns. There are a small number of base chemicals produced from a single, nonsustainable source: petroleum refining.

Separation of concerns arises because the production of the base chemicals is siloed, and, from a manufacturing point of view, the sheer volume of inexpensive materials provided by petroleum blocks most company's vision of an alternative. This led to the development of plastics that last forever, and at such low cost that they are disposable. This logic defies reality; it neglects the impact on health and the environment and ignores the simple truth that petroleum is not sustainable. Commerce based only on financial terms, ignoring long-term impact for short-term profit, favors the interchangeability, cost, and inventory of petroleum-based materials. The additional expenses of carbon dioxide released into the atmosphere and plastics into landfills are separate concerns, dollars for which the producer bears no responsibility. The separation of concerns paradigm, while efficient, breaks down when customers demand that the overall value includes sustainability and concern for the environment.

Valorization comes from the German *Verwertung*, meaning the productive use of a resource: maximizing the value of a product or process to realize greater earnings.[72,73] Waste valorization is the use of waste to create other valued products, thereby increasing the money generated from the starting materials. Minimization and reutilization are significant ways to create more value and reduce cost.

The petroleum industry valorizes a barrel of oil, and we should model our development of bio-based materials on that, optimizing the value of a kilogram of biomass and work to replace all the things generated from a barrel of crude. For many, a barrel of oil translates to "how much gasoline?" Interestingly, only 40% of a barrel of oil is used in the production of gasoline. The balance becomes other, higher-value products, increasing the valorization of the barrel; they include jet fuel, plastics, and industrial chemicals. This sentiment is reflected in the US Department of Energy's Biomass Program in 2011 on precisely this theme: *Replace the Whole Barrel, Supply the Whole Market—the New Horizons of Bioenergy.* At Biomass 2011, Secretary Steven Chu remarked,[74] "When oil prices rise, markets tend to panic; when oil prices stabilize, markets tend to hit the "snooze button." To effectively displace crude oil imports, biomass substitutes must be developed for gasoline, diesel, jet fuel, and other petroleum products."

Sapphire Energy[75] raised over 250 million dollars to make biofuels as a scale-up test bed. The sudden drop in petroleum prices in late 2008 eventually made the company insolvent as a biofuels manufacturer. The facility has since been repurposed and now isolates omega-3 fatty acids.[76] Market acceptance will come when producers demonstrate to the public that high quality, sustainable, biodegradable, biomass-based materials with no compromise in quality or performance can be produced. The challenge is in the form of education to create market awareness, acceptance, and eventual profitability. In the future, urban areas will adopt more efficient mass transit, and most remaining personal vehicles will be electric. Sustainable, renewable energy sources such as solar and wind may well handle this energy need. However, there will be a need for liquid fuels in (1) long-distance trucking, rail locomotion, and shipping (biodiesel and renewable diesel) and (2) commercial and military aviation (aviation gas from the saturated hydrocarbon stream). Algae biomass must be valorized in much the same way as petroleum. Extract the valuable omega-3 fatty acids, capture the carbohydrates and proteins for feed, convert the saturated fatty acids to fuel, and produce higher-value polymeric polyester polyurethane precursors from the remaining unsaturated fats.

Stumbling out the gate: Greenwashing and the eco-consumer

Greenwashing uses deceptive content and imagery in the marketing of a product or practice as environmentally friendly. Intentional greenwashing is simply fraud and is reminiscent of the pattern identified by Rachel Carson[77] and repeated with the topic of climate change (Oreskes and Conway, *Merchants of Doubt*).[78] The authors described the deliberate use of misinformation, the consumer's unquestioning acceptance of advertising claims, and the development of institutional policy that clouds the issue rather than provides transparency. How does this occur? The following quotation from Narayan, Doi, and Fukada[79] is informative:

The U.S. biodegradable's industry fumbled at the beginning by introducing starch-filled (6%–15%) polyolefins as true biodegradable materials. These, at best, were only bio disintegrable and not completely biodegradable. Data showed that only the surface starch biodegraded, leaving behind a resistant polyethylene material. Starch entrapped within the P.E. matrix did not appear to be degraded.

This connects a poorly crafted governmental definition to the abusive practice of greenwashing. Carefully written institutional definitions and science-based measurements combined with educating both product developers and consumers are key. It will allow analysts to decide for themselves which products and practices are truly eco-friendly. Lastly, a watchdog group should be established to help identify entities engaged in this unethical and misleading practice. The marketing materials must be linked to methods and tests qualified to evaluate the environmental impact of products. LCA is an excellent means of getting to the bottom line within an environmental assessment.

A new question: If we were going to start the polymer industry today, how would we do it? Why not take advantage of recent advances in genetics and biotechnology to create a new and remarkably diverse set of molecules to work with rather than being limited to those generated by petroleum refining. Bio-based polymers are not direct drop-ins to existing petroleum-based polymer formulations, which makes them not fit for purpose in the old paradigm. We will describe the use of bio-based polymers, formulations, and processes using bio-based sustainable monomers, with the long-term goal to create biopolymers that have performance characteristics similar to or better than petroleum polymers and that are renewable, sustainable, biodegradable, and recyclable.

Organization of the book

Part I is an overview.

Part II explores algae's relationship to petroleum and the use of algae as a renewable and, more importantly, sustainable feedstock. Feedstocks and methods of refining to convert biomass into the base materials for polyesters, polyurethanes, polyols, and diisocyanates are described.

Part III defines biodegradation and the biological and physical conditions that promote or inhibit the breakdown of polyester polyurethanes. Analytical techniques and challenges associated with providing the metrics to honestly evaluate the efficacy of algae-based polyester polyurethanes are described. The methodology and practices required to properly evaluate the economic and environmental impacts of polyurethanes are given.

Part IV describes the introductory chemistry and basic concepts around formulating two major types of polyurethanes, foams and thermoplastics. Coatings, adhesives, and sealants are introduced, one of the fastest growing sectors of the polyurethane industry. A chapter examines how the structural mechanics of materials can be enhanced by the creation of composites.

Part V examines the holy grail of plastics: a circular economy of forming, depolymerizing, and reassembling polyester polyurethanes. The real-world realities of bringing renewable polyester polyurethanes into today's marketplace are explored, and a big-picture view of the future science of algae-based materials is presented.

Close-up: Some thoughts about plastics for the future

Plastics have transformed our lives. They have changed the ways we live, work, and play. Plastics changed the ways we shop, prepare our food, eat, and store our food. Plastics changed the ways we get and drink many of our favorite beverages. Most of us live in houses that get our water and rid of our wastewater through plastic pipes. Because of plastics, we get increased mileage from our cars because they are lighter than the metals they replace. In hospitals, plastics have reduced infections and have even replaced defective body parts. In sports, plastics provided better protection against head injuries, allowed us to hit golf balls farther, pole vault higher, and hit tennis balls faster. Furthermore, they changed the way we fish as plastic nets replaced hemp nets.

The uses and benefits of plastics are plentiful; however, we are now aware that these miracle materials come at a cost. That cost comes primarily from the fact that most plastics are not recyclable and do not biodegrade but break down physically into smaller and smaller pieces that persist for centuries. The costs also come because we humans fail to dispose of or recycle plastics properly when we are done with them. Today, plastics are ubiquitous in the environment, the air we breathe, the water we drink, and the food we eat. Thank goodness, we will not get rid of plastics, but we need to rethink them and reengineer them.

Technology can solve the problem of creating recyclable plastics and biodegrade, but not the human behavioral problem. That requires education and public awareness. Aquariums have essential educational roles because many plastics end up in the ocean—the ultimate sink—killing marine life. The Pacific Garbage Patch in the northern Pacific has been a lightning rod for bringing attention to the problems of plastics in the ocean. Several aborted efforts to remove plastic debris from the Garbage Patch have failed. A better strategy is to keep plastics out of the ocean in the first place.

Bans that limit or prohibit certain plastics such as single-use plastics and Styrofoam can play a role with local and regional benefits. However, plastics are produced, used, and disposed of globally, and they degrade the environment and put wildlife at risk globally.

It is clear now that we need to create new materials that mimic many of the desirable qualities of plastics but that eliminate their undesirable properties. It is time for us to rethink plastics. Because plastics find so many diverse applications that are so widely used, they should be designed to be recycled and to biodegrade when appropriate. We should find sustainable feedstock to replace oil to produce them. Recycling allows the manufacture of a wide range of new products that benefit from the qualities that make plastics desirable.

California's offshore oil and gas industry will end within the next 10 to 15 years. California will lose the third most significant component of its ocean economy in dollars and jobs. These losses could be offset by growing vast amounts of algae in aquaculture farms as feedstock for plastics. If algae farms were located in areas of the ocean with high concentrations of nutrients (nitrogen and phosphorous), such as off the discharges from sewage treatment plants or at the mouths of large rivers, the algae would grow faster and contribute to better water quality. One of California's soon to be decommissioned offshore oil rigs and its 500-m radius restricted zone might be designated as a site to explore various options for growing marine algae to produce a new generation of plastics.

California is well positioned to play a leadership role in rethinking and reengineering the next generation of plastics. It should seize the opportunity.

J.R. Schubel
Former President, Aquarium of the Pacific, Long Beach, CA, United States

References

1. Mclendon R. What is the Great Pacific Ocean Garbage Patch? *Mother Nat Netw.* 2018. Published online.
2. Egger M, Sulu-Gambari F, Lebreton L. First evidence of plastic fallout from the North Pacific Garbage Patch. *Sci Rep.* 2020;10(1). https://doi.org/10.1038/s41598-020-64465-8.
3. Lau WWY, Shiran Y, Bailey RM, et al. Evaluating scenarios toward zero plastic pollution. *Science (80-).* 2020;369(6509). https://doi.org/10.1126/SCIENCE.ABA9475.
4. Rada EC, Ragazzi M, Torretta V, Castagna G, Adami L, Cioca LI. Circular economy and waste to energy. In: *AIP Conference Proceedings.* vol. 1968; 2018. https://doi.org/10.1063/1.5039237.
5. Gourmelon G, Mármol Z, Páez G, Rincón M, Araujo K, Aiello C. Global plastic production rises, recycling lags. *Rev Tcnocientifica URU.* April 2016;2015.
6. Kirby AJ. Hydrolysis and formation of esters of organic acids. *Compr Chem Kinet.* 1972;10 (C). https://doi.org/10.1016/S0069-8040(08)70344-3.
7. Geyer R, Jambeck JR, Law KL. Production, use, and fate of all plastics ever made. *Sci Adv.* 2017;3(7). https://doi.org/10.1126/sciadv.1700782.
8. Ebnesajjad S, ed. *Handbook of Biopolymers and Biodegradable Plastics;* 2013. https://doi.org/10.1016/c2011-0-07342-8.
9. Geyer R. A brief history of plastics. In: *Mare Plasticum—The Plastic Sea;* 2020. https://doi.org/10.1007/978-3-030-38945-1_2.
10. BPF. A history of plastics. *Br Plast Fed Website.* 2019. Published online.
11. Taylor DA. The dawn of the age of plastic? *Undark.* 2019;8.
12. Harvey JA. Handbook of Environmental Degradation of Materials; 2005.
13. Oxford University Press. Oxford Dictionaries. Oxford University Press.
14. Gibson RF. Principles of composite material mechanics; 2016. https://doi.org/10.1201/b19626.
15. Irving P, Soutis C. Polymer composites in the aerospace industry; 2019. https://doi.org/10.1016/C2017-0-03502-4.
16. Uddin I, Thomas S, Mishra RK, Asiri AM. Sustainable polymer composites and nanocomposites; 2019. https://doi.org/10.1007/978-3-030-05399-4.
17. Väntsi O, Kärki T. Environmental assessment of recycled mineral wool and polypropylene utilized in wood polymer composites. *Resour Conserv Recycl.* 2015;104. https://doi.org/10.1016/j.resconrec.2015.09.009.
18. Chaudhuri UR. Fundamentals of petroleum and petrochemical engineering; 2016. https://doi.org/10.1201/b10486.
19. ModorPlastics. Thermoset vs. thermoplastics. *ModorPlastics.* 2016. Published online.
20. Alauddin M, Choudhury IA, El Baradie MA, Hashmi MSJ. Plastics and their machining: a review. *J Mater Process Technol.* 1995;54(1–4). https://doi.org/10.1016/0924-0136(95)01917-0.
21. Baur E, Osswald TA, Rudolph N. Common acronyms in plastics technology. In: *Plastics Handbook;* 2019. https://doi.org/10.3139/9781569905609.001.
22. Tissot BP, Welte DH. Petroleum Formation and Occurrence, A New Approach to Oil and Gas Exploration. 2nd ed; 1984.
23. Philp RP. Petroleum formation and occurrence. *Eos, Trans Am Geophys Union.* 1985;66(37). https://doi.org/10.1029/eo066i037p00643.
24. Tissot BP, Welte DH. Petroleum Formation and Occurrence; 1978. https://doi.org/10.1007/978-3-642-96446-6.
25. Dembicki HJ. Practical Petroleum Geochemistry for Exploration and Production (Introduction); 2017.
26. Dembicki H. Practical Petroleum Geochemistry for Exploration and Production; 2016. https://doi.org/10.1016/c2014-0-03244-3.

27. Birx H. La Brea Tar Pits. In: *Encyclopedia of Time: Science, Philosophy, Theology, & Culture*; 2014. https://doi.org/10.4135/9781412963961.n301.
28. Speight JG. The Chemistry and Technology of Petroleum; 2014. https://doi.org/10.1201/b16559.
29. Hubbert M.K. Nuclear energy and the fossil fuels. In: Drilling and Production Practice, 1956. vol 1956, January 1956.
30. Jones TH, Willms NB. A critique of Hubbert's model for peak oil. *Facets*. 2018;3(1). https://doi.org/10.1139/facets-2017-0097.
31. Norouzi N, Fani M, Ziarani ZK. The fall of oil Age: A scenario planning approach over the last peak oil of human history by 2040. *J Petrol Sci Eng*. 2020;188. https://doi.org/10.1016/j.petrol.2019.106827.
32. Bardi U. Peak oil, 20 years later: Failed prediction or useful insight? *Energy Res Soc Sci*. 2019;48. https://doi.org/10.1016/j.erss.2018.09.022.
33. Outline C. Chapter 7—Petroleum Refining; 2016.
34. Speight JG. Handbook of Petroleum Refining; 2016. https://doi.org/10.1201/9781315374079.
35. Doble M, Kruthiventi AK. Green Chemistry and Engineering; 2007. https://doi.org/10.1016/B978-0-12-372532-5.X5000-7.
36. Archer D. The Global Carbon Cycle; 2010.
37. Lal R, Kimble JM, Follett RF, Stewart BA. Soil Processes and the Carbon Cycle; 2018. https://doi.org/10.1201/9780203739273.
38. Emerson S, Hedges J. Chemical Oceanography and the Marine Carbon Cycle; 2008. https://doi.org/10.1017/CBO9780511793202.
39. ACS. What is Green Chemistry? American Chemical Society; 2015. www.Acs.Org/Content/Acs/En/Greenchemistry/What-Is-Green-Chemistry.Html. Published online.
40. United States Environmental Protection Agency. Basics of Green Chemistry | Green Chemistry | US EPA. *Basics Green Chem*. 2015. Published online.
41. Jarvis EE, NREL EEJ. Aquatic species program (ASP): lessons learned. *AFOSR Work*. 2008;2008.
42. Sheehan J. A look back at the U.S. Department of Energy's Aquatic Species Program—Biodiesel from Algae. *Program*. 1998; (July. https://doi.org/10.2172/15003040.
43. Johnson DA, Sprague S. Liquid Fuels from Microalgae; 1987.
44. NSF CCI. https://www.nsfcci.org/. Accessed 10 May 2021.
45. Hauschild MZ, Rosenbaum RK, Olsen SI. Life Cycle Assessment: Theory and Practice; 2017. https://doi.org/10.1007/978-3-319-56475-3.
46. Acero AP, Rodríguez C, Ciroth A. LCIA Methods Impact Assessment Methods in Life Cycle Assessment and Their Impact Categories; 2014.
47. Fröhling M, Hiete M. The sustainability and life cycle assessments of industrial biotechnology: an introduction. In: *Advances in Biochemical Engineering/Biotechnology*. vol. 173; 2020. https://doi.org/10.1007/10_2020_123.
48. JRC, European Commission, Joint Research Centre, Institute for Environment and Sustainability. Characterisation Factors of the ILCD. Recommended Life Cycle Assessment Methods. Database and Supporting Information; 2012.
49. Sari YW, Kartikasari K, Widyarani SI, Lestari D. Techno-economic assessment of microalgae for biofuel, chemical, and bioplastic. In: *Microalgae*; 2021. https://doi.org/10.1016/b978-0-12-821218-9.00013-x.
50. Nanda S, Vo DVN, Sarangi PK. Biorefinery of Alternative Resources: Targeting Green Fuels and Platform Chemicals; 2020. https://doi.org/10.1007/978-981-15-1804-1.
51. Verma D, Fortunati E, Jain S, Zhang X. Biomass, Biopolymer-Based Materials, and Bioenergy: Construction, Biomedical, and Other Industrial Applications; 2019. https://doi.org/10.1016/C2017-0-00839-X.

52. Hytönen E, Stuart P. Techno-Economic Assessment and Risk Analysis of Biorefinery Processes. 29; 2011. https://doi.org/10.1016/B978-0-444-54298-4.50054-4.
53. Zubiate R. Patent leather: two case histories. In: *20th Century Materials, Testing and Textile Conservation: Harpers Ferry Regional Textile Group 9th Symposium, November 3–4, 1988*; 1988.
54. Lawson WD, Lynch CA, Richards JC. Corfam: research brings chemistry to footwear. *Respir Manage.* 1965;8(1). https://doi.org/10.1080/00345334.1965.11755739.
55. Anon. Third generation poromerics. *Rubber Plast Age.* 1969;50(8).
56. DU PONT. End of Corfam. *Chem Eng News Arch.* 1971;49(12). https://doi.org/10.1021/cen-v049n012.p019.
57. Littler DA, Pearson AW. Marketing a new industrial good: a case study. *Ind Mark Manage.* 1972;1(3). https://doi.org/10.1016/0019-8501(72)90022-3.
58. Li J, Liu H, Paul CJ. Microplastics in freshwater systems: a review on occurrence, environmental effects, and methods for microplastics detection. *Water Res.* 2018;137. https://doi.org/10.1016/j.watres.2017.12.056.
59. Hale RC, Seeley ME, La Guardia MJ, Mai L, Zeng EY. A global perspective on microplastics. *J Geophys Res Ocean.* 2020;125(1). https://doi.org/10.1029/2018JC014719.
60. Zhang Y, Kang S, Allen S, Allen D, Gao T, Sillanpää M. Atmospheric microplastics: A review on current status and perspectives. *Earth-Science Rev.* 2020;203. https://doi.org/10.1016/j.earscirev.2020.103118.
61. Wright SL, Thompson RC, Galloway TS. The physical impacts of microplastics on marine organisms: a review. *Environ Pollut.* 2013;178. https://doi.org/10.1016/j.envpol.2013.02.031.
62. Botterell ZLR, Beaumont N, Dorrington T, Steinke M, Thompson RC, Lindeque PK. Bioavailability and effects of microplastics on marine zooplankton: A review. *Environ Pollut.* 2019;245. https://doi.org/10.1016/j.envpol.2018.10.065.
63. Szycher M. Szycher's Handbook of Polyurethanes. 2nd ed; 2012. https://doi.org/10.1201/b12343.
64. Sonnenschein MF. Polyurethanes: Science, Technology, Markets, and Trends; 2014. https://doi.org/10.1002/9781118901274.
65. Urbánek T, Jäger E, Jäger A, Hrubý M. Selectively biodegradable polyesters: Nature-inspired construction materials for future biomedical applications. *Polymers (Basel).* 2019;11(6). https://doi.org/10.3390/POLYM11061061.
66. Diaz C, Mehrkhodavandi P. Strategies for the synthesis of block copolymers with biodegradable polyester segments. *Polym Chem.* 2021;12(6). https://doi.org/10.1039/d0py01534b.
67. Center for the Polyurethanes Industry. https://polyurethane.americanchemistry.com/. Accessed 10 May 2021.
68. PWMI. An introduction to plastic recycling. *Plast Waste Manag Inst.* 2019. Published online.
69. Hopewell J, Dvorak R, Kosior E. Plastics recycling: challenges and opportunities. *Philos Trans R Soc B Biol Sci.* 2009;364(1526). https://doi.org/10.1098/rstb.2008.0311.
70. Goodfriend M, McDermott J. The American System of economic growth. *J Econ Growth.* 2021;26(1). https://doi.org/10.1007/s10887-021-09186-x.
71. Dijkstra EW. On the role of scientific thought. In: *Selected Writings on Computing: A Personal Perspective*; 1982. https://doi.org/10.1007/978-1-4612-5695-3_12.
72. Garba MD, Galadima A. Catalytic hydrogenation of hydrocarbons for gasoline production. *J Physiol Sci.* 2018;29(2). https://doi.org/10.21315/jps2018.29.2.10.
73. Ben Tahar N, Mimoun H. Valorization of loads petroleum by thermal process. *Asian J Chem.* 2013;25(6). https://doi.org/10.14233/ajchem.2013.13535.
74. Biomass. Replace the Whole Barrel, Supply the Whole Market. Department of Energy; 2011. https://www.energy.gov/index.php/eere/bioenergy/biomass-2011-replace-whole-barrel-supply-whole-market. Accessed 10 May 2021.

75. Sapphire Energy—Wikipedia, n.d. Accessed 10 May 2021. https://en.wikipedia.org/wiki/Sapphire_Energy.
76. Qualitas Health. n.d. Accessed 10 May 2021. https://www.qualitas-health.com/.
77. Carson R. Silent spring. In: *Key Readings in Journalism*; 2012. https://doi.org/10.9774/gleaf.978-1-907643-44-6_4.
78. Oreskes N, Conway EM. Merchant of Doubt; 2010.
79. Doi Y. Biodegradable plastics and polymers. *J Pestic Sci.* 1994;19(1). https://doi.org/10.1584/jpestics.19.S11.

PART II

Re-evaluating the sources

CHAPTER

2

Why algae?

Ryan Simkovsky and Frances Carcellar
University of California San Diego, La Jolla, CA, United States

Introduction

Ancient algae and cyanobacteria are the original sources of petroleum

Modern plastics are almost exclusively derived from petroleum. Why do we use petroleum to make them? With the invention and patenting by Leo Baekeland of the first synthetic plastic, Bakelite, in 1909 from phenol and formaldehyde,[1] and the invention of a plasticized polyvinylchloride by Walter Semon while working for the B.F. Goodrich Company in 1926,[2] it became obvious that useful plastic products could be generated from petroleum-derived chemicals. Further inventions of novel synthetic plastics in the 1930s and early 1940s, including polyurethanes (PUs), polytetrafluoroethylene, nylon, neoprene, polyvinylidene chloride (more commonly known as Saran, as in Saran Wrap), and polyethylene terephthalate, allowed field testing and development of applications and durability during World War II.[3,4] Thus, by the time the petrochemical industry started to focus on development of long-lasting and mass produced plastics for private commercial applications in the 1950s, petroleum had become the primary source of raw materials due to these prior innovations and the low cost and high availability of petroleum, which in turn resulted from innovations in extraction and refinement technologies, ever-increasing demand for fuels for the growing private and military uses of the internal combustion engine, and geopolitical engagements to enable secure access to global reserves in the early 20th century.[5] Thus the primary reasons for using petroleum are based more on historical factors than the requirements of the chemistry.

What created these global oil reserves? Contrary to the popular belief that oil was the result of accumulation and geothermal conversion of dead

Rethinking Polyester Polyurethanes
https://doi.org/10.1016/B978-0-323-99982-3.00002-X

dinosaurs, an assumption promoted as part of publicity campaigns at the World's Fairs in 1933 and 1964 by the Sinclair Oil Company, which had a dinosaur as its mascot, much of the oil found in underground deposits is derived from microbial life that evolved billions of years before the dinosaurs.[6,7] Specifically, the critical developments that led to accumulation of these oils were the evolution of photosynthesis in cyanobacteria, which occurred over 2.4 billion years ago as evidenced by stromatolite fossils,[8] and the subsequent evolution of eukaryotic algae through the endosymbiotic evolution of an engulfed cyanobacterium into the chloroplast. Prior to the explosive evolution of phytoplankton, a collective term interchangeably used with algae to denote all photosynthetic microorganisms, the Earth's atmosphere during the early Archean period was composed primarily of nitrogen, carbon dioxide (CO_2), and inert gases, with no detectable traces of oxygen present according to the geological record.[9] As phytoplankton proliferated across the planet, long before plants or animals had evolved, their photosynthetic activities released free oxygen into the atmosphere, resulting in the Great Oxidation Event around 2.4 billion years ago and a subsequent rise in oxygen levels around 0.9 billion years ago as the oxygen they produced finally saturated the exposed reserves of reductive metals and other oxygen-absorbing materials.[10,11] This production of oxygen is the result of only the first part of oxygenic photosynthesis, in which the light reactions use solar energy to split water to generate oxygen and ultimately chemical energy in the form of adenosine triphosphate (ATP) and nicotinamide adenine dinucleotide phosphate (NADPH). The second part of photosynthesis proceeds through the light-independent reactions of photosynthesis, in which ATP and NADPH are used to fix CO_2 into glyceraldehyde 3-phosphate, which is ultimately converted into all other organic molecules in the organism's biomass.[12] Thus the secondary result of these 1.5 billion years of microorganismal photosynthesis was the conversion of massive amounts of atmospheric carbon into algal biomass. The accumulation, decomposition, burial, pressurization, and heating through geological events of this phytoplankton biomass in the oceans, freshwater systems, and land masses is what led to the creation of the oil deposits we now use for petroleum products, including plastics.

Modern algae as a sustainable alternative for petroleum products

Because phytoplankton is the original source for petroleum, it follows that modern algae, which make the same types of chemicals as their ancient ancestors, are a viable alternative source for producing the chemicals needed to manufacture plastics. Can plastics be produced directly from biological sources? In short, yes—some of the original polymer-based materials invented as late as the 19th century, which inspired the early 20th century invention of synthetic plastics, were made wholly or in part from biological sources.[13] Examples are natural rubber and latex harvested from the rubber tree; linoleum, whose name derives from the linseed oil first used to make it; shellac, derived from a resin secreted by the lac bug; polystyrene, which was originally derived from the resin of the oriental sweetgum tree[14]; and celluloid, which was made by mixing a waxy terpenoid from the camphor laurel with nitrated cellulose.[15] Although the development of these early bioplastics, an umbrella term for biologically derived polymers many of which are biodegradable, took a back seat to the easy and cheap to manufacture synthetics, the recent acknowledgment of the limits on petroleum reserves, the impact of releasing petroleum-derived carbon into the atmosphere, and the environmentally destructive accumulation of nondegradable synthetic plastics in the environment have caused a renewed interest in biodegradable bioplastics. These newer or newly rediscovered bioplastics are primarily derived from fermentation of plant-derived sugars, as with polylactic acid (PLA), which was first made from corn in a cost-efficient manner in 1992 by Dr. Patrick R. Gruber,[16] and polyhydroxyalkanoates (PHAs) including polyhydroxybutyrate (PHB), which was first reported in 1926 by Maurice Lemoigne as a polymer produced by the bacterium *Bacillus megaterium.*[17] Novel bioplastics have also been made from blends of dried plant biomass such as banana peel starch[18] or any other fruit and vegetable matter that provides lignocellulosic material that can act as the core component of the bioplastic,[19] or through extraction and derivatization of natural polymers from plants or seaweeds.[20]

Instead of directly generating novel plastics, biomass can also be converted into the raw materials for manufacturing the synthetic plastics. Because petroleum is composed primarily of deoxygenated hydrocarbons, production of biodegradable synthetic plastics, including polyglycolic acid, polybutylene succinate, polycaprolactone, and PUs, requires reoxygenation of petroleum to generate precursor chemicals, including glycolic acid, succinic acid, 1,4-butanediol, hexanoic acid, and polyols, which are hydrocarbon molecules with multiple alcohol groups. Many of these precursors are naturally produced by organisms, particularly as a result of fermentation processes that can be performed at industrially relevant scale, and can act as a drop-in replacement of petroleum-based plastic precursors.[21]

Although derived from renewable, biological sources, most of the feedstocks for the raw materials that go into producing these bio-derived plastics, such as the plant-derived sugars that are fed to fermentation processes, compete with food production resources through the use of agricultural land and potable water. Estimates suggest that 1 million km^2, or 7% of the planet's arable land, would be needed to replace the annual manufacturing of 300 million tons of petroleum-derived plastics with plant-based analogs.[22] In addition, life cycle analysis on the production of these plant-based bioplastics suggests that the energy and carbon costs associated with growing and manufacturing these materials are on par with those of petroleum-based plastics.[23] Finally, many of these novel biopolymers lack the desired physical properties found in the synthetic plastics.[24] In spite of growing demand for bio-based petroleum alternatives, these combined obstacles to large-scale production have limited the market penetration of bioplastics to less than 1% of the current global polymer market, as of 2016.[25,26] Thus although plant oils and plant-derived bioplastics and chemicals are a technically feasible and renewable alternative to petroleum-based plastics, a more sustainable option, such as algae-derived materials, is needed to viably replace petroleum-based plastics.

How can algae overcome these obstacles and provide a sustainable source of raw materials to replace petroleum in the production of biodegradable plastics, including polyester PUs? First, numerous algae are capable of directly producing polymers and polymer precursors, including PHAs,[27] chitin,[28] and cellulose,[29] as well as diacids and diols needed to produce PU polyols.[30,31] In addition, algal biomass can be converted to plastic precursors and bioplastics using the same processes as with plant-derived biomass, including plasticization[32] or through fermentation by heterotrophic microorganisms.[33] Second, algae can grow much more rapidly than plants, with some of the fastest doubling times reported as once every two hours[34]; can do so using nonpotable water in nonarable environments; and can naturally reach scales visible by satellite. Readers are likely familiar with news reports of this large-scale growth in the form of pond scum or harmful algal blooms, where specific species of algae that often produce toxins, such as microcystin, can develop into blooms in the ocean or in freshwater reservoirs that kill the local wildlife and shut down potable water supplies.[35] While toxic strains should obviously be excluded from agricultural production, the fact that multiple blooms in the past decade have been reported at over 5000 $mile^2$ in surface area, equivalent to 13,000 km^2, highlights algae's natural growth capacity and agricultural potential.[36,37] Third, algae are more efficient at converting light and CO_2 to biomass than plants, due to a combination of factors such as more efficient mechanisms of carbon concentration and their aquatic lifestyle allowing faster uptake of nutrients and inorganic carbon.[38–40] Estimates of the surface area, which can be nonarable for many strains of algae, required to supply the same amount of algae biomass to replace the world's annual plastic production, are approximately 5000–12,500 km^2, based on similar estimates for biofuel production that algae require 80- to 200-fold less land than plants to generate equivalent yields.[41–43] For visualization of these estimated sizes, the entire combined

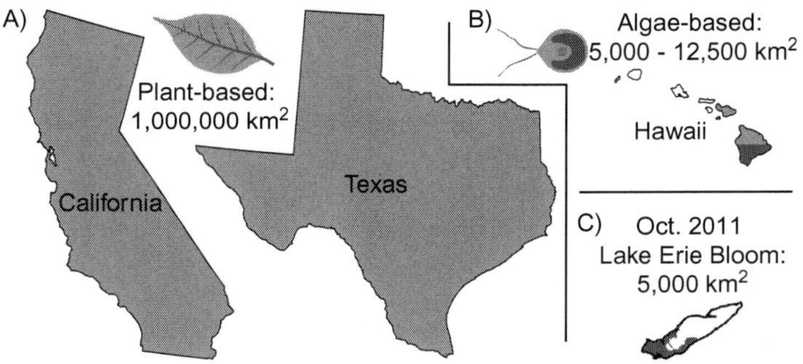

FIG. 2.1 Estimated land areas required to replace the world's annual plastic manufacturing using (A) plant-based or (B) algae-based production. (C) For comparison, the observed surface area of the October 2011 Lake Erie algae bloom is shown in *green (gray in the printed version).*

area of California and Texas would be needed for plant-based plastics, while only half the big island of Hawaii would be needed for algae-based plastics to replace the world's annual production of petroleum plastics (Fig. 2.1). Finally, the diversity of algae species and the varying environments in which they can thrive are extraordinary, meaning that for any given production location across the globe, a suitable strain can be found for robust, high production yields. Equally diverse is the set of natural products that algae can produce, and if a suitable natural producer of a desired chemical cannot be found for a specific production location, then it is likely that a suitable production strain can be genetically modified through direct biological engineering or more traditional methods of breeding to produce the desired chemicals at industrially relevant levels.

The foundations

Algae as a natural source of industrial products

Diversity of algae

Having described in broad strokes the industrial and agricultural potential of algae, we need to answer the question "What are algae?" to better understand how these organisms could be used to generate the raw materials to replace petroleum in products such as plastics. "Algae" is an umbrella word for the highly diverse group of photosynthetic organisms, life forms capable of converting light energy into chemical energy through the conversion of CO_2 and water into oxygen and sugars, that are not land plants. This grouping is polyphyletic, meaning that it includes multiple evolutionary lineages that do not share a single, direct common ancestor that is also not an ancestor of other more closely related nonalgal clades.[44] To better exemplify this, the term algae can include cyanobacteria, which are prokaryotic photosynthetic microbes that lack a nucleus and membrane-bound organelles, and the *Chlorophyta*, which are eukaryotic microbes that pack their genomes into a nucleus and have membrane-bound organelles such as chloroplasts and mitochondria, but excludes nonphotosynthetic bacteria, such as the more commonly known *Escherichia coli* or *Salmonella*, and eukaryotic microbes such as the protistan ciliates and Cercozoan amoeboids, that are more closely related to one of these algal clades but not the other.

One way to come to terms with the diversity of algae is in the context of their evolutionary history. As stated in the introduction, fossil and geological evidence indicate that the invention of oxygenic photosynthesis occurred only once, through the evolution of the cyanobacteria over 2.4 billion years ago (bya).[45,46] Approximately 1.2 billion years later, or 1.5bya, a likely biflagellated heterotrophic eukaryote engulfed a cyanobacteria (Fig. 2.2), which through the process of endosymbiotic gene transfer, in which genes from the cyanobacterium's genome were moved over time to the nuclear genome of the engulfing symbiotic host cell, evolved into the chloroplast.[46–48] The first experimental evidence for this endosymbiotic event was in the form of morphological comparisons between cyanobacteria and chloroplasts, as well as the existence of DNA in the chloroplast.[49] Later, DNA sequences of cyanobacteria and eukaryotic algae provided further evidence for this evolutionary relationship,[50–52] while studies on amoebal predation of bacteria have shown some of the missing link steps toward this endosymbiosis. Examples include the development of codependence through survival of the ingested bacteria and its conferral of resistance to external stresses, such as antibiotics, to the host amoeba[53]; or how imperfect mechanisms of predatory amoebal digestion of engulfed bacteria have allowed the evolution of primitive farming, where engulfed bacteria are maintained alive within the amoeba until it finds itself in a low resource environment, at which point the bacteria are released to seed a new population of food.[54]

The result of this endosymbiosis of cyanobacteria was the evolution of the eukaryotic algae, which then diverged into three main groups: glaucophytes, green algae, and red algae.[48] Glaucophytes are unicellular freshwater eukaryotes that can undergo only asexual reproduction, and whose chloroplasts retain the peptidoglycan cell wall that cyanobacteria use to hold their shape but other algal and plant chloroplasts do not make, thus providing more evidence for the endosymbiotic theory. The green algae include subgroups such as the *Chlorophyta*, which are predominantly unicellular aquatic organisms, and *Charophyta*, a group of freshwater species that gave rise to terrestrial plants[55,56] and include complex, multicellular species such as stoneworts. Red algae, or *Rhodophyta*, are predominantly multicellular marine seaweeds falling in the class

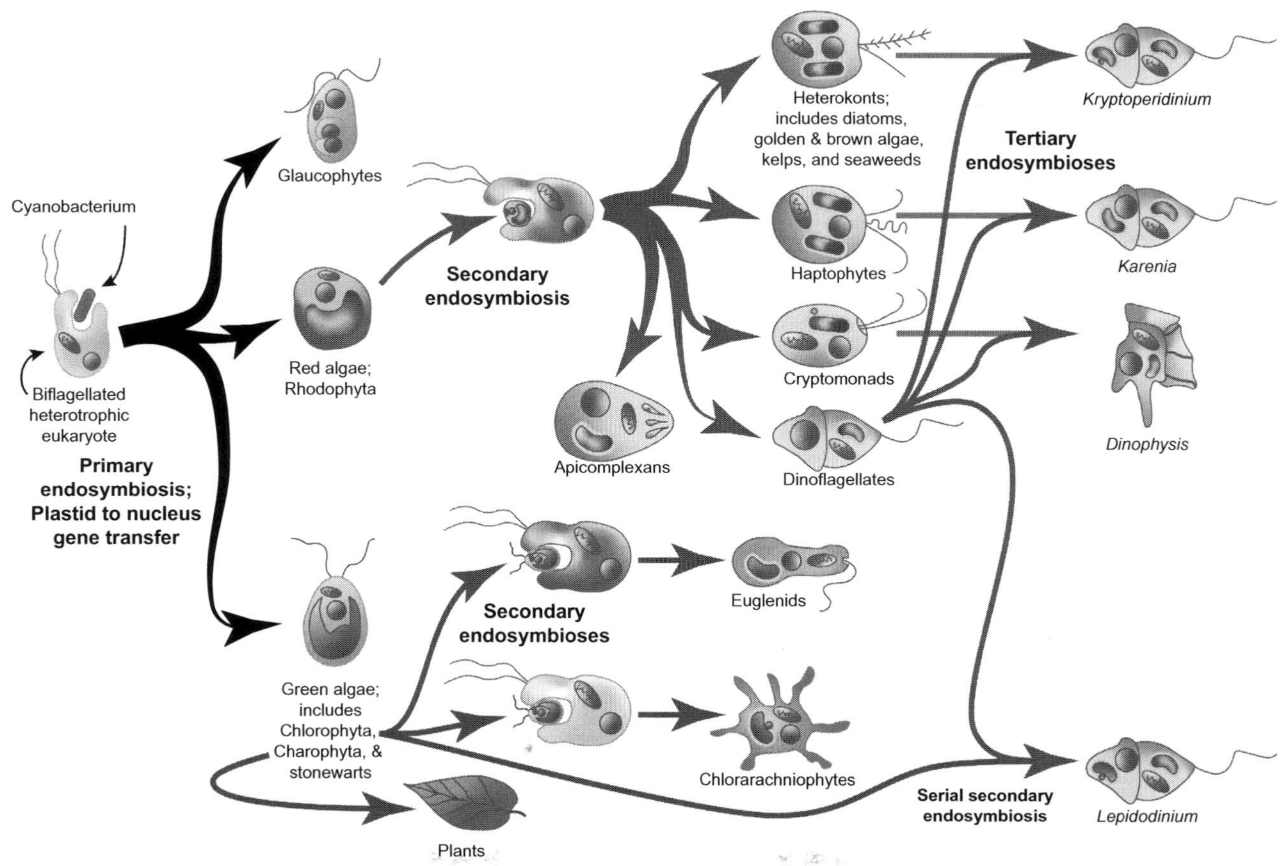

FIG. 2.2 Evolutionary history of endosymbiosis leading to the diversification of eukaryotic algae.

Florideophyceae.[47] A number of secondary and tertiary endosymbiotic events of green algae or red algae by heterotrophic eukaryotes led to further diversification of these groups of algae.[56,57] Two separate secondary endosymbioses of green algae led to the evolution of the euglenids, unicellular flagellates that can obtain energy from both photosynthesis and predatory engulfment of bacteria and smaller flagellates, and the *Chlorarachniophytes,* amoeboid unicellular marine mixotrophic algae.[48] Secondary endosymbiosis of a red alga followed by diversification led to groups such as the *haptophytes; cryptomonads;* dinoflagellates, unicellular phytoplankton that are the cause of red tides and shellfish poisoning[58]; apicomplexans; and the heterokonts, a group which includes diatoms, intricate unicellular plankton that have a cell wall composed of silica called a frustule[59]; *Chrysophyceae,* also called golden algae, that are predominantly multicellular freshwater organisms; and *Phaeophyceae,* brown algae that include kelps and seaweeds generally found in temperate and polar marine environments, such as the commonly consumed *Saccharina japonica.*[60] Serial secondary endosymbiosis of a green algae by a dinoflagellate red algae gave rise to *Lepidodinium,* while tertiary endosymbiosis events of red algae led to the evolution of *Dinophysis, Karenia,* and *Kryptoperidinium.*[48,56] While this evolutionary history, which fossil evidence suggests spans from the development of the oxygenic atmosphere to the early Jurassic approximately 200 million years ago,[61] is by no means exhaustive, it demonstrates the vast diversity and complexity of algae that exist. Over 33,000 species of algae have been described and classified, while estimates of the total number of algal species range from 72,000 to nearly 300,000, meaning that the number of algae yet to be discovered and the kinds of chemical products they can produce vastly outweigh what is known.[62,63]

With so many species, how can we begin to understand the potential for harnessing algae as a source for petroleum replacements? A secondary method to their evolutionary history is to classify these organisms based upon their physical and metabolic characteristics. For example, within most of the phylogenetic groups of algae described before, species can further be divided into groups of macroalgae, such as seaweeds that are visible to the naked eye and can reach over 60 m in length, and microalgae, which can be as small as 0.5 μm.[57] Although much of the current interest in growing petroleum replacements is on microalgae because of their robust high growth rates, ability to be genetically manipulated, and the generally more simple systems required to cultivate them,[64] it is important to keep in mind the potential of seaweed farming for both traditional food products, such as nori made from the red algae *Pyropia,*[65] and chemicals for fuels and materials, such as alginate and cellulose.[31,32]

A common misunderstanding of this purely size-based classification is that macroalgae are multicellular while microalgae are unicellular. Looking at the cyanobacteria alone disproves this misconception, along with

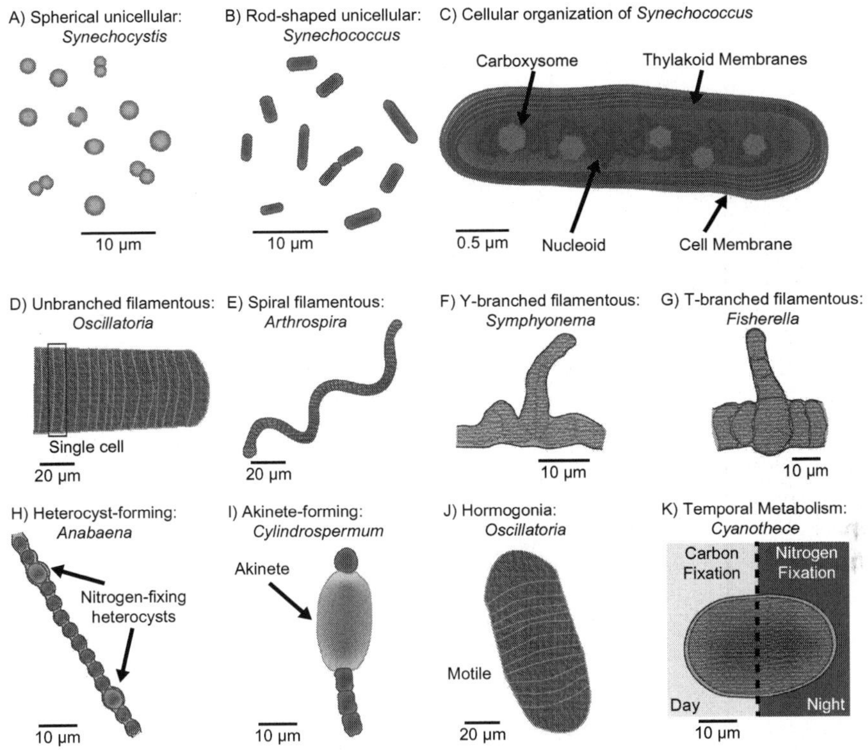

FIG. 2.3 Representative species of cyanobacteria demonstrating diversity in morphologies, internal structures, and metabolisms.

the common misconception that bacteria in general are simple unicellular pill-shaped bags of chemicals. Although numerous cyanobacteria are indeed unicellular, and many of these are elongated, others are coccoid or spherical, as with the freshwater *Synechocystis* and the marine *Prochlorococcus*, and none of these are as simple as most believe[66,67] (Fig. 2.3). Although prokaryotes lack membrane-bound organelles, they do not lack internal structure and organization. Photosynthetic production of chemical energy depends upon the accumulation of protons released during the water splitting and electron transport chain steps of photosynthesis into the thylakoid lumen, out of which the accumulated protons move to the cytoplasm through an ATPase embedded in the thylakoid membranes to generate ATP, the primary energy currency of the cell.[68] In addition to membrane-based compartmentalization, cyanobacteria produce self-assembling protein-based microcompartments[69] called carboxysomes that are filled with the enzymes carbonic anhydrase and ribulose-1,5-bisphosphate carboxylase/oxygenase (RuBisCO), the most abundant enzyme on the planet, which is primarily responsible for carbon fixation.[70]

The carbonic anhydrase converts carbonate (HCO_3^-) into aqueous CO_2 that cannot leak out of the carboxysome, thus concentrating CO_2 locally to enhance RuBisCO's carbon fixation efficiency while also excluding oxygen, which reduces RuBisCO's enzymatic capacity by outcompeting CO_2 for binding to the enzyme active site.[68]

In addition to this internal complexity, many cyanobacteria exist as multicellular filaments with diverse morphologies—some can be straight and unbranched, as with the freshwater *Oscillatoria* and *Ulothrix moniliformis*; unbranched and spiral shaped, as with *Arthrospira platensis*, commonly known as the edible spirulina; or branched, as with the Y-branched *Symphyonema* and T-branched *Fischerella*.[57,67,71] Some filamentous cyanobacteria are further capable of cellular differentiation, where individual cells morph from the more common vegetative cell into cells dedicated to particular functions, including heterocysts, which degrade their photosynthetic complexes and terminally dedicate their cellular structure to the exclusion and elimination of oxygen to enable nitrogen fixation by oxygen-inhibited nitrogenases[72]; akinetes, which are dormant thick-walled storage cells resistant to cold and desiccation, formed during times of environmental stress and revived in a more favorable environment decades after forming[73]; and hormogonia, motile filaments that allow the cells to find a more favorable environment.[66]

These morphological intricacies often reflect metabolic complexities. For example, in the nitrogen-fixing species *Anabaena* sp. PCC 7120, cells in the filament are separated from each other by their own inner membranes but share a continuous periplasm and single outer membrane to aid in nutrient exchange between nitrogen-fixing heterocysts and carbon-fixing vegetative cells.[74,75] During the early stages of heterocyst development, a spacing of ~10 cells between each heterocyst is established through a diffusible gradient of an inhibitory 5 amino acid long peptide PatS,[76] while the late stages of heterocyst differentiation require the irreversible excision of three DNA segments from the genome.[77,78] Such complex behaviors, often observed only in eukaryotes, are just one solution that cyanobacteria have evolved to perform nitrogen fixation. A second is to temporally separate photosynthesis and nitrogen fixation by expressing nitrogenases only at night when oxygen-producing photosynthesis is not active,[72] as occurs in *Cyanothece*[79] and *Crocosphaera*.[80] Thus it is important to remember that even what is considered the simplest of microalgae, the cyanobacteria, can be immensely complex in the diversity of morphologies and metabolic processes that they can perform. This, of course, holds true for the eukaryotic algae as well, where multicellularity allows extreme differences in size and shape, or multiple life cycle phases allow sexual mating that can result in microscopic diploid or macroscopic haploid stages of development, as occurs with *Pyropia*.[81] Various forms of motility and organelle-dependent metabolic processes can drastically alter the behavior of eukaryotic algae in response to the environment, including the accumulation of protective chemicals, as occurs when *Haematococcus pluvialis* accumulates the antioxidant astaxanthin at levels up to 5% of dry biomass to protect its cyst form from UV damage,[82,83] or rapid energy storage, as occurs in diatoms under nutrient stress in the form of neutral lipid droplets.[84]

Finally, over the course of these 2.5 billion years of diversification and natural selection, algae have evolved to survive and thrive in nearly every biome, including extreme habitats. Beyond the well-known tropical and temperate freshwater and marine habitats, algae survive in the freezing waters, ices, tundra, and permafrost of the Arctic and Antarctic polar regions.[56] In the other temperature extreme, algae species are found growing in hydrothermal pools and near hydrothermal vents.[85] Algae, including diatoms, cyanobacteria, and green algae such as *Dunaliella* and *Picocystis,* have been found to survive in acidic (pH 2.3–3.2), alkaline (>pH 10), and/or hypersaline (> 35 g/L salt) lakes, on the highest altitude volcanoes, airborne in water vapor, and in the jet streams of the stratosphere,[85–88] which likely has contributed to the cosmopolitan nature, or ubiquitous distribution, of many groups of algae. Algae are found in microbial mats in desert crusts and desert-like tidal regions where nonphotosynthetic organisms exist in a mutualistic relationship involving nutrient exchange between spatially distinct layers of nonphotosynthetic microbes and algae. To survive in these arid regions, these algae and cyanobacteria have evolved the molecular and cellular mechanisms to resist desiccation, including sporulation as exhibited by hormogonia in cyanobacteria, that also allow tolerance to high light, heat, salinity, and desiccation, as observed with the green alga *Chlorella ohadii,* until they can rapidly bloom during brief periods of rain.[66,85,89–91] Algae have become symbiotic with numerous other creatures, typically through an exchange of nutrients, including lichen[92]; coral[93,94]; sea sponges[95]; the egg capsules of the North American spotted salamander where algae raise the oxygen levels of the eggs and remove waste while the eggs provide the algae with ammonia[96]; and sloths, where algae gain access to moisture pockets in the fur and the sloths benefit from the green camouflage.[97] Adaptation to these various niches has allowed a diversification of metabolic acquisition of energy. While many cyanobacteria and eukaryotic algae are obligate photoautotrophs, meaning their only means of acquiring carbon is through photosynthesis, many have developed photomixotrophic lifestyles where the cell can also acquire carbon and energy through the uptake and processing of sugars or other organic compounds, such as acetate, from their environment.[57,98] Some, such as the *Chlorarachniophytes* and *euglenids,* are predatory, ingesting bacteria and small protists.

While this chapter cannot serve as an exhaustive encyclopedia of all known algae species and their capabilities, the lesson to be learned here is that the term algae represents an extraordinarily diverse and complex group of typically photosynthetic organisms that include species that can thrive nearly anywhere on the planet. There is still a considerable amount to learn about algae, particularly given the high estimates of species that have yet to be discovered, let alone studied, in order to take advantage of and optimize their productivity for making petroleum replacements.

Diversity of current and potential products

What kinds of natural products do algae make? Algae have been used for food, animal feed, pharmaceuticals, and nutraceuticals across the world since ancient times.[99] Macroalgae continue to play a major role in the diet of many cultures and are used extensively in fish farming. Various green algae, red algae, and brown algae are raised to produce a variety of bioactive compounds, including antiinflammatory agents, nutritional supplements, and phenolic compounds for antiaging creams, sun protection, and hair care.[100–105] Use as a nutritional supplement has been particularly successful, where either the whole cell is consumed, as is the case for species like *Arthrospira platensis* (spirulina), *Chlorella vulgaris*, *Chlamydomonas reinhardtii*, and three others that have received generally regarded as safe (GRAS) status by the US Food and Drug Administration,[106] or species that can convert most of their biomass into a product of interest, as with *Dunaliella salina* which accumulates beta-carotene in its chloroplasts when under stress conditions to a degree that beta-carotene accounts for 95% of the dry mass of harvests[107] (Fig. 2.4). High compound accumulations have also allowed the commercial production of omega-3 fatty acids from the diatom *Phaeodactylum tricornutum*, which accumulates neutral lipids to ~61% of its cell mass, and the green alga *Nannochloropsis oculata*.[108] The diversity of secondary metabolites produced by algae, particularly marine red and brown macroalgae as well as marine cyanobacteria that live symbiotically with corals, is only now being appreciated for their uses as pharmaceuticals that prevent a host of infectious or alleviate noncommunicable diseases, including cancer.[109–111] Although these products and uses are in many cases not direct replacements for petroleum products, they do represent alternatives that can displace their petroleum-derived counterparts in health and beauty markets, while further valorizing algal biomass that is otherwise grown to produce other components designated for direct petroleum replacements.

What petroleum replacements do algae make? In recent decades, algae have become of specific interest for their potential to produce biofuels, including biodiesel, biogas in the form of methane or hydrogen, bioethanol, biobutanol, and jet fuel.[7,41,42,64,105] Strains of interest have varied over time, depending upon their ease of cultivation, nutritional requirements, and production yields of polysaccharides, lipids, and overall composition

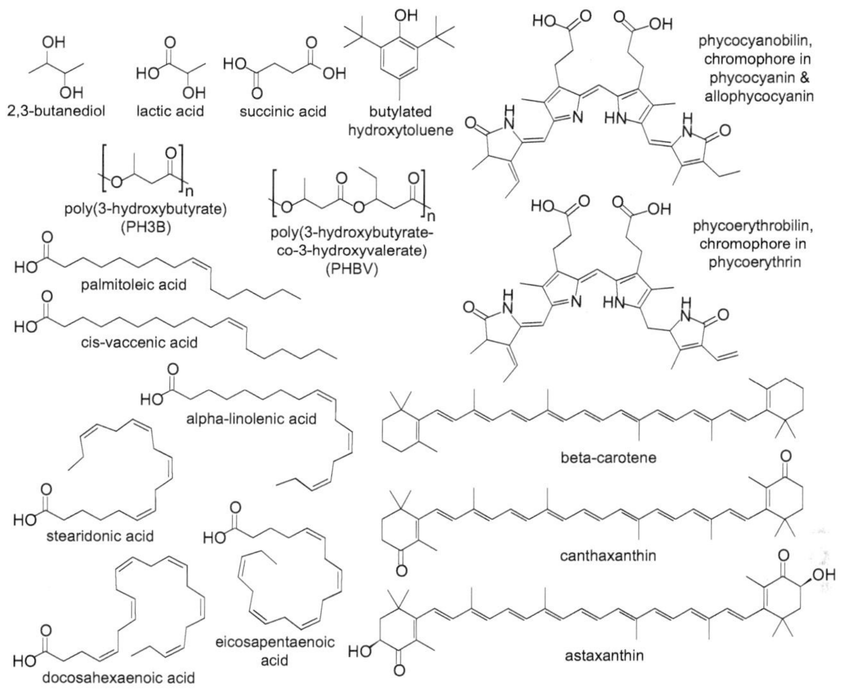

FIG. 2.4 Chemical structures of representative natural algae products of interest to replace petroleum-derived goods.

of carbon and nitrogen.[112] Beyond commodity markets of chemicals that would simply be burned, microalgae are also used to produce direct replacements of molecules otherwise derived from petroleum. One example is butylated hydroxytoluene, a food additive and antioxidant that has been more commonly produced as a petrochemical but can be made directly by the green algae *Botryococcus braunii* and at least three different cyanobacteria—*Oscillatoria sp., Cylindrospermopsis raciborskii,* and *Microcystis aeruginosa.*[113] Additionally, algae pigments are being used as replacements for inks and artificial food dyes, many of which could be used in coloring plastics.[114–117] Examples are canthaxanthin, an orange to dark pink keto-carotenoid derived from green algae like *Chlorella zofingiensis* and *Nannochloropsis gaditana*[118,119]; light harvesting chromophores from the photosystems of cyanobacteria, including the red protein-pigment complex called phycoerythrin and the blue protein-pigment complexes of phycocyanin and allophycocyanin[115]; and astaxanthin, an orange-red pigment that gives salmon meat, crustaceans, and flamingos their pink color when they ingest astaxanthin-producing algae, such as the green alga *Haematococcus.*[82]

What products do algae make that can replace petroleum-derived plastics? As noted previously, a number of algae have been used to make bioplastics or are capable of making bioplastics naturally.[27,30] In several cases, algae such as brown algae or green algae grown on wastewater were used as the raw material for enzymatic or microbial fermentation processes to produce bioplastics like PHB.[120,121] Alternatively, compounds commonly produced by algae, such as proteins, cellulose, or starches, can be plasticized with water or glycerol-based thermal processing to generate bioplastics.[32] Many algae are capable of producing PHAs, which needs only to be extracted from the algae biomass.[122] At least 137 cyanobacteria species that produce PHA have been identified,[123] while only a few green algae, such as *Chlorella minutissima,*[124] or their microbial symbionts, as with the microorganism UMI-21 from the green alga Ulva,[125] have been shown to accumulate PHA. Some of these PHA-producing cyanobacteria make specific types of PHAs that have material uses, including PHB produced by *Synechocystis,* and *Calothrix scytonemicola,*[126–129] poly-β-hydroxybutyrate, also known as poly(3-hydroxybutyrate) (P3HB), produced by *Synechocystis salina, Nostoc muscorum, and Arthrospira platensis,*[32,130,131] and at least 23 different species that produce poly(3-hydroxybutyrate-co-3-hydroxyvalerate) (PHBV),[132] including *Aulosira fertilissima,* which accumulate PHB or PHBV, depending upon the specific strain investigating, to 85% and 77% of its dry cell weight, respectively.[133]

Are algae able to naturally produce precursor chemicals for the manufacture of polyester PUs? Recent advances have demonstrated that polyester polyols can be derived from the ozonolysis of purified algal unsaturated fatty acids, as in the conversion of palmitoleic acid to azelaic acid, which can then be esterified into a polyol monomer for PU production.[134] Many green algae and red algae, such as the haptophytes, have membranes rich in mono- and poly-unsaturated fatty acids, including cis-vaccenic acid that is produced in *Schizochytrium* and a variety of omega-3 fatty acids that can be analogously converted into diacids for polyol production, including docosahexaenoic acid, eicosapentaenoic acid, stearidonic acid, and α-linolenic acid.[105,135,136] Alternatively, polyester polyols can be generated through the reaction of diacids with diols. A few *Nannochloropsis* species, such as *N. gaditana*, *N. oculata*, and *N. salina*, naturally produce long-chain (C29 to C32) diols.[137,138] The marine green alga *Chlamydomonas perigranulata* is capable of producing the short-chain diol 2,3-butanediol.[139] All algae naturally produce the small diacid succinate as part of their central metabolism, specifically in the tricarboxylic acid (TCA) cycle. The marine cyanobacterium *Synechococcus* sp. PCC 7002 naturally secretes succinate, along with lactate, acetate, alanine, and hydrogen, into its environment upon dark anoxic incubation.[140] Similar dark anaerobic conditions lead to secretion of succinate by the cyanobacterium *Synechocystis* sp. PCC 6803, whose yields have been further increased to 0.4 g/L with a combination of higher temperatures and bicarbonate addition.[127,141] Because of recent developments in flow chemistry methods, a diversity of diacids can be chemically converted into diisocyanates, the chemicals that crosslink polyols into PUs.[142] Thus, while some of these chemicals are produced by only a handful of known species, the basic chemicals required for PU production can be generated from algae through photosynthesis.

Algae as an engineered source of industrial production

There are many reasons for which the diversity of natural species and their chemical products may not be sufficient for industrial production of petroleum and plastic replacements. The first is that few if any organisms have evolved to focus the majority of their carbon resources on production of chemicals that humans deem interesting or desirable, especially for applications generating synthetic polymers not found in nature. Second, even if such a species exists, it may not grow well under the required conditions at the desired production location. Finally, the desired chemical may not be produced by any known algae, even if biosynthetic pathways are known for their production in other organisms, as is the case for the unusual branched long-chain diacid diabolic acid, a potential diacid polyol precursor, produced in the extremophile bacteria *Thermotoga*.[143] If algae are not capable of making a desired molecule, or the ones that do are not compatible with the growth environment or do not produce it at economically viable yields, then what options exist to obtain that chemical in a sustainable manner with algae?

The answer can be summed up simply: genetic engineering. An ever-growing number of species of algae have been genetically altered,[144] including ones never previously cultured in the laboratory[145] and ones that have been notoriously difficult to alter, as in the case of *Arthrospira platensis*.[146] Recent advances in synthetic biology have generated a large library of tools that allow genetic engineering to introduce genes that enable heterologous expression of foreign proteins and production of novel products.[147,148] With the application of genomic sequencing, metabolic modeling, genome editing tools such as CRISPR, and systems-based approaches, including metabolic engineering, breeding, and large-scale mutagenesis combined with high-throughput screening, significant improvements in yields of heterologous and natural products or the adaptation of a species to a desired cultivation conditions can be achieved.[144,149–151]

Genetically manipulatable species

What species of algae are genetically manipulatable? Traditionally, the answer was relegated to model organisms—a few rare species that had been brought into the lab and are well studied. For cyanobacteria, this list includes one marine species, *Synechococcus* sp. PCC 7002, and a few freshwater species, such as *Synechococcus elongatus* PCC 7942 and UTEX 2973, *Anabaena* sp. PCC 7120, and *Synechocystis* sp. PCC 6803.[150] For green algae, this was predominantly *Chlamydomonas reinhardtii*.[152] However, in recent decades, there has been a revolution in genetic engineering which has allowed novel species to have DNA introduced into them, a process referred to as transformation, and to have that DNA remain inside of the cell as either a self-replicating plasmid, a circular piece of DNA, or through a process called recombination in which the cell inserts the DNA into its own.[152,153] Introduction of DNA can occur through a number of different mechanisms, in part dependent upon the characteristics of the species to be transformed. For example, many cyanobacteria are naturally competent, meaning that they naturally take up DNA from their environment.[152,154] This fact has allowed genetic engineering of several newly isolated strains,[34,150,155–157] and even the introduction of plasmids into cyanobacteria that were not previously isolated from their environment except during the course of the transformation procedure.[145] Alternatively, cyanobacteria can be transformed through mechanical means, as with electroporation, or through conjugation, a mechanism of horizontal gene transfer in which a conjugal strain, typically *E. coli*, transfers DNA through contact via a structure called a pilus, independent of the competence of the receiving strain.[145,158,159]

For eukaryotic microalgae, various methods are available for introduction of DNA, including bacterial conjugation, Agrobacterium-mediated gene transfer, electroporation; chemical-mediated methods including lipofection and PEG-mediated transformation; and physical methods including glass bead abrasion, microinjection, and biolistic microparticle bombardment.[144,160–163] Eukaryotic algal species that have been transformed, either transiently or stably in either the nucleus or the chloroplast, now include representatives from the green, red, and brown algal clades, including multiple species of *Chlorella, Haematococcus pluvialis, Ostreococcus, Picochlorum celeri, Nannochloropsis oceanica, Porphyridium, Cyanidioschyzon merolae, Gracilaria changii, Laminaria japonica*; diatoms such as *Phaeodactylum tricornutum*; and euglenids such as *Euglena gracilis*.[144,148,151,152,163–169]

Genetic tools

What types of genetic tools are used to alter algae for production of relevant chemicals? To make a new product, heterologous expression of a protein of interest requires the stable presence of the DNA encoding the foreign protein and regulatory elements, such as promoters and ribosome binding sites, that the host cell can recognize to transcribe the DNA code into RNA and translate the RNA into protein.[150] Foreign DNA can be maintained in a host cell through either its presence on a self-replicating plasmid or integration into the host's genome, so that when the cell divides, the foreign DNA will get replicated and passed on to both daughter cells.[147] For certain genomes, including those of many cyanobacteria, chloroplasts, and some eukaryotic nuclei such as *Nannochloropsis*,[170] native homologous recombination complexes allow the DNA to be integrated into a predetermined location if the introduced DNA is flanked by the sequence of the upstream and downstream sequences of the integration site.[153] In other cases, as with the nuclear genome of *Chlamydomonas reinhardtii*,[148] integration occurs in random locations using nonhomologous end-joining mechanisms.[171] Because random integration results in an inhomogeneous population of cells, where some integration sites express more poorly than others due to chromosomal compaction or epigenetic repression, cells must be screened or selected for the desired phenotype or production capabilities.[172,173] A third alternative, developed in recent years for targeted integration into nuclei, chloroplasts, and cyanobacteria for an ever-growing number of algae hosts, is CRISPR-based genome editing, which can either specify the insertion location or aid in selecting for its successful integration.[151,174]

Because the maintenance of foreign DNA can be an energetic burden, result in the production of a protein or chemical that damages the cell, or be recognized as foreign by host defense mechanisms, transgene stability is typically maintained through the use of a selective marker.[147,150,175,176] Most often, selection is for the presence of a coencoded antibiotic resistance gene that allows the transformed cell, but not the wild-type host cell, to survive in the presence of the antibiotic of choice, for which many options exist for both prokaryotic and eukaryotic hosts.[147,150,175,176] Alternatives to antibiotic selection include selection based on desired growth conditions, for example tolerance to high light and high temperature[177] or auxotrophic growth on alternative nutrients, such as melamine and phosphite.[178,179]

What is required to express a new pathway or gene(s) of interest (GOI)? First, the encoding of the protein coding sequence itself must be compatible with the host and must be prefaced by regulatory elements that allow the host cell to express the protein. Depending upon the algal host, the GOI may require codon optimization to best match the gene sequence's codon usage to that of the host cell's codon usage and/or guanine-cytosine content.[144,147] For eukaryotic systems, the inclusion of introns must also be considered to optimize expression.[180] The choice of promoter, operators, ribosome binding sites, riboswitches, terminators, and the relative distances between these elements and the GOI's start or stop codons can all affect expression levels or the condition under which the protein is expressed.[147,175,176] Depending upon the chosen promoter, expression can be constitutive, meaning always turned on, or induced by an external stimulus, typically a chemical such as IPTG (isopropylthio-β-galactoside), metals, or nutrients, as with the nitrate-responsive promoter developed for *Nannochloropsis gaditana.*[181] Alternatively, light-induced or wavelength-specific responsive promoter systems have been developed,[182,183] either of which can be particularly useful for outdoor production where the diel cycle of day and night directly affects the production state of photosynthetic algal systems.[144,184]

To rapidly generate vectors to encode these expression systems, decades of research have led to synthetic biology tool kits and DNA assembly methods that allow the engineering of a diversity of algal production species. Standardized modular assembly toolkits comprising the necessary components to clone and express novel proteins or to otherwise alter the genome have been developed, including Cyano-Vector and various cyanobacteria and green algae toolkits based on the MoClo or BioBricks standards.[147,175,176,185–189] In some cases, these toolkits can assemble vectors that can be used with multiple species, including nonalgal species such as *E. coli.*[176,183] These toolkits include reporter proteins, such as Green Fluorescent Protein, to allow the visualization, localization, and quantitation of protein expression, as well as high-throughput screening for high expressing clones.[190,191]

To improve the productivity of a strain in producing either a native or a heterologous product, the expression of native genes can be upregulated, downregulated, or abolished using these toolkits or other synthetic biology tools, including CRISPR-based gene editing or expression interference systems such as CRISPRi.[151,192,193] This type of metabolic engineering for enhancements in biomass or product yield is typically driven by hypotheses resulting from analysis of physiologically relevant datasets, including comparative genomics, transcriptomics, proteomics, and metabolomics, all of which can feed into a metabolic model to predict alterations to the host that improve its ability to grow and produce the molecule of interest under the desired conditions.[175,194–198] This kind of metabolic engineering, and some of the genetic engineering described before, critically relies on the recent explosion in available algal genome sequences, which now total over 130 cyanobacterial and 220 eukaryotic algal genome assemblies.[199–201] In spite of all of this knowledge, however, there are still many unknowns that can affect productivity and cannot be captured by these models without further basic research, particularly when attempting to engineer a recently isolated or characterized strain. To overcome this deficit, strain improvements can be made through systematic, discovery-based methods of generating diversity, combined with high-throughput screening or selection, a process otherwise known as directed or adaptive laboratory evolution.[202] Diversity can be generated through mutagenic methods, such as UV, chemical-induced, or barcoded or nonbarcoded transposon insertion mutagenesis, or breeding techniques for species that are capable of sexual reproduction that produce combinatorial diversity through genome shuffling.[153,202–205] Combining mutagenesis, breeding, and high-throughput screening of microalgae has proven to rapidly and significantly increase productivity, as compared to when these individual methods are used separately.[204]

Engineered products from algae

What products have been generated using these genetic and metabolic engineering techniques in algae? Both algae and cyanobacteria have been genetically engineered to make novel proteins, including vaccine antigens, and to increase production of native and nonnative pigments, terpenes, vitamins, and accumulated and secreted sugars.[114,144,146,187,206–208] In the realm of bioplastics, algae that do not naturally produce bioplastics have been engineered with pathways from bacteria to produce bioplastics such as PHB.[209] In the case of the diatom *Phaeodactylum tricornutum*, expression of the heterologous PHB synthesis pathway led to accumulation of PHB up to 10.6% of algal dry weight.[210] Cyanobacteria that do naturally produce PHB have been metabolically engineered to improve production yields. *Synechocystis* sp. PCC 6803, whose natural accumulation of PHB is on the order of 10%–20% of the dry cell weight, that has had its regulatory protein PirC genetically removed and is grown in nitrogen and phosphorus depleted medium with acetate added, accumulates PHB up to 81% of its dry cell weight.[211]

More closely related to PU precursor chemical production, both eukaryotic algae and cyanobacteria have been widely used to improve production of lipids, diacids, and diols.[212–214] For eukaryotic algae, including green algae and diatoms, a primary target of genetic engineering efforts has been in yield and content improvements of fatty acids and their energy-rich storage molecule, triacylglycerol (TAG).[144] For example, starch-deficient strains of *Chlamydomonas* and *Chlorella* show increased production of TAG and polyunsaturated fatty acid content, while deletion of a phospholipase in *Chlamydomonas* results in a 64% increase in lipid accumulation.[144,212] Alternatively, overexpression of different transcription factors or fatty acid synthesis genes results in changes in chain lengths of fatty acids or enhanced accumulation or secretion of lipids in green algae and diatoms.[212] Although most of these improvements were performed in the context of producing biofuels, they can improve algae strains for the production of fatty acids that can be precursors for PU materials.

Cyanobacteria in particular have become a well-developed system for genetic engineering of cell factories for chemical production.[213,214] A short list of the chemicals that cyanobacteria have been engineered to produce includes sucrose, ethanol, ethylene, terpenoids, 3-hydroxybutyrate, isoprene, isopropanol, limonene, isobutyraldehyde, and hydrogen gas. Of particular interest for PU production are the novel production of

2,3-butanediol, 1,2-propanediol, saturated and unsaturated free fatty acids, and the natively produced succinate.[213,214] Many strategies have been implemented that each provide incremental increases in production of these chemicals, including rational manipulations to enhance the efficiency of carbon fixation, metabolic engineering of carbon flux through central metabolism, secretion of products to create energy sinks that drive photosynthesis more efficiently, mixotrophic production by enabling cyanobacteria to take up endogenously added sugars, and implementation of production pathways in novel fast growing strains.[175,215,216] As an example of the improvements this generates, overexpression of the *pepC* gene involved in carbon flux toward the TCA cycle based on metabolomics data and metabolic modeling of *Synechocystis* sp. PCC 6803 increased the photosynthetically produced titer of succinate from 0.4 g/L in the nonengineered strain to 1.8 g/L.[141] Alternatively, moving succinate production from a model cyanobacterial strain (PCC 7942) to a newly isolated fast-growing strain (PCC 11801) doubled the titer of succinate secreted from the cells after 5 days of production, and further metabolic engineering resulted in another 50% increase in titer above that.[197]

Summary: Algae at agricultural and industrial scales

The promise of algae: Scalability and sustainability

The proliferation and diversification of algae during much of the geological record not only converted the ancient CO_2-abundant atmosphere into biomass that ultimately became the oil reserves from which we get all our petroleum, but also generated the vast diversity in morphology, metabolisms, behaviors, and environmental tolerances that the polyphyletic term algae now encompasses. Why is this diversity so important for the production of petroleum replacements and polyester PUs in particular?

To produce commodity and high-value products at industrial scale through agricultural means that do not compete with food products, algae must be able to grow rapidly in nonpotable water on nonarable land in environmental conditions across the world that will vary by location, season, and even time of day. Eons of algal evolution have resulted in a vast collection of species and metabolisms that can thrive under nearly any desired production condition, including locations with extreme or fluctuating temperatures, humidities, salinities, acidities or alkalinities, and light intensities. The proliferation of harmful microalgal blooms at satellite-visible scales exceeding 13,000 km^2 accentuates the fact that species exist that, under the right conditions, can proliferate to the necessary scales for industrial production, which are estimated at 5000–12,500 km^2 to meet annual plastic production demands. In addition, evolution has resulted in algal species that naturally generate, accumulate, or secrete products that are useful to humanity, including bioplastics and PU precursors. Because a large portion of algae species have yet to be characterized, biodiscovery and screening of novel species continues to be a critical set of tools for determining if the perfect species that does it all already exists—grows rapidly under the desired conditions and efficiently produces high yields of the desired product.

Even if such a species cannot be found, the promise of genetic and metabolic engineering is that a species can be rapidly designed, altered, or bred that can accomplish these goals. With the development of novel genetic tool sets that can be used across species, this potential can be extended to newly identified species that may grow or produce better in a specific environment than other strains. The combination of genome editing technology with those species that can undergo sexual breeding and selection for desired traits has the potential to rapidly generate ideal production strains in a time frame relevant to our current climate crisis, as opposed to the thousands of years it took for traditional agricultural breeding of vegetable crops to accomplish similar goals of altering species, such as the wild mustard plant *Brassica oleracea*, to generate a diversity of variant strains (broccoli, cauliflower, cabbage, kale, etc.) that can be resiliently and robustly grown in varying environments.[217]

What are the current practices for agricultural scale growth and harvesting of algae? In many cases, natural blooms of microalgae and macroalgae are harvested or farmed from both freshwater and marine systems, as is the case for traditional seaweed farming practices and the modern aquaculture industries that produce edible algae products.[60,218] Industrial methods that include chemical or electroflocculation, filtration, and sedimentation have been developed to harvest microalgae from naturally infested waterways for material and fuel uses, though it is unclear if the products from such harvests include toxic species or harmful chemicals and whether their chemical composition is consistent enough to be of high value for replacement of petroleum products.[219,220] For purposefully growing specific species of aquatic microalgae for industrial or agricultural purposes, several growth systems have been developed, which include open or enclosed ponds, tubular or flat panel photobioreactors, hanging bags, revolving algal biofilm systems, and fermentation reactors.[43,221–223] While fermenters and photobioreactors allow high levels of control over conditions such as pH, temperature, and light to optimize productivity, these systems require large investments to build, clean, and maintain in appropriate conditions such as temperature range in outdoor, exposed environments, ultimately making them difficult to commercialize.[43,224,225] In contrast, while open pond systems have less control over environmental conditions and can suffer more from evaporative loss and contamination, their economics improve with larger scales.[226] The most prevalent open pond design, the shallow outdoor raceway pond driven by a paddle wheel, derives from designs used for wastewater treatment and became popular due to its international use in the scaled growth and commercialization of spirulina.[225,227,228] In addition to environmental fluctuations, other considerations that affect the productivity and cost of these large-scale growth systems include the choice of media, which in part depends upon the culture strain but can include municipal and agricultural wastewater, brackish water, fresh water, ocean water, fertilizer-rich media, defined media, and recycled media, and flow rates that allow proper mixing to avoid settling or biofouling and to ensure aeration and gas exchange.[43,105,225,226] Independent of the growth system, algal biomass must be harvested, often using similar methods described for harvesting microalgae from natural environments, and processed for the products of interest. Ideally, this processing generates numerous product streams from the lipids, sugars, proteins, and high-value coproducts derived from the algal cells in order to fully valorize the biomass.[21,229]

Future challenges

To achieve robust and market-competitive scales and production yields, there are still technical challenges to overcome.[226] They include strain, cultivation, and harvesting optimization, all of which will be affected by continued bioprospecting, genetic engineering, and breeding of production strains. However, whether legal entities will allow the growth of genetically modified organisms at commercially viable scale is still an open question. Although GMO food crops are grown worldwide,[230] the evolutionary diversity that allows algae to grow rapidly at large scale under many environmental conditions raises concerns that GMO algae could become invasive species and that design controls must be developed to prevent spread into natural environments. Though few studies have investigated whether GMO species can become invasive and persist in the natural ecosystems, recent experiments that gained US Environmental Protection Agency approval for outdoor trials through a Toxic Substances Control Act Environmental Release Application demonstrated that while GMO strains could spread to new water bodies, both the wild-type and GMO strain failed to dominate these ecosystems or displace their native diversity.[231]

The fact that the GMO strain could not outcompete the native species highlights another consequence of the evolved diversity of algae: competitors and contaminants in algal biomass growth systems pose a large challenge to the robustness of these production cultures. Although more typically seen as a threat, native algae species that outcompete the desired strain can represent an opportunity for improvements, especially in the context of genetic engineering tools that enable alteration of these native strains as new production strains that are better adapted to the culture location and conditions. That said, contamination by predators or infectious species, such as zooplanktonic ciliates, rotifers, bacteria, and fungi, that can wipe out entire acres of algae crops in a few days, does represent a major challenge to robust production schedules.[98,232] While contamination more frequently affects open pond systems, even closed photobioreactors are susceptible to infestation and infection.[225] Preventative measures against contamination include new technologies to monitor for deleterious species[233,234]; polycultures where biological diversity is chosen to prevent contamination[235]; biological and chemical controls that attack potential predators; genetic engineering to use alternative nutrients that contaminants cannot, such as phosphite[179,236]; and media optimizations to prevent the survival of undesired organisms without killing the algae crop, as occurs naturally when spirulina cultures raise the pH of their culture media to alkaline levels.[234,237]

In spite of these challenges, algae are the clear best hope for sustainable and renewable production of petroleum replacements, including bioplastics and the precursor chemicals needed to generate bio-based polyester PUs.

Close-up: Algae for the future

Algae are incredibly diverse, with members across both the eukaryotic and prokaryotic phylogenetic trees. With this great species diversity, there is diversity in genetics, in biochemistry and photochemistry, in multicellularity, in organelle structure and organization, in cultivation and environmental resilience, and much more. In addition to genetic variation, diverse algal genera are characterized by phenotypic differences such as size, shape, and cell structure; pigments involved in light capture and conversion; molecules utilized for carbon storage; and biochemical pathways for nutrient utilization. This wealth of chemistry that varies across species is a treasure trove of valuable bioproducts. Algae are grown industrially for the production of novel lipids and dietary supplements such as eicosapentaenoic acid (an omega-3 fatty acid) and astaxanthin; colorful pigments such as phycocyanin; whole foods such as spirulina; animal feed; and biofuels. Scientists are only beginning to fully characterize, engineer, and exploit algae, with tool development advances varying greatly between species. Safe and secure genetic engineering tools, or tools that modify algae without escaping or transferring to other organisms in the environment, are in their infancy, with CRISPR-Cas toolboxes and gene editing technologies available for only a handful of algal species.

While algae lipid polymers have long been cultivated for the production of biofuels, the isolation of petroleum polymer replacements from algae has only recently become commercial, with the production of polyester PUs from algae in everything from surfboards to flip-flops to skis. To date, these polymers have been harvested from whole wild-type (nongenetically modified) algae cultivated as single species or collected as part of mixed algal biomass from waterways. As more safe and secure genetic engineering tools are developed for algae species of interest in the production of petroleum polymers, there's the potential for a revolution in greener plastic production.

As one of the originators of petroleum, algae have always been the precursors of petroleum products. By tapping into how to develop greener plastics from growing algae, it could be said that scientists are contributing to a cleaner planet by bypassing a few hundred million years of fossilization.

Amanda N. Barry
Sandia National Laboratories, Albuquerque, NM, United States

References

1. Baekeland LH. Process of Making Condensation Products of phenols And Formaldehyde; 1916. Published online June 13 https://patents.google.com/patent/US1187229A/en?q=Baekeland+bakelite&before=priority:19100101&after=priority:19000101.
2. Semon WL. Synthetic Rubber-Like Composition and Method of Making Same; 1933. Published online October 10 https://patents.google.com/patent/US1929453A/en?oq=US1929453A.
3. Smith PH. Plastics come of age. *Sci Am.* 1935;152(1):5–7.
4. Geyer R. A brief history of plastics. In: Streit-Bianchi M, Cimadevila M, Trettnak W, eds. *Mare Plasticum—The Plastic Sea: Combatting Plastic Pollution Through Science and Art.* Springer International Publishing; 2020:31–47. https://doi.org/10.1007/978-3-030-38945-1_2.
5. Molchanov M. Petroleum geopolitics. In: *Encycl Glob Stud*; 2012:1317–1320. Published online March.
6. Dinosaur Fever - Sinclair's Icon. American Oil & Gas Historical Society; 2021. Published April 25 https://aoghs.org/oil-almanac/sinclair-dinosaur/. Accessed 5 April 2022.
7. Nersesian RL. Energy for the 21st Century: A Comprehensive Guide to Conventional and Alternative Sources. p; 2015.
8. Demoulin CF, Lara YJ, Cornet L, et al. Cyanobacteria evolution: insight from the fossil record. *Free Radic Biol Med.* 2019;140:206–223. https://doi.org/10.1016/j.freeradbiomed.2019.05.007.
9. Catling DC, Zahnle KJ. The Archean atmosphere. *Sci Adv.* 2020;6(9), eaax1420. https://doi.org/10.1126/sciadv.aax1420.
10. Holland HD. The oxygenation of the atmosphere and oceans. *Philos Trans R Soc B Biol Sci.* 2006;361(1470):903–915. https://doi.org/10.1098/rstb.2006.1838.
11. Lyons TW, Reinhard CT, Planavsky NJ. The rise of oxygen in Earth's early ocean and atmosphere. *Nature.* 2014;506(7488):307–315. https://doi.org/10.1038/nature13068.
12. Junge W. Oxygenic photosynthesis: history, status and perspective. *Q Rev Biophys.* 2019;52, e1. https://doi.org/10.1017/S0033583518000112.
13. Meikle JL. American Plastic: A Cultural History. Rutgers University Press; 1995.
14. Simon E. Ueber den flüssigen Storax (Styrax liquidus). *Ann Pharm.* 1839;31(3):265–277.

15. Painter PC, Coleman MM. Essentials of Polymer Science and Engineering. DEStech Publications; 2009.
16. Drumright RE, Gruber PR, Henton DE. Polylactic acid technology. *Adv Mater.* 2000;12 (23):1841–1846. https://doi.org/10.1002/1521-4095(200012)12:23<1841::AID-ADMA1841>3.0.CO;2-E.
17. Lemoigne M. Products of dehydration and of polymerization of β-hydroxybutyric acid. *Bull Soc Chem Biol.* 1926;8:770–782.
18. Shafqat A, Al-Zaqri N, Tahir A, Alsalme A. Synthesis and characterization of starch based bioplatics using varying plant-based ingredients, plasticizers and natural fillers. *Saudi J Biol Sci.* 2021;28(3):1739–1749. https://doi.org/10.1016/j.sjbs.2020.12.015.
19. Mekonnen T, Mussone P, Khalil H, Bressler D. Progress in bio-based plastics and plasticizing modifications. *J Mater Chem A.* 2013;1(43):13379–13398. https://doi.org/10.1039/C3TA12555F.
20. Reddy MM, Vivekanandhan S, Misra M, Bhatia SK, Mohanty AK. Biobased plastics and bionanocomposites: current status and future opportunities. *Prog Polym Sci.* 2013;38 (10):1653–1689. https://doi.org/10.1016/j.progpolymsci.2013.05.006.
21. Sardon H, Mecerreyes D, Basterretxea A, Avérous L, Jehanno C. From lab to market: current strategies for the production of biobased polyols. *ACS Sustain Chem Eng.* 2021;9(32):10664–10677. https://doi.org/10.1021/acssuschemeng.1c02361.
22. Raschka A, Carus M, Piotrowski S. Renewable raw materials and feedstock for bioplastics. In: Kabasci S, ed. *Bio-Based Plastics.* John Wiley & Sons Ltd; 2013:331–345. https://doi.org/10.1002/9781118676646.ch13.
23. Walker S, Rothman R. Life cycle assessment of bio-based and fossil-based plastic: a review. *J Clean Prod.* 2020;261:121158. https://doi.org/10.1016/j.jclepro.2020.121158.
24. Machmud MN, Fahmi R, Abdullah R, Kokarkin C. Characteristics of red algae bioplastics/latex blends under tension. *Int J Sci Eng.* 2013;5(2):81–88. https://doi.org/10.12777/ijse.5.2.81-88.
25. Ready to grow: the biodegradable polymers market. *Plast Eng.* 2016;72(3):1–4. https://doi.org/10.1002/j.1941-9635.2016.tb01489.x.
26. Künkel A, Becker J, Börger L, et al. Polymers, biodegradable. In: Wiley-VCH Verlag GmbH & Co. KGaA, ed. *Ullmann's Encyclopedia of Industrial Chemistry.* Wiley-VCH Verlag GmbH & Co. KGaA; 2016:1–29. https://doi.org/10.1002/14356007.n21_n01.pub2.
27. Rahman A, Miller CD. Chapter 6—microalgae as a source of bioplastics. In: Rastogi RP, Madamwar D, Pandey A, eds. *Algal Green Chemistry.* Elsevier; 2017:121–138. https://doi.org/10.1016/B978-0-444-63784-0.00006-0.
28. Pearlmutter NL, Lembi CA. Localization of chitin in algal and fungal cell walls by light and electron microscopy. *J Histochem Cytochem Off J Histochem Soc.* 1978;26(10):782–791. https://doi.org/10.1177/26.10.722047.
29. Samiee S, Ahmadzadeh H, Hosseini M, Lyon S. Chapter 17—algae as a source of microcrystalline cellulose. In: Hosseini M, ed. *Advanced Bioprocessing for Alternative Fuels, Biobased Chemicals, and Bioproducts.* Woodhead Publishing; 2019:331–350. Woodhead Publishing Series in Energy; https://doi.org/10.1016/B978-0-12-817941-3.00017-6.
30. Chia WY, Ying Tang DY, Khoo KS, Kay Lup AN, Chew KW. Nature's fight against plastic pollution: algae for plastic biodegradation and bioplastics production. *Environ Sci Ecotechnol.* 2020;4:100065. https://doi.org/10.1016/j.ese.2020.100065.
31. Karan H, Funk C, Grabert M, Oey M, Hankamer B. Green bioplastics as part of a circular bioeconomy. *Trends Plant Sci.* 2019;24(3):237–249. https://doi.org/10.1016/j.tplants.2018.11.010.
32. Dang BT, Bui XT, Tran DPH, et al. Current application of algae derivatives for bioplastic production: a review. *Bioresour Technol.* 2022;347:126698. https://doi.org/10.1016/j.biortech.2022.126698.
33. Sathish A, Glaittli K, Sims RC, Miller CD. Algae biomass based media for Poly(3-hydroxybutyrate) (PHB) production by Escherichia coli. *J Polym Environ.* 2014;22 (2):272–277. https://doi.org/10.1007/s10924-014-0647-x.

34. Włodarczyk A, Selão TT, Norling B, Nixon PJ. Newly discovered Synechococcus sp. PCC 11901 is a robust cyanobacterial strain for high biomass production. *Commun Biol.* 2020;3(1):1–14. https://doi.org/10.1038/s42003-020-0910-8.
35. Griffith AW, Gobler CJ. Harmful algal blooms: a climate change co-stressor in marine and freshwater ecosystems. *Harmful Algae.* 2020;91:101590. https://doi.org/10.1016/j.hal.2019.03.008.
36. Lin J, Miller PI, Jönsson BF, Bedington M. Early warning of harmful algal bloom risk using satellite ocean color and Lagrangian particle trajectories. *Front Mar Sci.* 2021;8:736262. https://doi.org/10.3389/fmars.2021.736262.
37. Michalak AM, Anderson EJ, Beletsky D, et al. Record-setting algal bloom in Lake Erie caused by agricultural and meteorological trends consistent with expected future conditions. *Proc Natl Acad Sci.* 2013;110(16):6448–6452. https://doi.org/10.1073/pnas.1216006110.
38. Tsai DDW, Chen PH, Ramaraj R. The potential of carbon dioxide capture and sequestration with algae. *Ecol Eng.* 2017;98:17–23. https://doi.org/10.1016/j.ecoleng.2016.10.049.
39. Moroney JV, Somanchi A. How do algae concentrate CO2 to increase the efficiency of photosynthetic carbon fixation? *Plant Physiol.* 1999;119(1):9–16. https://doi.org/10.1104/pp.119.1.9.
40. Creed JC, Vieira VMNCS, Norton TA, Caetano D. A meta-analysis shows that seaweeds surpass plants, setting life-on-Earth's limit for biomass packing. *BMC Ecol.* 2019;19(1):6. https://doi.org/10.1186/s12898-019-0218-z.
41. Hannon M, Gimpel J, Tran M, Rasala B, Mayfield S. Biofuels from algae: challenges and potential. *Biofuels.* 2010;1(5):763–784.
42. Ullah K, Ahmad M, Sofia, et al. Algal biomass as a global source of transport fuels: overview and development perspectives. *Prog Nat Sci: Mater Int.* 2014;24(4):329–339. https://doi.org/10.1016/j.pnsc.2014.06.008.
43. Georgianna DR, Mayfield SP. Exploiting diversity and synthetic biology for the production of algal biofuels. *Nature.* 2012;488(7411):329–335. https://doi.org/10.1038/nature11479.
44. Grigoriev IV, Hayes RD, Calhoun S, et al. PhycoCosm, a comparative algal genomics resource. *Nucleic Acids Res.* 2021;49(D1):D1004–D1011. https://doi.org/10.1093/nar/gkaa898.
45. Baumgartner RJ, Van Kranendonk MJ, Wacey D, et al. Nano−porous pyrite and organic matter in 3.5-billion-year-old stromatolites record primordial life. *Geology.* 2019;47(11):1039–1043. https://doi.org/10.1130/G46365.1.
46. Hohmann-Marriott MF, Blankenship RE. Evolution of photosynthesis. *Annu Rev Plant Biol.* 2011;62(1):515–548. https://doi.org/10.1146/annurev-arplant-042110-103811.
47. De Clerck O, Bogaert KA, Leliaert F. Chapter two—diversity and evolution of algae: primary endosymbiosis. In: Piganeau G, ed. *Genomic Insights into the Biology of Algae.* Academic Press; 2012:55–86. Advances in Botanical Research; Vol 64. https://doi.org/10.1016/B978-0-12-391499-6.00002-5.
48. Keeling PJ. The endosymbiotic origin, diversification and fate of plastids. *Philos Trans R Soc B Biol Sci.* 2010;365(1541):729–748. https://doi.org/10.1098/rstb.2009.0103.
49. Sagan L. On the origin of mitosing cells. *J Theor Biol.* 1967;14(3):225–IN6. https://doi.org/10.1016/0022-5193(67)90079-3.
50. Moore KR, Magnabosco C, Momper L, Gold DA, Bosak T, Fournier GP. An expanded ribosomal phylogeny of cyanobacteria supports a deep placement of plastids. *Front Microbiol.* 2019;10:1612. https://doi.org/10.3389/fmicb.2019.01612.
51. Schwartz RM, Dayhoff MO. Origins of prokaryotes, eukaryotes, mitochondria, and chloroplasts: a perspective is derived from protein and nucleic acid sequence data. *Science.* 1978;199(4327):395–403. https://doi.org/10.1126/science.202030.

52. Rodríguez-Ezpeleta N, Brinkmann H, Burey SC, et al. Monophyly of primary photosynthetic eukaryotes: green plants, red algae, and glaucophytes. *Curr Biol.* 2005;15 (14):1325–1330. https://doi.org/10.1016/j.cub.2005.06.040.
53. Jeon KW. Bacterial endosymbiosis in amoebae. *Trends Cell Biol.* 1995;5(3):137–140. https://doi.org/10.1016/S0962-8924(00)88966-7.
54. Brock DA, Douglas TE, Queller DC, Strassmann JE. Primitive agriculture in a social amoeba. *Nature.* 2011;469(7330):393–396. https://doi.org/10.1038/nature09668.
55. Strother PK, Foster C. A fossil record of land plant origins from charophyte algae. *Science.* 2021;373(6556):792–796. https://doi.org/10.1126/science.abj2927.
56. Hopes A, Mock T. Evolution of microalgae and their adaptations in different marine ecosystems. In: John Wiley & Sons, Ltd, ed. *ELS.* 1st ed. Wiley; 2015:1–9. https://doi.org/10.1002/9780470015902.a0023744.
57. Andersen RA. The Microalgal Cell. In: Richmond A, Hu Q, eds. *Handbook of Microalgal Culture.* John Wiley & Sons, Ltd; 2013:1–20. https://doi.org/10.1002/9781118567166.ch1.
58. Anderson DM, Glibert PM, Burkholder JM. Harmful algal blooms and eutrophication: nutrient sources, composition, and consequences. *Estuaries.* 2002;25(4):704–726. https://doi.org/10.1007/BF02804901.
59. Mishra M, Arukha AP, Bashir T, Yadav D, Prasad GBKS. All new faces of diatoms: potential source of nanomaterials and beyond. *Front Microbiol.* 2017;8:1239. https://doi.org/10.3389/fmicb.2017.01239.
60. Abbott IA. Food and food products from seaweeds. In: *Algae Hum Aff Ed Carole Lembi J Robert Waaland Spons Phycol Soc Am Inc;* 1988. Published online https://scholar.google.com/scholar_lookup?title=Food+and+food+products+from+seaweeds&author=Abbott%2C+I.A.&publication_year=1988. Accessed 5 April 2022.
61. Girard V, Saint Martin S, Buffetaut E, et al. Thai amber: insights into early diatom history? In: Saint Martin JP, Saint Martin S, eds. *BSGF—Earth Sci Bull.* Vol. 191; 2020:23. https://doi.org/10.1051/bsgf/2020028.
62. Pachiadaki MG, Brown JM, Brown J, et al. Charting the complexity of the marine microbiome through single-cell genomics. *Cell.* 2019;179(7):1623–1635.e11. https://doi.org/10.1016/j.cell.2019.11.017.
63. Guiry MD. How many species of algae are there? *J Phycol.* 2012;48(5):1057–1063. https://doi.org/10.1111/j.1529-8817.2012.01222.x.
64. Rasala BA, Mayfield SP. Photosynthetic biomanufacturing in green algae; production of recombinant proteins for industrial, nutritional, and medical uses. *Photosynth Res.* 2015;123(3):227–239. https://doi.org/10.1007/s11120-014-9994-7.
65. Cao J, Wang J, Wang S, Xu X. *Porphyra* species: a mini-review of its pharmacological and nutritional properties. *J Med Food.* 2016;19(2):111–119. https://doi.org/10.1089/jmf.2015.3426.
66. Singh SP, Montgomery BL. Determining cell shape: adaptive regulation of cyanobacterial cellular differentiation and morphology. *Trends Microbiol.* 2011;19(6):278–285. https://doi.org/10.1016/j.tim.2011.03.001.
67. Stanier RY, Cohen-Bazire G. PHOTOTROPHIC PROKARYOTES: THE CYANOBACTERIA. *Annu Rev Microbiol.* 1977;31(1):225–274. https://doi.org/10.1146/annurev.mi.31.100177.001301.
68. Johnson MP. Photosynthesis. *Essays Biochem.* 2016;60(3):255–273. https://doi.org/10.1042/EBC20160016.
69. Kerfeld CA, Aussignargues C, Zarzycki J, Cai F, Sutter M. Bacterial microcompartments. *Nat Rev Microbiol.* 2018;16(5):277–290. https://doi.org/10.1038/nrmicro.2018.10.
70. Hayer-Hartl M, Hartl FU. Chaperone machineries of Rubisco—the Most abundant enzyme. *Trends Biochem Sci.* 2020;45(9):748–763. https://doi.org/10.1016/j.tibs.2020.05.001.

71. Gugger MF, Hoffmann L. Polyphyly of true branching cyanobacteria (Stigonematales). *Int J Syst Evol Microbiol.* 2004;54(2):349–357. https://doi.org/10.1099/ijs.0.02744-0.
72. Stal LJ. Nitrogen fixation in cyanobacteria. In: *ELS.* John Wiley & Sons, Ltd; 2015:1–9. https://doi.org/10.1002/9780470015902.a0021159.pub2.
73. Livingstone D, Jaworski GHM. The viability of akinetes of blue-green algae recovered from the sediments of Rostherne Mere. *Br Phycol J.* 1980;15(4):357–364. https://doi.org/10.1080/00071618000650361.
74. Mariscal V, Herrero A, Flores E. Continuous periplasm in a filamentous, heterocyst-forming cyanobacterium. *Mol Microbiol.* 2007;65(4):1139–1145. https://doi.org/10.1111/j.1365-2958.2007.05856.x.
75. Wilk L, Strauss M, Rudolf M, et al. Outer membrane continuity and septosome formation between vegetative cells in the filaments of Anabaena sp. PCC 7120: intercellular septum structure of Anabaena sp. *Cell Microbiol.* 2011;13(11):1744–1754. https://doi.org/10.1111/j.1462-5822.2011.01655.x.
76. Wu X, Liu D, Lee MH, Golden JW. patS minigenes inhibit heterocyst development of Anabaena sp. strain PCC 7120. *J Bacteriol.* 2004;186(19):6422–6429. https://doi.org/10.1128/JB.186.19.6422-6429.2004.
77. Golden JW, Whorff LL, Wiest DR. Independent regulation of nifHDK operon transcription and DNA rearrangement during heterocyst differentiation in the cyanobacterium Anabaena sp. strain PCC 7120. *J Bacteriol.* 1991;173(22):7098–7105.
78. Golden JW. Programmed DNA rearrangements in cyanobacteria. In: de Bruijn FJ, Lupski JR, Weinstock GM, eds. *Bacterial Genomes.* Springer US; 1998:162–173. https://doi.org/10.1007/978-1-4615-6369-3_16.
79. Toepel J, Welsh E, Summerfield TC, Pakrasi HB, Sherman LA. Differential transcriptional analysis of the cyanobacterium Cyanothece sp. strain ATCC 51142 during light-dark and continuous-light growth. *J Bacteriol.* 2008;190(11):3904–3913. https://doi.org/10.1128/JB.00206-08.
80. Shi T, Ilikchyan I, Rabouille S, Zehr JP. Genome-wide analysis of diel gene expression in the unicellular N2-fixing cyanobacterium Crocosphaera watsonii WH 8501. *ISME J.* 2010;4(5):621–632. https://doi.org/10.1038/ismej.2009.148.
81. Drew KM. Conchocelis-phase in the life-history of Porphyra umbilicalis (L.) Kütz. *Nature.* 1949;164(4174):748–749. https://doi.org/10.1038/164748a0.
82. MdMR S, Liang Y, Cheng JJ, Daroch M. Astaxanthin-producing green microalga *Haematococcus pluvialis*: from single cell to high value commercial products. *Front Plant Sci.* 2016;7:531. https://doi.org/10.3389/fpls.2016.00531.
83. Li X, Wang X, Duan C, et al. Biotechnological production of astaxanthin from the microalga Haematococcus pluvialis. *Biotechnol Adv.* 2020;43:107602. https://doi.org/10.1016/j.biotechadv.2020.107602.
84. Yu ET, Zendejas FJ, Lane PD, Gaucher S, Simmons BA, Lane TW. Triacylglycerol accumulation and profiling in the model diatoms Thalassiosira pseudonana and Phaeodactylum tricornutum (Baccilariophyceae) during starvation. *J Appl Phycol.* 2009;21(6):669. https://doi.org/10.1007/s10811-008-9400-y.
85. Malavasi V, Soru S, Cao G. Extremophile microalgae: the potential for biotechnological application. *J Phycol.* 2020;56(3):559–573. https://doi.org/10.1111/jpy.12965.
86. Spijkerman E. Phosphorus limitation of algae living in iron-rich, acidic lakes. *Aquat Microb Ecol.* 2008;53:201–210. https://doi.org/10.3354/ame01244.
87. Sahu N, Tangutur AD. Airborne algae: overview of the current status and its implications on the environment. *Aerobiologia.* 2015;31(1):89–97. https://doi.org/10.1007/s10453-014-9349-z.
88. Smith DJ, Ravichandar JD, Jain S, et al. Airborne bacteria in earth's lower stratosphere resemble taxa detected in the troposphere: results from a new NASA aircraft bioaerosol collector (ABC). *Front Microbiol.* 2018;9. https://www.frontiersin.org/article/10.3389/fmicb.2018.01752. Accessed 10 April 2022.

89. Holzinger A, Karsten U. Desiccation stress and tolerance in green algae: consequences for ultrastructure, physiological and molecular mechanisms. *Front Plant Sci.* 2013;4:327. https://doi.org/10.3389/fpls.2013.00327.
90. Lewis LA, Flechtner VR. Green algae (Chlorophyta) of desert microbiotic crusts: diversity of north American taxa. *Taxon.* 2002;51(3):443–451. https://doi.org/10.2307/1554857.
91. Treves H, Raanan H, Kedem I, et al. The mechanisms whereby the green alga Chlorella ohadii, isolated from desert soil crust, exhibits unparalleled photodamage resistance. *New Phytol.* 2016;210(4):1229–1243. https://doi.org/10.1111/nph.13870.
92. Sanders WB, Masumoto H. Lichen algae: the photosynthetic partners in lichen symbioses. *Lichenologist.* 2021;53(5):347–393. https://doi.org/10.1017/S0024282921000335.
93. Charpy L, Casareto BE, Langlade MJ, Suzuki Y. Cyanobacteria in coral reef ecosystems: a review. *J Mar Biol.* 2012;2012, e259571. https://doi.org/10.1155/2012/259571.
94. Roth MS. The engine of the reef: photobiology of the coral-algal symbiosis. *Front Microbiol.* 2014;5:422. https://doi.org/10.3389/fmicb.2014.00422.
95. Venn AA, Loram JE, Douglas AE. Photosynthetic symbioses in animals. *J Exp Bot.* 2008;59(5):1069–1080. https://doi.org/10.1093/jxb/erm328.
96. Kerney R, Kim E, Hangarter RP, Heiss AA, Bishop CD, Hall BK. Intracellular invasion of green algae in a salamander host. *Proc Natl Acad Sci U S A.* 2011;108(16):6497–6502. https://doi.org/10.1073/pnas.1018259108.
97. Fountain ED, Pauli JN, Mendoza JE, Carlson J, Peery MZ. Cophylogenetics and biogeography reveal a coevolved relationship between sloths and their symbiont algae. *Mol Phylogenet Evol.* 2017;110:73–80. https://doi.org/10.1016/j.ympev.2017.03.003.
98. Mata TM, Martins AA, Caetano NS. Microalgae for biodiesel production and other applications: a review. *Renew Sust Energ Rev.* 2010;14(1):217–232. https://doi.org/10.1016/j.rser.2009.07.020.
99. Dillehay TD, Ramírez C, Pino M, Collins MB, Rossen J, Pino-Navarro JD. Monte Verde: seaweed, food, medicine, and the peopling of South America. *Science.* 2008;320(5877):784–786. https://doi.org/10.1126/science.1156533.
100. Singh S, Kate BN, Banerjee UC. Bioactive compounds from cyanobacteria and microalgae: an overview. *Crit Rev Biotechnol.* 2005;25(3):73–95. https://doi.org/10.1080/07388550500248498.
101. Kiran BR, Venkata MS. Microalgal cell biofactory-therapeutic, nutraceutical and functional food applications. *Plants Basel Switz.* 2021;10(5):836. https://doi.org/10.3390/plants10050836.
102. Kim SK, Wijesekara I. Marine-derived nutraceuticals: trends and prospects. In: *Marine Nutraceuticals.* CRC Press; 2013.
103. Mahmoud AM, Bin-Jumah M, Abukhalil MH. 9—Antiinflammatory natural products from marine algae. In: Gopi S, Amalraj A, Kunnumakkara A, Thomas S, eds. *Inflammation and Natural Products.* Academic Press; 2021:175–203. https://doi.org/10.1016/B978-0-12-819218-4.00012-2.
104. Dittmann E, Gugger M, Sivonen K, Fewer DP. Natural product biosynthetic diversity and comparative genomics of the cyanobacteria. *Trends Microbiol.* 2015;23(10):642–652. https://doi.org/10.1016/j.tim.2015.07.008.
105. Mathimani T, Pugazhendhi A. Utilization of algae for biofuel, bio-products and bioremediation. *Biocatal Agric Biotechnol.* 2019;17:326–330. https://doi.org/10.1016/j.bcab.2018.12.007.
106. Torres-Tiji Y, Fields FJ, Mayfield SP. Microalgae as a future food source. *Biotechnol Adv.* 2020;41:107536. https://doi.org/10.1016/j.biotechadv.2020.107536.
107. Paniagua-Michel J. Microalgal nutraceuticals. In: *Handbook of Marine Microalgae.* Elsevier; 2015:255–267. https://doi.org/10.1016/B978-0-12-800776-1.00016-9.
108. Song M, Pei H, Hu W, Ma G. Evaluation of the potential of 10 microalgal strains for biodiesel production. *Bioresour Technol.* 2013;141:245–251. https://doi.org/10.1016/j.biortech.2013.02.024.

109. Jones AC, Gu L, Sorrels CM, Sherman DH, Gerwick WH. New tricks from ancient algae: natural products biosynthesis in marine cyanobacteria. *Curr Opin Chem Biol.* 2009;13 (2):216–223. https://doi.org/10.1016/j.cbpa.2009.02.019.
110. Teixeira VL, Lima JCR, Lechuga GC, et al. Natural products from marine red and brown algae against Trypanosoma cruzi. *f.* 2019;29(6):735–738. https://doi.org/10.1016/j.bjp.2019.08.003.
111. Swain SS, Padhy RN, Singh PK. Anticancer compounds from cyanobacterium Lyngbya species: a review. *Antonie Van Leeuwenhoek.* 2015;108(2):223–265. https://doi.org/10.1007/s10482-015-0487-2.
112. Sánchez J, Curt MD, Robert N, Fernández J. Biomass resources. In: *The Role of Bioenergy in the Bioeconomy.* Elsevier; 2019:25–111. https://doi.org/10.1016/B978-0-12-813056-8.00002-9.
113. Babu B, Wu JT. Production of natural butylated hydroxytoluene as an antioxidant by freshwater phytoplankton. *J Phycol.* 2008;44(6):1447–1454. https://doi.org/10.1111/j.1529-8817.2008.00596.x.
114. Sen T, Barrow CJ, Deshmukh SK. Microbial pigments in the food industry—challenges and the Way forward. *Front Nutr.* 2019;6:7. https://doi.org/10.3389/fnut.2019.00007.
115. Rana B, Bhattacharyya M, Patni B, Arya M, Joshi GK. The realm of microbial pigments in the food color market. *Front Sustain Food Syst.* 2021;5. https://www.frontiersin.org/article/10.3389/fsufs.2021.603892. Accessed 10 April 2022.
116. Gonçalves BRP, Machado BAS, Hanna SA, Umsza-Guez MA. Prospective study of microbial colorants under the focus of patent documents. *Recent Pat Biotechnol.* 2020;14(3):184–193. https://doi.org/10.2174/1872208313666191002125035.
117. Nigam PS, Luke JS. Food additives: production of microbial pigments and their antioxidant properties. *Curr Opin Food Sci.* 2016;7:93–100. https://doi.org/10.1016/j.cofs.2016.02.004.
118. Millao S, Uquiche E. Extraction of oil and carotenoids from pelletized microalgae using supercritical carbon dioxide. *J Supercrit Fluids.* 2016;116:223–231. https://doi.org/10.1016/j.supflu.2016.05.049.
119. Hua-Bin L, Fan KW, Chen F. Isolation and purification of canthaxanthin from the microalga Chlorella zofingiensis by high-speed counter-current chromatography. *J Sep Sci.* 2006;29(5):699–703. https://doi.org/10.1002/jssc.200500365.
120. Jeong DW, Hyeon JE, Lee ME, Ko YJ, Kim M, Han SO. Efficient utilization of brown algae for the production of polyhydroxybutyrate (PHB) by using an enzyme complex immobilized on Ralstonia eutropha. *Int J Biol Macromol.* 2021;189:819–825. https://doi.org/10.1016/j.ijbiomac.2021.08.149.
121. Rahman A, Putman RJ, Inan K, et al. Polyhydroxybutyrate production using a wastewater microalgae based media. *Algal Res.* 2015;8:95–98. https://doi.org/10.1016/j.algal.2015.01.009.
122. Hong K, Beld J, Davis TD, Burkart MD, Palenik B. Screening and characterization of polyhydroxyalkanoate granules, and phylogenetic analysis of polyhydroxyalkanoate synthase gene PhaC in cyanobacteria. *J Phycol.* 2021;57(3):754–765. https://doi.org/10.1111/jpy.13123.
123. Asada Y, Miyake M, Miyake J, Kurane R, Tokiwa Y. Photosynthetic accumulation of poly-(hydroxybutyrate) by cyanobacteria—the metabolism and potential for CO2 recycling. *Int J Biol Macromol.* 1999;25(1):37–42. https://doi.org/10.1016/S0141-8130(99)00013-6.
124. Costa SS, Miranda AL, Andrade BB, et al. Influence of nitrogen on growth, biomass composition, production, and properties of polyhydroxyalkanoates (PHAs) by microalgae. *Int J Biol Macromol.* 2018;116:552–562. https://doi.org/10.1016/j.ijbiomac.2018.05.064.
125. Han X, Satoh Y, Kuriki Y, et al. Polyhydroxyalkanoate production by a novel bacterium Massilia sp. UMI-21 isolated from seaweed, and molecular cloning of its

polyhydroxyalkanoate synthase gene. *J Biosci Bioeng.* 2014;118(5):514–519. https://doi.org/10.1016/j.jbiosc.2014.04.022.

126. Ansari S, Fatma T. Cyanobacterial polyhydroxybutyrate (PHB): screening, optimization and characterization. *PLoS One.* 2016;11(6), e0158168. https://doi.org/10.1371/journal.pone.0158168.
127. Osanai T, Oikawa A, Shirai T, et al. Capillary electrophoresis-mass spectrometry reveals the distribution of carbon metabolites during nitrogen starvation in Synechocystis sp. PCC 6803. *Environ Microbiol.* 2014;16(2):512–524. https://doi.org/10.1111/1462-2920.12170.
128. Koch M, Berendzen KW, Forchhammer AK. On the role and production of polyhydroxybutyrate (PHB) in the cyanobacterium Synechocystis sp. PCC 6803. *Life Basel Switz.* 2020;10(4):E47. https://doi.org/10.3390/life10040047.
129. Kaewbai-ngam A, Incharoensakdi A, Monshupanee T. Increased accumulation of polyhydroxybutyrate in divergent cyanobacteria under nutrient-deprived photoautotrophy: an efficient conversion of solar energy and carbon dioxide to polyhydroxybutyrate by Calothrix scytonemicola TISTR 8095. *Bioresour Technol.* 2016;212:342–347. https://doi.org/10.1016/j.biortech.2016.04.035.
130. Bhati R, Mallick N. Production and characterization of poly(3-hydroxybutyrate-co-3-hydroxyvalerate) co-polymer by a N2-fixing cyanobacterium, Nostoc muscorum Agardh. *J Chem Technol Biotechnol.* 2012;87(4):505–512. https://doi.org/10.1002/jctb.2737.
131. Bhati R, Samantaray S, Sharma L, Mallick N. Poly-β-hydroxybutyrate accumulation in cyanobacteria under photoautotrophy. *Biotechnol J.* 2010;5(11):1181–1185. https://doi.org/10.1002/biot.201000252.
132. Chotchindakun K, Pathom-Aree W, Dumri K, Ruangsuriya J, Pumas C, Pekkoh J. Low crystallinity of poly(3-hydroxybutyrate-co-3-hydroxyvalerate) bioproduction by hot spring cyanobacterium Cyanosarcina sp. AARL T020. *Plants.* 2021;10(3):503. https://doi.org/10.3390/plants10030503.
133. Singh AK, Sharma L, Mallick N, Mala J. Progress and challenges in producing polyhydroxyalkanoate biopolymers from cyanobacteria. *J Appl Phycol.* 2017;29(3):1213–1232. https://doi.org/10.1007/s10811-016-1006-1.
134. Phung Hai TA, Neelakantan N, Tessman M, et al. Flexible polyurethanes, renewable fuels, and flavorings from a microalgae oil waste stream. *Green Chem.* 2020;22(10):3088–3094. https://doi.org/10.1039/D0GC00852D.
135. Hauvermale A, Kuner J, Rosenzweig B, Guerra D, Diltz S, Metz JG. Fatty acid production in Schizochytrium sp.: involvement of a polyunsaturated fatty acid synthase and a type I fatty acid synthase. *Lipids.* 2006;41(8):739–747. https://doi.org/10.1007/s11745-006-5025-6.
136. Renaud SM, Zhou HC, Parry DL, Thinh LV, Woo KC. Effect of temperature on the growth, total lipid content and fatty acid composition of recently isolated tropical microalgae Isochrysis sp., Nitzschia closterium, Nitzschia paleacea, and commercial species Isochrysis sp. (clone T.ISO). *J Appl Phycol.* 1995;7(6):595–602. https://doi.org/10.1007/BF00003948.
137. Méjanelle L, Sanchez-Gargallo A, Bentaleb I, Grimalt JO. Long chain n-alkyl diols, hydroxy ketones and sterols in a marine eustigmatophyte, Nannochloropsis gaditana, and in Brachionus plicatilis feeding on the algae. *Org Geochem.* 2003;34(4):527–538. https://doi.org/10.1016/S0146-6380(02)00246-2.
138. Volkman JK, Barrett SM, Dunstan GA, Jeffrey SW. C30-C32 alkyl diols and unsaturated alcohols in microalgae of the class Eustigmatophyceae. *Org Geochem.* 1992;18(1):131–138. https://doi.org/10.1016/0146-6380(92)90150-V.

139. Hon-Nami K. A unique feature of hydrogen recovery in endogenous starch-to-alcohol fermentation of the marine microalga, Chlamydomonas perigranulata. *Appl Biochem Biotechnol.* 2006;129–132:808–828.
140. McNeely K, Xu Y, Bennette N, Bryant DA, Dismukes GC. Redirecting reductant flux into hydrogen production via metabolic engineering of fermentative carbon metabolism in a cyanobacterium. *Appl Environ Microbiol.* 2010;76(15):5032–5038. https://doi.org/10.1128/AEM.00862-10.
141. Hasunuma T, Matsuda M, Kato Y, Vavricka CJ, Kondo A. Temperature enhanced succinate production concurrent with increased central metabolism turnover in the cyanobacterium Synechocystis sp. PCC 6803. *Metab Eng.* 2018;48:109–120. https://doi.org/10.1016/j.ymben.2018.05.013.
142. Phung Hai TA, De Backer LJS, Cosford NDP, Burkart MD. Preparation of mono- and diisocyanates in flow from renewable carboxylic acids. *Org Process Res Dev.* 2020;24(10):2342–2346. https://doi.org/10.1021/acs.oprd.0c00167.
143. Damsté JSS, Rijpstra WIC, Hopmans EC, Schouten S, Balk M, Stams AJM. Structural characterization of diabolic acid-based tetraester, tetraether and mixed ether/ester, membrane-spanning lipids of bacteria from the order Thermotogales. *Arch Microbiol.* 2007;188(6):629–641. https://doi.org/10.1007/s00203-007-0284-z.
144. Radakovits R, Jinkerson RE, Darzins A, Posewitz MC. Genetic engineering of algae for enhanced biofuel production. *Eukaryot Cell.* 2010;9(4):486–501. https://doi.org/10.1128/EC.00364-09.
145. Bishé B, Taton A, Golden JW. Modification of RSF1010-based broad-host-range plasmids for improved conjugation and cyanobacterial bioprospecting. *iScience.* 2019;20:216–228. https://doi.org/10.1016/j.isci.2019.09.002.
146. Jester BW, Zhao H, Gewe M, et al. Development of spirulina for the manufacture and oral delivery of protein therapeutics. *Nat Biotechnol.* 2022;1–9. https://doi.org/10.1038/s41587-022-01249-7. Published online March 21.
147. Wang B, Wang J, Zhang W, Meldrum DR. Application of synthetic biology in cyanobacteria and algae. *Front Microbiol.* 2012;3:344. https://doi.org/10.3389/fmicb.2012.00344.
148. Specht E, Miyake-Stoner S, Mayfield S. Micro-algae come of age as a platform for recombinant protein production. *Biotechnol Lett.* 2010;32(10):1373–1383. https://doi.org/10.1007/s10529-010-0326-5.
149. Polle JEW, Calhoun S, McKie-Krisberg Z, et al. Genomic adaptations of the green alga Dunaliella salina to life under high salinity. *Algal Res.* 2020;50:101990. https://doi.org/10.1016/j.algal.2020.101990.
150. Mukherjee B, Madhu S, Wangikar PP. The role of systems biology in developing non-model cyanobacteria as hosts for chemical production. *Curr Opin Biotechnol.* 2020;64:62–69. https://doi.org/10.1016/j.copbio.2019.10.003.
151. Patel VK, Soni N, Prasad V, Sapre A, Dasgupta S, Bhadra B. CRISPR-Cas9 system for genome engineering of photosynthetic microalgae. *Mol Biotechnol.* 2019;61(8):541–561. https://doi.org/10.1007/s12033-019-00185-3.
152. León R, Cejudo AG, Fernández E. Transgenic Microalgae as Green Cell Factories. Springer Science & Business Media; 2008.
153. Clerico EM, Ditty JL, Golden SS. Specialized techniques for site-directed mutagenesis in cyanobacteria. *Methods Mol Biol Clifton NJ.* 2007;362:155–171. https://doi.org/10.1007/978-1-59745-257-1_11.
154. Stucken K, Ilhan J, Roettger M, Dagan T, Martin WF. Transformation and conjugal transfer of foreign genes into the filamentous multicellular cyanobacteria (subsection V) Fischerella and Chlorogloeopsis. *Curr Microbiol.* 2012;65(5):552–560. https://doi.org/10.1007/s00284-012-0193-5.
155. Schirmacher AM, Hanamghar SS, Zedler JAZ. Function and benefits of natural competence in cyanobacteria: from ecology to targeted manipulation. *Life.* 2020;10(11):249. https://doi.org/10.3390/life10110249.

156. Jaiswal D, Sengupta A, Sohoni S, et al. Genome features and biochemical characteristics of a robust, fast growing and naturally transformable cyanobacterium Synechococcus elongatus PCC 11801 isolated from India. *Sci Rep.* 2018;8:16632. https://doi.org/10.1038/s41598-018-34872-z.
157. Jaiswal D, Sengupta A, Sengupta S, Madhu S, Pakrasi HB, Wangikar PP. A novel cyanobacterium Synechococcus elongatus PCC 11802 has distinct genomic and metabolomic characteristics compared to its neighbor PCC 11801. *Sci Rep.* 2020;10:191. https://doi.org/10.1038/s41598-019-57051-0.
158. Elhai J, Vepritskiy A, Muro-Pastor AM, Flores E, Wolk CP. Reduction of conjugal transfer efficiency by three restriction activities of Anabaena sp. strain PCC 7120. *J Bacteriol.* 1997;179(6):1998–2005.
159. Thiel T, Poo H. Transformation of a filamentous cyanobacterium by electroporation. *J Bacteriol.* 1989;171(10):5743–5746.
160. Karas BJ, Diner RE, Lefebvre SC, et al. Designer diatom episomes delivered by bacterial conjugation. *Nat Commun.* 2015;6:6925. https://doi.org/10.1038/ncomms7925.
161. Rathod JP, Prakash G, Pandit R, Lali AM. Agrobacterium-mediated transformation of promising oil-bearing marine algae Parachlorella kessleri. *Photosynth Res.* 2013;118(1–2):141–146. https://doi.org/10.1007/s11120-013-9930-2.
162. Thomy J, Sanchez F, Gut M, et al. Combining nanopore and illumina sequencing permits detailed analysis of insertion mutations and structural variations produced by PEG-mediated transformation in Ostreococcus tauri. *Cell.* 2021;10(3):664. https://doi.org/10.3390/cells10030664.
163. Faktorová D, Nisbet RER, Fernández Robledo JA, et al. Genetic tool development in marine protists: emerging model organisms for experimental cell biology. *Nat Methods.* 2020;17(5):481–494. https://doi.org/10.1038/s41592-020-0796-x.
164. Han X, Song X, Li F, Lu Y. Improving lipid productivity by engineering a control-knob gene in the oleaginous microalga Nannochloropsis oceanica. *Metab Eng Commun.* 2020;11, e00142. https://doi.org/10.1016/j.mec.2020.e00142.
165. Sanchez F, Geffroy S, Norest M, Yau S, Moreau H, Grimsley N. Simplified transformation of Ostreococcus tauri using polyethylene glycol. *Genes.* 2019;10(5), E399. https://doi.org/10.3390/genes10050399.
166. Krishnan A, Likhogrud M, Cano M, et al. Picochlorum celeri as a model system for robust outdoor algal growth in seawater. *Sci Rep.* 2021;11(1):11649. https://doi.org/10.1038/s41598-021-91106-5.
167. Apt KE, Grossman AR, Kroth-Pancic PG. Stable nuclear transformation of the diatom Phaeodactylum tricornutum. *Mol Gen Genet MGG.* 1996;252(5):572–579. https://doi.org/10.1007/BF02172403.
168. Wang K, Cui Y, Wang Y, et al. Chloroplast genetic engineering of a unicellular green alga Haematococcus pluvialis with expression of an antimicrobial peptide. *Mar Biotechnol (NY).* 2020;22(4):572–580. https://doi.org/10.1007/s10126-020-09978-z.
169. Lapidot M, Raveh D, Sivan A, Arad SM, Shapira M. Stable chloroplast transformation of the unicellular red alga Porphyridium species. *Plant Physiol.* 2002;129(1):7–12.
170. Kilian O, Benemann CSE, Niyogi KK, Vick B. High-efficiency homologous recombination in the oil-producing alga Nannochloropsis sp. *Proc Natl Acad Sci.* 2011;108(52):21265–21269. https://doi.org/10.1073/pnas.1105861108.
171. Yamano T, Iguchi H, Fukuzawa H. Rapid transformation of Chlamydomonas reinhardtii without cell-wall removal. *J Biosci Bioeng.* 2013;115(6):691–694. https://doi.org/10.1016/j.jbiosc.2012.12.020.
172. Franklin SE, Mayfield SP. Prospects for molecular farming in the green alga Chlamydomonas reinhardtii. *Curr Opin Plant Biol.* 2004;7(2):159–165. https://doi.org/10.1016/j.pbi.2004.01.012.
173. Zhang MP, Wang M, Wang C. Nuclear transformation of Chlamydomonas reinhardtii: a review. *Biochimie.* 2021;181:1–11. https://doi.org/10.1016/j.biochi.2020.11.016.

174. Nymark M, Sharma AK, Sparstad T, Bones AM, Winge P. A CRISPR/Cas9 system adapted for gene editing in marine algae. *Sci Rep.* 2016;6(1):24951. https://doi.org/10.1038/srep24951.
175. Santos-Merino M, Singh AK, Ducat DC. New applications of synthetic biology tools for cyanobacterial metabolic engineering. *Front Bioeng Biotechnol.* 2019;7:33. https://doi.org/10.3389/fbioe.2019.00033.
176. Taton A, Unglaub F, Wright NE, et al. Broad-host-range vector system for synthetic biology and biotechnology in cyanobacteria. *Nucleic Acids Res.* 2014;42(17), e136. https://doi.org/10.1093/nar/gku673.
177. Zhang S, Zheng S, Sun J, et al. Rapidly improving high light and high temperature tolerances of cyanobacterial cell factories through the convenient introduction of an AtpA-C252F mutation. *Front Microbiol.* 2021;12:647164. https://doi.org/10.3389/fmicb.2021.647164.
178. Selão TT, Włodarczyk A, Nixon PJ, Norling B. Growth and selection of the cyanobacterium Synechococcus sp. PCC 7002 using alternative nitrogen and phosphorus sources. *Metab Eng.* 2019;54:255–263. https://doi.org/10.1016/j.ymben.2019.04.013.
179. Loera-Quezada MM, Leyva-González MA, Velázquez-Juárez G, et al. A novel genetic engineering platform for the effective management of biological contaminants for the production of microalgae. *Plant Biotechnol J.* 2016;14(10):2066–2076. https://doi.org/10.1111/pbi.12564.
180. Shahar N, Landman S, Weiner I, et al. The integration of multiple nuclear-encoded transgenes in the green alga Chlamydomonas reinhardtii results in higher transcription levels. *Front Plant Sci.* 2020;10:1784. https://doi.org/10.3389/fpls.2019.01784.
181. Jackson HO, Berepiki A, Baylay AJ, Terry MJ, Moore CM, Bibby TS. An inducible expression system in the alga Nannochloropsis gaditana controlled by the nitrate reductase promoter. *J Appl Phycol.* 2019;31(1):269–279. https://doi.org/10.1007/s10811-018-1510-6.
182. Kobayashi S, Atsumi S, Ikebukuro K, Sode K, Asano R. Light-induced production of isobutanol and 3-methyl-1-butanol by metabolically engineered cyanobacteria. *Microb Cell Factories.* 2022;21(1):7. https://doi.org/10.1186/s12934-021-01732-x.
183. Sengupta A, Madhu S, Wangikar PP. A library of tunable, portable, and inducer-free promoters derived from cyanobacteria. *ACS Synth Biol.* 2020;9(7):1790–1801. https://doi.org/10.1021/acssynbio.0c00152.
184. Welkie DG, Rubin BE, Diamond S, Hood RD, Savage DF, Golden SS. A hard day's night: cyanobacteria in diel cycles. *Trends Microbiol.* 2019;27(3):231–242. https://doi.org/10.1016/j.tim.2018.11.002.
185. Vasudevan R, Gale GAR, Schiavon AA, et al. CyanoGate: a modular cloning suite for engineering cyanobacteria based on the plant MoClo syntax. *Plant Physiol.* 2019;180(1):39–55. https://doi.org/10.1104/pp.18.01401.
186. Kim WJ, Lee SM, Um Y, Sim SJ, Woo HM. Development of SyneBrick vectors as a synthetic biology platform for gene expression in Synechococcus elongatus PCC 7942. *Front Plant Sci.* 2017;8:293. https://doi.org/10.3389/fpls.2017.00293.
187. Jackson HO, Taunt HN, Mordaka PM, Smith AG, Purton S. The algal chloroplast as a testbed for synthetic biology designs aimed at radically rewiring plant metabolism. *Front Plant Sci.* 2021;12:708370. https://doi.org/10.3389/fpls.2021.708370.
188. Crozet P, Navarro FJ, Willmund F, et al. Birth of a photosynthetic chassis: a MoClo toolkit enabling synthetic biology in the microalga *Chlamydomonas reinhardtii*. *ACS Synth Biol.* 2018;7(9):2074–2086. https://doi.org/10.1021/acssynbio.8b00251.
189. Behle A, Axmann IM. pSHDY: a new tool for genetic engineering of cyanobacteria. *Methods Mol Biol Clifton NJ.* 2022;2379:67–79. https://doi.org/10.1007/978-1-0716-1791-5_4.
190. Sproles AE, Berndt A, Fields FJ, Mayfield SP. Improved high-throughput screening technique to rapidly isolate Chlamydomonas transformants expressing recombinant

proteins. *Appl Microbiol Biotechnol.* 2022;106(4):1677–1689. https://doi.org/10.1007/s00253-022-11790-9.

191. Rasala BA, Chao SS, Pier M, Barrera DJ, Mayfield SP. Enhanced genetic tools for engineering multigene traits into green algae. Chen WN, ed. *PLoS One.* 2014;9(4):e94028. https://doi.org/10.1371/journal.pone.0094028.
192. Kao PH, Ng IS. CRISPRi mediated phosphoenolpyruvate carboxylase regulation to enhance the production of lipid in Chlamydomonas reinhardtii. *Bioresour Technol.* 2017;245:1527–1537. https://doi.org/10.1016/j.biortech.2017.04.111.
193. Yao L, Shabestary K, Björk SM, et al. Pooled CRISPRi screening of the cyanobacterium Synechocystis sp PCC 6803 for enhanced industrial phenotypes. *Nat Commun.* 2020;11 (1):1666. https://doi.org/10.1038/s41467-020-15491-7.
194. Mora Salguero DA, Fernández-Niño M, Serrano-Bermúdez LM, et al. Development of a Chlamydomonas reinhardtii metabolic network dynamic model to describe distinct phenotypes occurring at different CO2 levels. *PeerJ.* 2018;6, e5528. https://doi.org/10.7717/peerj.5528.
195. Broddrick JT, Welkie DG, Jallet D, Golden SS, Peers G, Palsson BO. Predicting the metabolic capabilities of Synechococcus elongatus PCC 7942 adapted to different light regimes. *Metab Eng.* 2019;52:42–56. https://doi.org/10.1016/j.ymben.2018.11.001.
196. Hendry JI, Prasannan CB, Joshi A, Dasgupta S, Wangikar PP. Metabolic model of Synechococcus sp. PCC 7002: prediction of flux distribution and network modification for enhanced biofuel production. *Bioresour Technol.* 2016;213:190–197. https://doi.org/10.1016/j.biortech.2016.02.128.
197. Sengupta S, Jaiswal D, Sengupta A, Shah S, Gadagkar S, Wangikar PP. Metabolic engineering of a fast-growing cyanobacterium Synechococcus elongatus PCC 11801 for photoautotrophic production of succinic acid. *Biotechnol Biofuels.* 2020;13:89. https://doi.org/10.1186/s13068-020-01727-7.
198. Vijay D, Akhtar MK, Hess WR. Genetic and metabolic advances in the engineering of cyanobacteria. *Curr Opin Biotechnol.* 2019;59:150–156. https://doi.org/10.1016/j.copbio.2019.05.012.
199. Shih PM, Wu D, Latifi A, et al. Improving the coverage of the cyanobacterial phylum using diversity-driven genome sequencing. *Proc Natl Acad Sci U S A.* 2013;110 (3):1053–1058. https://doi.org/10.1073/pnas.1217107110.
200. Hirose Y, Ohtsubo Y, Misawa N, et al. Genome sequencing of the NIES Cyanobacteria collection with a focus on the heterocyst-forming clade. *DNA Res Int J Rapid Publ Rep Genes Genomes.* 2021;28(6), dsab024. https://doi.org/10.1093/dnares/dsab024.
201. Hanschen ER, Starkenburg SR. The state of algal genome quality and diversity. *Algal Res.* 2020;50:101968. https://doi.org/10.1016/j.algal.2020.101968.
202. LaPanse AJ, Krishnan A, Posewitz MC. Adaptive laboratory evolution for algal strain improvement: methodologies and applications. *Algal Res.* 2021;53:102122. https://doi.org/10.1016/j.algal.2020.102122.
203. Rubin BE, Wetmore KM, Price MN, et al. The essential gene set of a photosynthetic organism. *Proc Natl Acad Sci.* 2015;112(48). https://doi.org/10.1073/pnas.1519220112.
204. Fields FJ, Ostrand JT, Tran M, Mayfield SP. Nuclear genome shuffling significantly increases production of chloroplast-based recombinant protein in Chlamydomonas reinhardtii. *Algal Res.* 2019;41:101523. https://doi.org/10.1016/j.algal.2019.101523.
205. Shan T, Pang S. Breeding in the economically important brown alga Undaria pinnatifida: a concise review and future prospects. *Front Genet.* 2021;12. https://www.frontiersin.org/article/10.3389/fgene.2021.801937. Accessed 14 April 2022.
206. Specht EA, Mayfield SP. Algae-based oral recombinant vaccines. *Front Microbiol.* 2014;5:60. https://doi.org/10.3389/fmicb.2014.00060.
207. Berndt AJ, Smalley TN, Ren B, et al. Recombinant production of a functional SARS-CoV-2 spike receptor binding domain in the green algae Chlamydomonas reinhardtii. *PLoS One.* 2021;16(11), e0257089. https://doi.org/10.1371/journal.pone.0257089.

208. Ducat DC, Avelar-Rivas JA, Way JC, Silver PA. Rerouting carbon flux to enhance photosynthetic productivity. *Appl Environ Microbiol.* 2012;78(8):2660–2668. https://doi.org/10.1128/AEM.07901-11.
209. Chaogang W, Zhangli H, Anping L, Baohui J. Biosynthesis of poly-3-hydroxybutyrate (phb) in the transgenic green alga Chlamydomonas Reinhardtii1. *J Phycol.* 2010;46(2):396–402. https://doi.org/10.1111/j.1529-8817.2009.00789.x.
210. Hempel F, Bozarth AS, Lindenkamp N, et al. Microalgae as bioreactors for bioplastic production. *Microb Cell Factories.* 2011;10:81. https://doi.org/10.1186/1475-2859-10-81.
211. Koch M, Bruckmoser J, Scholl J, Hauf W, Rieger B, Forchhammer K. Maximizing PHB content in Synechocystis sp. PCC 6803: a new metabolic engineering strategy based on the regulator PirC. *Microb Cell Factories.* 2020;19(1):231. https://doi.org/10.1186/s12934-020-01491-1.
212. Fayyaz M, Chew KW, Show PL, Ling TC, Ng IS, Chang JS. Genetic engineering of microalgae for enhanced biorefinery capabilities. *Biotechnol Adv.* 2020;43:107554. https://doi.org/10.1016/j.biotechadv.2020.107554.
213. Ducat DC, Way JC, Silver PA. Engineering cyanobacteria to generate high-value products. *Trends Biotechnol.* 2011;29(2):95–103. https://doi.org/10.1016/j.tibtech.2010.12.003.
214. Pathania R, Srivastava A, Srivastava S, Shukla P. Metabolic systems biology and multi-omics of cyanobacteria: perspectives and future directions. *Bioresour Technol.* 2022;343:126007. https://doi.org/10.1016/j.biortech.2021.126007.
215. Treece TR, Gonzales JN, Pressley JR, Atsumi S. Synthetic biology approaches for improving chemical production in cyanobacteria. *Front Bioeng Biotechnol.* 2022;10. https://www.frontiersin.org/article/10.3389/fbioe.2022.869195. Accessed 15 April 2022.
216. Carroll AL, Case AE, Zhang A, Atsumi S. Metabolic engineering tools in model cyanobacteria. *Metab Eng.* 2018;50:47–56. https://doi.org/10.1016/j.ymben.2018.03.014.
217. Stansell Z, Hyma K, Fresnedo-Ramírez J, et al. Genotyping-by-sequencing of Brassica oleracea vegetables reveals unique phylogenetic patterns, population structure and domestication footprints. *Hortic Res.* 2018;5:38. https://doi.org/10.1038/s41438-018-0040-3.
218. Duarte CM, Marbá N, Holmer M. Rapid domestication of marine species. *Science.* 2007;316(5823):382–383. https://doi.org/10.1126/science.1138042.
219. Valero E, Álvarez X, Cancela Á, Sánchez Á. Harvesting green algae from eutrophic reservoir by electroflocculation and post-use for biodiesel production. *Bioresour Technol.* 2015;187:255–262. https://doi.org/10.1016/j.biortech.2015.03.138.
220. Corcoran AA, Hunt RW. Capitalizing on harmful algal blooms: from problems to products. *Algal Res.* 2021;55:102265. https://doi.org/10.1016/j.algal.2021.102265.
221. Zhao X, Kumar K, Gross MA, Kunetz TE, Wen Z. Evaluation of revolving algae biofilm reactors for nutrients and metals removal from sludge thickening supernatant in a municipal wastewater treatment facility. *Water Res.* 2018;143:467–478. https://doi.org/10.1016/j.watres.2018.07.001.
222. Perez-Garcia O, Bashan Y. Microalgal heterotrophic and mixotrophic culturing for biorefining: from metabolic routes to techno-economics. In: Prokop A, Bajpai RK, Zappi ME, eds. *Algal Biorefineries.* Springer International Publishing; 2015:61–131. https://doi.org/10.1007/978-3-319-20200-6_3.
223. Schoepp NG, Stewart RL, Sun V, et al. System and method for research-scale outdoor production of microalgae and cyanobacteria. *Bioresour Technol.* 2014;166:273–281. https://doi.org/10.1016/j.biortech.2014.05.046.
224. Gupta PL, Lee SM, Choi HJ. A mini review: photobioreactors for large scale algal cultivation. *World J Microbiol Biotechnol.* 2015;31(9):1409–1417. https://doi.org/10.1007/s11274-015-1892-4.

225. Borowitzka MA. Energy from microalgae: a short history. In: Borowitzka MA, Moheimani NR, eds. *Algae for Biofuels and Energy.* Developments in Applied Phycology. Springer Netherlands; 2013:1–15. https://doi.org/10.1007/978-94-007-5479-9_1.
226. Davis R, Aden A, Pienkos PT. Techno-economic analysis of autotrophic microalgae for fuel production. *Appl Energy.* 2011;88(10):3524–3531. https://doi.org/10.1016/j.apenergy.2011.04.018.
227. Oswald WJ, Gotaas HB. Photosynthesis in sewage treatment. *Trans Am Soc Civ Eng.* 1957;122(1):73–97. https://doi.org/10.1061/TACEAT.0007483.
228. Pushparaj B, Pelosi E, Tredici MR, Pinzani E, Materassi R. An integrated culture system for outdoor production of microalgae and cyanobacteria. *J Appl Phycol.* 1997;9(2):113–119. https://doi.org/10.1023/A:1007988924153.
229. Prabha S, Vijay AK, Paul RR, George B. Cyanobacterial biorefinery: towards economic feasibility through the maximum valorization of biomass. *Sci Total Environ.* 2022;814:152795. https://doi.org/10.1016/j.scitotenv.2021.152795.
230. Committee on Genetically Engineered Crops: Past Experience and Future Prospects, Board on Agriculture and Natural Resources, Division on Earth and Life Studies, National Academies of Sciences, Engineering, and Medicine. Genetically Engineered Crops: Experiences and Prospects. National Academies Press; 2016, https://doi.org/10.17226/23395.
231. Szyjka SJ, Mandal S, Schoepp NG, et al. Evaluation of phenotype stability and ecological risk of a genetically engineered alga in open pond production. *Algal Res.* 2017;24:378–386. https://doi.org/10.1016/j.algal.2017.04.006.
232. Zhu Z, Jiang J, Fa Y. Overcoming the biological contamination in microalgae and cyanobacteria mass cultivations for photosynthetic biofuel production. *Molecules.* 2020;25(22):5220. https://doi.org/10.3390/molecules25225220.
233. Sauer JS, Simkovsky R, Moore AN, et al. Continuous measurements of volatile gases as detection of algae crop health. *Proc Natl Acad Sci.* 2021;118(40), e2106882118. https://doi.org/10.1073/pnas.2106882118.
234. Deore P, Beardall J, Noronha S. A perspective on the current status of approaches for early detection of microalgal grazing. *J Appl Phycol.* 2020;32(6):3723–3733. https://doi.org/10.1007/s10811-020-02241-x.
235. Newby DT, Mathews TJ, Pate RC, et al. Assessing the potential of polyculture to accelerate algal biofuel production. *Algal Res.* 2016;19:264–277. https://doi.org/10.1016/j.algal.2016.09.004.
236. González-Morales SI, Pacheco-Gutiérrez NB, Ramírez-Rodríguez CA, et al. Metabolic engineering of phosphite metabolism in Synechococcus elongatus PCC 7942 as an effective measure to control biological contaminants in outdoor raceway ponds. *Biotechnol Biofuels.* 2020;13:119. https://doi.org/10.1186/s13068-020-01759-z.
237. Soni RA, Sudhakar K, Rana RS. Spirulina—from growth to nutritional product: a review. *Trends Food Sci Technol.* 2017;69:157–171. https://doi.org/10.1016/j.tifs.2017.09.010.

CHAPTER

3

Renewable, sustainable sources and bio-based monomers

Bhausaheb S. Rajput, Anton A. Samoylov, and Thien An Phung Hai

University of California San Diego, La Jolla, CA, United States

Introduction

Over the last several decades, the world has witnessed a revolution inducing a new age of heightened demand in production and consumption. As a result, CO_2 emissions have increased by 1%–2% per year since 2010, linked to the rising use of fossil fuels. Investment in biodiversity resources and environmental sustainability is a priority to maintain the ecology of the planet. The use of petrochemical-based polymers has been a global concern for many decades. Renewable resources such as cellulose, hemicellulose, lignin, starch, chitosan, and natural oils hold high potential as alternatives,[1–3] and much effort is being devoted to finding substitutes for fossil resources in the polymer field.[4,5]

Among the renewable resources currently studied, vegetable oils are the most promising and can be directly subjected to chemical modification and functionalization.[6] Their biodegradability is also a characteristic of interest. They consist of triglycerides, a three-pronged molecule containing a triester of higher fatty acids. Each vegetable oil has a similar triglyceride structure with different fatty acid chains. Saponification and transesterification with methanol is a standard procedure to purify fatty acid methyl esters from plant-based oils. Most of these common fatty acids contain a chain of 14–20 carbons with a degree of unsaturation between 0 and 3. Unsaturation indicates the number of double bonds in a molecule, often used as a substrate for functionalization. For example, sunflower oil and soybean oil contain primarily oleic acid (C18:1) and linoleic acid (C18:2). Apart from their chemical or structural features, these oils and

https://doi.org/10.1016/B978-0-323-99982-3.00016-X

fatty acids can be used to synthesize various polymers including the polyols for polyurethane (PU) applications.[1]

Recent research has also focused on converting inexpensive algae into valuable bioproducts using natural or genetic manipulation. Converting algal biomass into a value-added chemical for synthesizing monomers for polymers applications is an area of high interest. This biomass is divided into microalgal and macroalgal. Microalgae are composed of lipids, proteins, and carbohydrates. Because of algae's high growth rate and oil content, much attention has been given to using them for biodiesel production, and engineered microalgae technology is showing rapid progress.[7]

Microalgae have many advantages, such as low production cost, rapid growth, and noncompetition with food resources, unlike vegetable crops that are grown mainly for food.[8] Algae can also utilize nonarable land and brackish water, making them more versatile than food crops, which are more energy and resource intensive.[9,10] Algae can grow at a much heavier density than vegetables and produce more unsaturated fatty acids per acre, saving valuable space.[11] Microalgae are unicellular photosynthetic organisms with relatively high tolerance and can be grown without arable land or fresh water, unlike oil-producing crops. They are diverse, with an estimated 200,000–800,000 species. The biomass is harvested and processed to obtain unsaturated lipids and fatty acids following biosynthetic pathways.[12] Production is not limited to the generation of lipids for biodiesel, but can be valorized to different biocomponents. Unlike vegetable oils, the fatty acid composition is unfortunately dependent on the species' climate conditions. However, research to commercialize algae production economically continues because of its utility in conversion to PU precursors similar to vegetable oils.

In PU chemistry, polyols, diisocyanates, and polyisocyanates play a crucial role in synthesizing various PU grades, making up most formulations by their combined weight. Applications in thermoplastic PUs (TPUs) and thermosetting PUs rely on the properties of the two components (polyols and diisocyanates) to produce optimal elastomers, coatings, sealants, adhesives, foams, solvent-borne PUs, and waterborne PUs.[1,13] TPUs can be used for high-performance applications, and more importantly can be reprocessed into sophisticated shapes and molds, depending on the final application. With increasing environmental consciousness and demand, bio-based TPUs are necessary for developing sustainable alternatives to petroleum-derived plastics. Bio-based polyester polyols can be obtained from renewable resources through either chemical pathways (azelaic and sebacic acids) or biotechnological processes (succinic and adipic acids). A slight difference in mechanical properties such as hardness and tensile and tear strength between renewable- and petroleum-based TPUs[14] shows promise for future renewable TPUs. Hence, the first part of this chapter is focused on polyester polyols based on oil from plants and algae, while the second part is focused on recent developments in bio-based isocyanate.

Different types of polyols

Polyols are a diverse category of molecules made from a variety of organic compounds which have multiple hydroxyl groups. They are of three main types: short (less than 12 carbons in their chains), aliphatic, and aromatic. They are also categorized into monomeric and polymeric, depending on molecular weight: monomeric exist as unique, organic molecules with no identifiable repeating units, while polymeric have high molecular weight. The structures and properties of the subsequent PUs can be modified by choosing suitable diisocyanates and polyols and then configuring the mass ratio between these two. Polymeric polyols act as soft segments, while diisocyanates act as hard segments in the synthesis of PUs. Few diisocyanates are commercially available compared to the variety of polyols designed and supplied for various applications. The design and production of novel polyols determines the performance of the PUs; a wide variety of molecules are available for formulation. Most of the polyols used in industry are derived from petroleum feedstocks, so there are challenges in moving production to renewable alternatives. The world's crude oil feedstock is being depleted at a time of increasing environmental concern to find renewable alternatives such as natural oils and cellulose to synthesize novel polyols.[15] The polyols are classified into polyethers and polyesters, depending on the application. They are widely used to make PUs by various industries.

There are three essential properties of polyols: hydroxyl (–OH) value, functionality, and average molecular weight (M_n) range. Hydroxyl value is a standard metric for quickly assessing how many groups are present per gram of polyol, which correlates with final PU properties. There is a strong correlation between polyol properties and chemical structure, and the behavior and type of the corresponding PU foam.[16] These results determine the final properties of the PU. The following sections describe the primary uses, sources, and synthesis of each introduced polyol classification. This chapter excludes discussion of rarer polyols such as polycarbonates, focusing on more available and biodegradation-friendly alternatives.

Monomeric polyols

These are generally low-molecular-weight organic compounds such as ethylene glycol, glycerol, diethylene glycol, propylene glycol, and

1,4-butanediol. They are usually added into formulations on their own or coupled with other agents such as dicarboxylic acids to create a polymeric polyol chain. Depending on the number of hydroxyl groups in one molecule, they are also classified as diols, triols, or tetraols, that is, having two, three, or four hydroxyl groups in one molecule. Some short-chain polyols such as xylitol, xylose, sorbitol, and mannitol are added to PU foams to improve mechanical properties. These short-chain monomeric polyols are widely used in rigid foaming processes because of their high reactivity (high hydroxyl functionalization) and commercial availability.

Diols are useful chain extenders in producing thermoplastic and thermosetting PUs because of their linear structure, which allows them to link up. Monomeric polyols such as triols and tetraols are extensively used as initiators to produce polyether polyols. They also serve as crosslinkers to produce a variety of harder PUs, which can link multiple chains together because of their higher functionalization.

Polyether polyols

These are characterized by ether linkages within the molecules. They are an option alongside polyester-based polymeric polyols. They are usually present in lower viscosity and have higher weight dispersity than polyester polyols. They are mainly used to improve mechanical properties in PU-based applications, especially foams.[16] Compared to polyester polyols, they have better hydrolytic and aging stability, at the expense of decreased biodegradation. Generally, they are obtained from ring-opening anionic polymerization of oxiranes (ethylene oxide, propylene oxide), called alkoxylation. They produce high-quality PU foams and elastomers. This reaction is completed under basic catalytic conditions at high temperature (110–200°C) in a closed pressure reactor.[16] A multihydroxyl alcohol initiator such as glycerol and pentaerythritol and a base catalyst such as alkali metal hydroxides are commonly used in the alkoxylation process.[17] Functionality can be tailored by using suitable initiators during alkoxylation.[18]

In Fig. 3.1A, poly(propylene oxide)polyols produced with propylene oxide and glycerol as initiators have three functions, meaning an average of three hydroxyl groups per molecule. A variety of them can be obtained by controlling the initiator's functionality and the degree of polymerization to produce various PUs (Fig. 3.1B). They have functionality in the range of 3–8 and a low molecular weight between 150 and 1000 $g\,mol^{-1}$. High hydroxyl numbers between 250 and 1000 mg KOH g^{-1} are used for producing rigid coatings and rigid foams in PU applications. On the other hand, high molecular weight (1000–6500 $g\,mol^{-1}$), low functionality (2–3), and low hydroxyl number (28–160 mg KOH g^{-1}) are used to

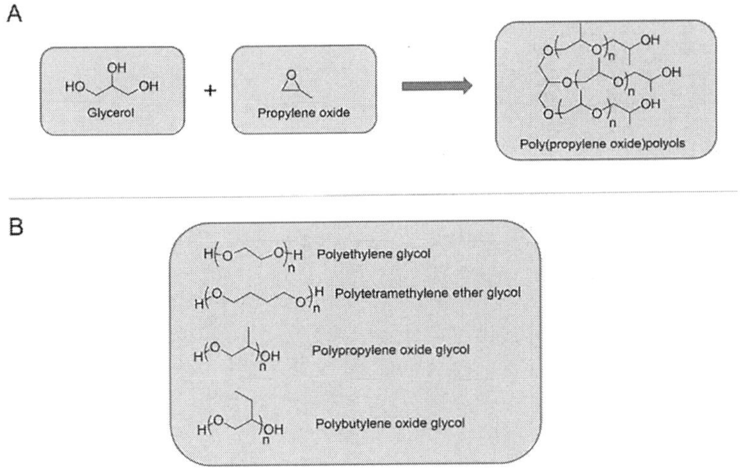

FIG. 3.1 Production of poly(propylene oxide)polyols (A) and examples of polyether polyols (B).

produce elastomers and flexible foams. Another well-known example is poly(tetramethylene ether)glycols, which are made by acid-catalyzed polymerization. They are used to produce thermoplastic PUs, PU elastomers, and PU fibers.

Polyester polyols

Biodegradable polyester polyols are produced via a condensation reaction between the aliphatic dicarboxylic acids and diols (Fig. 3.2A). They contain internal ester linkages, which are more prone to cleavage by various biological pathways. Carothers first demonstrated the step-

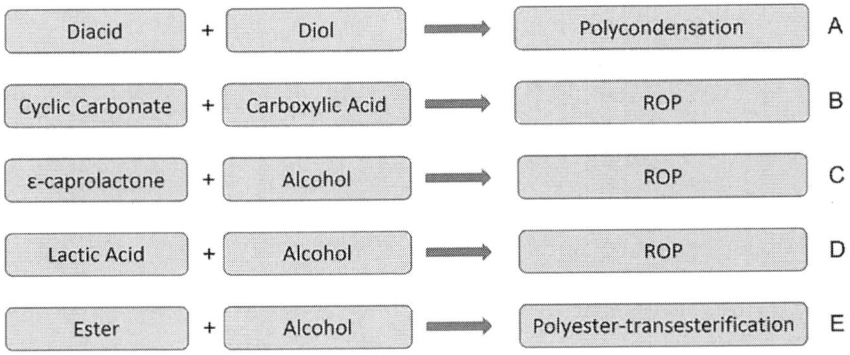

FIG. 3.2 Different pathways to obtain polyester polyols: (A) polycondensation of diols and diacids, (B) ROP of cyclic carbonate and a carboxylic acid, (C) ROP of ε-caprolactone and alcohol, (D) ROP of lactic acid and alcohol, (E) polyester transesterification.

growth mechanism of polyester synthesis in 1929.[19–21] Several synthetic routes exist, such as polycondensation reactions, transesterification, and ring-opening polymerization (ROP). All are used to make polyester polyols for PUs (Fig. 3.2). The linear structures are produced mainly by a reaction between a select dicarboxylic acid, ester, or anhydride, coupled with monomeric diols. The branched polyester polyol can be obtained by the addition of triols, such as glycerol and 1,1,1-trimethylolpropane. Polyester polyols are classified into aliphatic and aromatic, based on the structures. The most common examples of aliphatics are succinic acid, adipic acid, and low-molecular-weight diols such as 1,2-ethanediol and 1,4-butanediol. The aromatics are based on terephthalic acid with 1,2-ethanediol or 1,4-butanediol. Phthalic anhydride is also commonly used as an aromatic dicarboxylic acid to produce aromatic polyester polyol.

One of the best-known polymerization routes for cyclic monomers is ROP of lactones such as ε-caprolactone or lactides to synthesize polyester polyol.[16,22] ROP of five-membered alkylene carbonates or cyclic acid anhydrides is used to obtain a mixture of hydroxyalkyl esters (Fig. 3.2B). Fig. 3.2C is an interesting example for preparing aliphatic polyester polyols from ROP of lactone or other cyclic monomers. Polycaprolactone polyols are synthesized by ring-opening polymerization of ε-caprolactone using various multihydroxyl alcohols as initiators, such as pentaerythritol, glucose, methyl-α-glucoside, castor oil, ethylene glycol, glycerol, and 1,4-butanediol.[16,17,23,24] They have lower viscosity and better water resistance than adipic acid-derived polyester polyols with similar molecular weights. Polycaprolactone-based polyols are well known for use in elastomers, coatings, foams, and adhesives.

The low-hydroxyl-number aliphatic polyester polyols are generally used to produce PU elastomers, flexible foams, and adhesives. In comparison with aliphatic members, aromatics show higher rigidity and are usually used to produce rigid PU foam flame retardants. Polyester polyols are generally a waxy solid or highly viscous liquid. The presence of an ester group makes them more susceptible to water than polyether polyols. This property allows them to be biodegradable. Apart from their properties, PU foams were frequently prepared from polyester polyols derived from adipic acid, phthalic anhydride, and glycol, such as trimethylolpropane.[16] Because of the higher cost, higher rigidity, lower hydrolysis resistance, and lower functionality of polyester polyols, they are less extensively developed than polyether polyols. Nowadays, strong polyester polyol development is linked to the fast development of polyisocyanurate foams, which present better mechanical properties and lower flammability.[25]

Apart from petroleum-based polyester polyols, bio-based polyester polyols are of current interest to many researchers and industries oriented toward a sustainable future. On the other hand, most large-scale manufacturers are focusing on using natural oils to produce polyols, considering the current global concern about the use of petroleum-derived chemicals.[26]

There are plenty of examples in the literature that discuss the use of natural oil-based polyols for PU applications. Vegetable oils contain an ester group on the glycerol backbone of triglycerides. Glycerides are produced by combining glycerol and fatty acids containing 8–18 carbon atoms. The fatty acids can be either poly- or monounsaturated depending on the plant category and climatic growing conditions. Some vegetable oils have built-in reactive functional groups in their fatty acid chains; for example, castor oil contains a hydroxyl group, and vernonia oil contains an epoxide functional group. However, most vegetable oils require introducing hydroxyl groups on the reactive sites (e.g., carbon-carbon double bonds) before their use as polyester polyols for PU applications. Some of the well-known methods were explored in the literature to convert the vegetable oil into the polyols, such as (a) epoxidation/ring opening of the epoxide, (b) hydroformylation/reduction, (c) transesterification or amidation, (d) thiol-ene addition, and (e) ozonolysis/reduction. The following sections discuss the use of natural oils for the synthesis of various polyester polyols for PU applications. Natural oils from vegetable sources predominate in the market; they may provide a template for adapting algae to produce some of the target molecules.

Polyester polyols derived from natural oils

Soybean oil is rich in unsaturated fatty acids and is most susceptible to chemical modification into preferred polymeric materials. Soy-based polyols were used as precursors for the synthesis of PUs. Soybean oil consists of naturally built fatty acid components such as the polyunsaturates alpha-linolenic acid (C-18:3), 7%–10%; linoleic acid (C-18:2), 51%; and the monounsaturated oleic acid (C-18:1), 23%. It also contains the saturated fatty acids stearic acid (C-18:0), 4%, and palmitic acid (C-16:0), 10%. Soybean oil was used mostly for food, but recent research has been done for other applications such as biodiesel, thermosetting resins, polyols for PU, bactericidal PU coatings, coating for lithium batteries, and waterborne PU coatings.

Once the fractions of the different kinds of fatty acids are separated, they become much more useful for functionalization into monomers. There are a few approaches to convert inexpensive natural oil into polyols to synthesize grades of PUs. For example, the presence of carbon-carbon double bonds in natural oil first undergoes an epoxidation reaction to create an epoxide ring that can be opened to produce a hydroxyl group.

The ring opening is done afterward with nucleophilic reagents such as alcohols, amines, and halides, and is an effective way to generate polyols. The resulting hydroxyl groups are generally secondary, known for being less reactive than primary groups. Another well-known method is hydroformylation, in which generating a primary alcohol is possible, producing a hydroxyl group that is more reactive than a secondary alcohol.[15] However, during the hydroformylation process, the use of hazardous gases such as CO and H_2 and expensive rhodium catalysts is needed to make the polyols. Transesterifying natural oils with amines or alcohols is also a known pathway, but it produces a mixture of unreactive mono- and di-glycerides due to the partial transesterification. Ozonolysis is another method to convert natural oils into polyols with terminal primary hydroxyl groups, but with this method, the maximum hydroxyl number per molecule is limited to 3. Apart from these methods, Yang and coworkers used the thiolene click reaction to produce polyols with a solvent-free and scalable method.[15] Their research group focused on this reaction to develop a new mode of polyol synthesis. It gives high yields, simple reaction conditions, short reaction time, and easy purification. These novel polyols can make various PUs along the same line using aliphatic, cycloaliphatic, and aromatic diisocyanates. The prepared PUs show good resulting thermal and mechanical properties.

Similar to soybean oil, sunflower oil is abundant because of its consumption in the food industry. Hyperbranched primary hydroxyl functionality polyols were derived from sunflower oil using a facile photochemical thiolene coupling method, and their applications were tested for the synthesis of thermoplastic PU films. The authors synthesized the sunflower oil-based polyols with high hydroxyl numbers and prepared the thermoplastic PU with a broad range of crosslinking densities. They also demonstrated that these prepared PUs could be useful as coatings.[27]

Palm oil has desirable fatty acids for polyol production. It is high yielding and more efficient than any other vegetable oil. Abundant supply and low cost make it an eco-friendly resource for bio-based materials or polymers. It consists of 46.6% saturated fatty acids and 53.4% unsaturated fatty acids. The primary fatty acids are oleic (approximately 40%) and palmitic (approximately 44%).

Epoxidation of palm oil and its subsequent epoxide ring-opening reaction has been intensively studied, as it is the most feasible route to generate polyols. However, typical palm oil-based polyols generated by this method usually have hydroxyl values of less than 200 mg KOH g^{-1}, which limits their suitability in specific applications such as rigid PU foams. Also, saturated fatty acids in palm oil are nonreactive moieties that behave as plasticizers and reduce the suitability for rigid PU foam formation. To address these issues, Arniza and coworkers performed the transesterification of palm olein with glycerol.[28] This study aimed to increase the functionality by introducing additional hydroxyl groups to the triglyceride molecule. This transformation has an advantage compared to using palm olein directly as feedstock for producing palm-based polyol. The transesterified palm olein-based polyester polyol exhibited hydroxyl values between 300 and 330 mg KOH g^{-1}. Furthermore, thiolene chemistry can be used to generate natural polyols. Metathesis is also used to produce palm oil-based polyols.[29]

Polyester polyols derived from algae oil

Microalgae oil contains longer unsaturated fatty acid chains than conventional vegetable oils. These acids can be converted into epoxy and subsequently ring opened using alcohol or hydroxyl acids into polyester polyols for PU applications.[12,30–32] Although the behavior of microalgae-based polyester polyols is promising, only a handful of research papers have been published in the last decade. In 2016 Peyton and coworkers synthesized a microalgae oil-based polyester polyol and tested it by formulating a PU foam. The objective was to study the effect of microalgae oil derivatives on PU foam closed cells. During the synthesis, carbon-carbon double bonds in the microalgae oil were converted into the corresponding epoxidized microalgae oil.

Increasing the lipid content, determining a high enough growth rate, and scaling up production are significant challenges. However, algae oil-derived omega-3 fatty acids were a high-value nutraceutical that helped balance out the cost of creating algae farms at commercial scale. After separating out the omega-3 fatty acids, the residual oil can be converted into a valuable renewable feedstock for petrochemical substitutions. It contains palmitic and palmitoleic acid as the two primary components of oils found in waste-stream algal biomass. Palmitoleic acid can be further transformed into azelaic acid and heptanoic acid by an ozonolysis process.[33] Purified azelaic acid is a valuable dicarboxylic acid that polymerizes with glycols such as ethylene glycol to generate the corresponding polyester polyol. A study from a University of California San Diego group[33] aimed to develop a scalable methodology to produce azelaic acid and polyester polyols for PU applications.

Ring-opening reactions of microalgae oil-based epoxides are known using acetic acid, diethylamine, ethanol, or hydrochloric acid to generate the polyester polyols.[12] These synthesized polyols were analyzed by determining molar mass, viscosity, functionality, hydroxyl, and acid number. For demonstrating PU foam performance, the authors used the bio-based polyols in combination with fossil-based polyether polyol. In their experiments, 25, 35, 50, and 75 wt% of polyether polyols were entirely substituted by one microalgae oil-derived polyol. The properties of bio-based foams were evaluated using the mechanical, thermal conductivity, and flammability tests for various applications. Yemul and coworkers synthesized rigid PU foams using algae oil derived from chlorella microalgae having an iodine value of 120 g I_2 100 g^{-1} of oil.[30] Oxidation of algae oil was done using environmentally friendly reagents such as hydrogen peroxide and acetic acid, followed by a ring-opening reaction of epoxy using ethylene glycol or lactic acid to produced bio-based polyols. The formation of epoxidized algae oil was confirmed by ^{1}H NMR and FT-IR analyses.

Ring opening of epoxidized algae oil with lactic acid showed the ability to build a crosslinked PU network rapidly. The incorporation of lactic acid increased the amount of the hydroxyl group to a fatty acid moiety and increased its overall hydroxyl functionality, making it more reactive. The ring opening of epoxidized algae oil with ethylene glycol shows both primary and secondary hydroxyl groups. The authors evaluated the properties of rigid PU foams using differential scanning calorimetric and thermogravimetric analyses; these foams were comparable with petroleum-based PU foam. Algae oil can also be used in combination with other diacids to make alkyd and polyester amide polyols to synthesize PU coatings.[34] These algae oil-derived coatings showed increased chemical stability and antimicrobial, water repellent, anticorrosive, thermal, and mechanical properties.

Bio-based diacids for polyester polyols

Diacid monomers from renewable resources are materials for polyester polyol synthesis in industry—lactic, adipic, succinic, sebacic, azelaic, and other acids.[1] Polyester polyols from four-carbon succinic acid are well known for producing PU with improved durability, solvent resistance, scratch resistance, and other mechanical properties. The improvement is mostly due to higher ester density, which makes higher polar interactions in PU chains. Catalytic hydrogenation of petroleum-based maleic anhydride has been replaced by a fermentation method to produce succinic acid.[35]

According to the International Energy Agency, adipic acid is the most important platform for applications such as plasticizers, cosmetics, lubricants, pharmaceuticals, and paper additives.[36] Similar to succinic acid, it is made by either indirect fermentation combined with chemical conversion or direct fermentation. Azelaic acid was made primarily from oleic acid originated from vegetable oil; now it can be synthesized from algae, a non-food biomass.[33] Sebacic acid, a natural dicarboxylic acid, is produced from castor oil by cleavage of ricinoleic acid. Sebacic and azelaic acid can be used as monomers for polyester polyol, nylon, plasticizers, lubricants, cosmetics, and other applications.[1]

Bio-based diisocyanates and polyisocyanates

Isocyanates are the other component of a PU formulation, reacting with the polyol. Those used in PU contain at least two functional groups and are manufactured at an industrial scale using phosgenation. Alternative methods such as that of Curtius, Hofmann, and Lossen's rearrangements have been explored, which aim to avoid the use of toxic phosgene.[37,38] A drawback is the highly exothermic decomposition of carbonyl nitrene intermediates, which poses a significant barrier for safe application at industrial scale.[39] Flow chemistry has provided safe alternatives to conducting nitrogen-carbon rearrangements thanks to the development of lab and industrial scale setups.[40] Advancements in this technology have allowed alternative rearrangement at kilogram scale.[3,5–7] Further development of flow synthesis of isocyanates to increase throughput is crucial for eliminating the need for phosgenation in future industrial production.[41] Several types of biomass, including wood, agriculture products, solid waste, and algae, have been used to synthesize bio-based isocyanates (Fig. 3.3).[42]

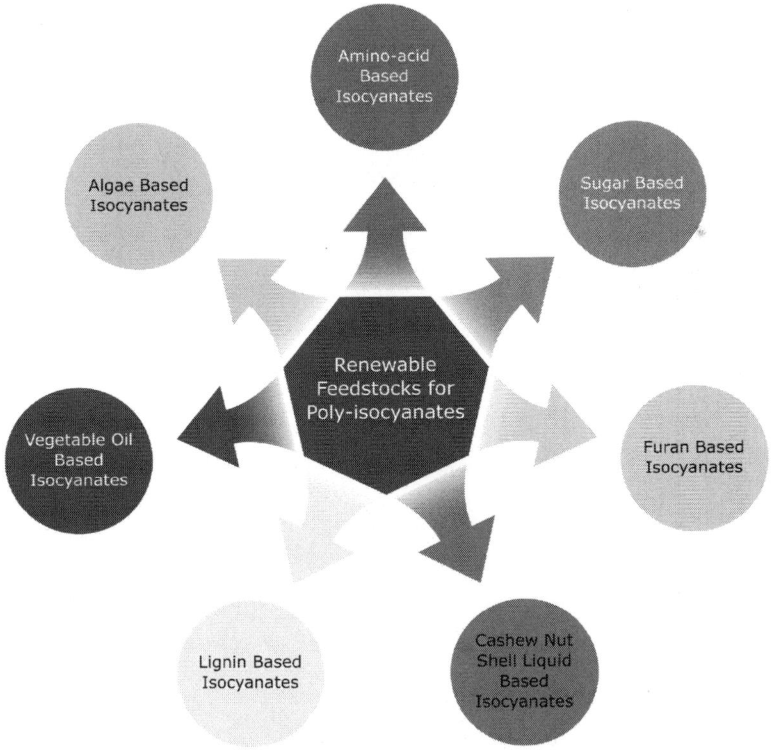

FIG. 3.3 Seven major renewable feedstocks that can be used for the synthesis of isocyanates. Most isocyanates are still made from petroleum and have no direct drop-in renewable replacements save for a couple of aliphatic varieties.

Most isocyanates are synthesized from petrochemical precursors, limiting the degree to which PU materials can be formulated from bio-based materials, including algae.[16] Bio-based sources have been researched for the production of isocyanates. An advantage of researching bio-based sources is the potential to discover new monomers with different mechanical properties and improve the overall biodegradability of PU formulations. While algae have not been extensively studied for their potential to produce functionalized molecules, plenty of other biomass sources ranging from wood to vegetable oils have been.[43–45] These molecules would directly substitute for petroleum-based monomers or provide entirely novel structures to redirect chemical and energy demands away from petroleum and toward environmentally friendlier alternatives. Bio-sourced isocyanates are grouped into seven classes (Fig. 3.3).[42] These seven were sourced from algae, wood, waste, and crop biomass,[42] all of which produced platform chemicals deemed suitable for the manufacture of monomer precursors for polymer synthesis.[46,47] During biodegradation, cleavage of the urethane bonds by hydrolysis, alcoholysis, acidolysis, glycolysis, or aminolysis results in free amine and the respective polyol, which was bonded to the isocyanate.[48] Enzymatic hydrolysis has also been reported to break down renewable polyester PU foams into the original polyol components and the amine derivative of the isocyanate.[49]

The α-amino acid L-lysine is a common nutrient in meat, fish, dairy, eggs, and plant sources such as soy. Its most common synthetic pathways are the diaminopimelate [50] and the α-aminoadipate.[51] It has been used as a precursor for methyl and ethyl ester L-lysine diisocyanates in PU applications.[52–55] Synthesis of diisocyanate is achieved by phosgenation of amine-terminated lysine ester.[54] Lysine-derived isocyanates have low vapor pressure, making them easier to adapt to manufacturing processes than other aliphatic diisocyanates.[56] Lysine-based diisocyanate is very promising for biomedical applications because of its ability to generate nontoxic breakdown products upon biodegradation.[57–60] These compounds include diamines, amino acid lysine, and carboxylic acid-terminated PU fragments.[61–64] The asymmetry in lysine-based isocyanates results in the generation of amorphous PU regions and disproportionately affects the reactivity of each isocyanate group.[53] The lysine-based diisocyanates have two NCO (nitrogen, carbon, oxygen) groups, α-NCO and ε-NCO, which are secondary and primary, respectively. The reactivity of ε-NCO is slightly higher because it is a primary functional group. α-NCO is also affected by the ethoxycarbonyl group, which introduces both steric hindrance and electron withdrawing, further reducing its reactivity.[52,65,66] The average reactivity of these two groups makes lysine-based diisocyanates on par with commercially available hexamethylene diisocyanate.[67,68] The average reactivity is also higher than that of commercial isophorone diisocyanate, but still less than toluene diisocyanate.[65,69] The increased reactivity can be explained because isophorone diisocyanate has high steric hindrance and poor electrophilicity. Steric hindrance makes it slower to react than aromatic diisocyanates such as toluene diisocyanate, which use delocalized electrons in the π-bonds and their planar structure to produce highly reactive NCO groups.[65,70]

Apart from animal-based sources, plant-based sources rich in starches or oils such as sugar, purified starch, cashew, or a variety of vegetable oils can be easily converted into polymer precursors. Starch or glucose can produce heterocyclic compounds called dianhydrohexitols, synthesized by hydrogenolysis of hydrolyzed starch or by double dehydration of sorbitol, which can be obtained from reduced glucose.[71] There are three possible configurations for dianhydrohexitols, which possess two reconfigurable hydroxyl groups, making it possible to convert diisocyanate via phosgenation.[71,72] Instead of dehydrating sorbitol, it can be turned into isosorbide by hydration, another polymer precursor highly valued for its thermal stability and rigid structure.[73] Both isomers of isosorbide can be converted into diisocyanates by first converting to their diacid chloride derivatives, followed by a multistep Curtius rearrangement. However, the yield is only approximately 55%.[73] The difference in the C2 position of isosorbide and its isomer isomannide-derived isocyanates produces stereochemically well-defined PU, also dependent on the ratio of the isomers.[73] Sugar-based diisocyanates were first placed on the market by Covestro under the name DESMODUR eco N7300, which is pentamethylene diisocyanate. The biomass-derived hydrocarbon backbone of pentamethylene diisocyanate gives it a renewable content of 70%, with the NCO groups added via a series of biotechnological and phosgenation-based procedures.[41,42] This approach gives a market advantage to pentamethylene diisocyanate when competing against other aliphatic diisocyanates such as hexamethylene diisocyanate because of its high energy efficiency and extremely low carbon footprint.[41,43]

The furan-based compounds furfural, 5-hydroxymethylfurfural, and 2,5-furan dicarboxylic acid are high-value bio-based chemicals by virtue of their versatility and ease of use in converting to renewable precursors for polymer formulation.[46,74–77] 2,5-Furandicarboxylic acid can be converted into various precursors for polymers, including polyamides, polyester, and PU.[74,78] Garber patented the first furan-based diisocyanate in 1962.[79] Soon after, 2,5-diisocyanatofuran was synthesized from methyl pyromucate (2-carbomethoxyfuran) through a five-step procedure including chloromethylation and oxidation, followed by the formation and Curtius rearrangement of acyl azide from acid chloride.[80] Furan 2,5-diisocyanate shows promise as a renewable diisocyanate due to its heteroatomic structure, as renewable alternatives are scarce. However, it is difficult to store because of high reactivity with moisture and oxygen.[81] In addition, a series of furan-based diisocyanates were developed for PU formulation, including methylenebis(2,5-furandiylmethylene) diisocyanates and ethylidene bis(2,5-furandiylmethylene) diisocyanate.[82] Those furan-derived chemicals hold high research interest because of structural elements analogous to well-known petroleum-based isocyanates such as 1,4-phenylene diisocyanate and 4,4′-methylenediphenyl diisocyanate. Furan-based diisocyanates also have a significant advantage due to the hypothesized generation of nontoxic by-products upon breakdown, furan diamines. These diamines can be further broken down into ammonia and erythritol. Petrochemical aromatic diisocyanates produce toxic aromatic amine degradation products, reported to be potent carcinogens and mutagens, which hinders their use.[83–86] Decomposition of the urethane group by chemical or biological degradation, including hydrolysis, alcoholysis, acidolysis, glycolysis, and aminolysis, results in the free amine and polyol.[48] Enzymatic hydrolysis has been reported to break down renewable PU foams into the original monomers.[49] The effects of furan-derived precursors on polymer synthesis and mechanical properties have been reported in detail.[87–90] The presence of furan rings does not chemically affect the structure or the final molecular weight of the resulting polymer.[88] Some authors claimed that the introduction of furan moieties increases free volume in a polymer matrix, resulting in lower glass transition temperature.[87] Thermal degradation was also affected significantly with the higher percentage of furan rings in the polymer structure, which is advantageous for applications where the polymer is intended to biodegrade.[88]

For some diisocyanates such as furan-derived 2,5-diisocyanatofuran, high reactivity can pose a challenge toward practical application in polymer formulation. The instability of 2,5-diisocyanate can be explained by the electron-withdrawing nature of the aromatic ring and the steric hindrance because of the close placement of the NCO groups.[65] The oxygen on the furan ring is sp^2 hybridized, while its lone pair is involved in an sp^2 orbital, producing an electronic induction effect in the aromatic ring. These two factors make the furan ring electron rich and planar,[91,92] which increases the reactivity of the isocyanate groups due to high electrophilicity. Highly reactive groups can be protected by blocking agents, which cap the ends of the functional groups until needed in a polymerizing reaction. The technique of blocked isocyanates is beneficial in applications such as coatings or adhesives.[93–96] The blocked isocyanate remains inactive until a suitable processing temperature is reached, which uncaps the blocking agents, reverting the isocyanate to its natural form and opening it up to reacting with the intended polyol or prepolymer blend. The deblock temperature can be manipulated through the chemical structure of the blocking agent.[93,94] p-Cyanophenol is one blocking agent particularly popular for the aforementioned furan-based diisocyanates, but unfortunately it requires high temperatures or loaded catalysts to uncap the isocyanate groups during processing.[97]

One of the largest sources of aromatic molecules comes from cellulosic biomass in the form of lignin. This helps make up the vascular tissue of plants and some algae, giving them their rigid forms.[98] The highly cross-linked phenolic structure of lignins makes them a good resource for producing chemicals and polymers.[99] Lignins are naturally formed polymers themselves. Lignin processing is already well developed, being a crucial step for industries such as paper mills. Most common by far is the lignosulfonate process, accounting for about 88% of total lignin extraction.[100,101] The remaining 12% is made up of the kraft, alkaline, and organosolv processes.[100,101] While the lignosulfonate and kraft processes use sulfur, the other two do not, which avoids incorporating functionalized sulfur into the resulting lignin mix.[100,101] Given its sulfur-free nature, the organosolv process will have the highest share growth in upcoming years.[102,103] With the global production of 100 million tonnes per year in 2015, the market's valuation is expected to rise by approximately 25% by 2025.[102,104] Several lignin-derived components such as vanillin, phenols, carbon fiber, oils, and other aromatic compounds are heavily researched for further incorporation into polymer production. The expected growth is needed to satisfy increasing demand as more lignin-based products enter the market.[42,101] As for isocyanates, several aromatic diisocyanates, including bis(4-isocyanato-2-methoxyphenoxy)alkanes, bis(4-isocyanato-2,6-dimethoxyphenoxy)alkanes, (E)-1,2-bis(4-cyanato-3-methoxyphenyl)ethane, and 1,2-bis(4-cyanato-3-methoxyphenyl)ethane, were produced from lignin-derived vanillin and vanillic acid.[70,71] Renewable PU was synthesized using the aforementioned lignin-based isocyanates via polycondensation reactions with a bio-based polyol to test thermoplastic and thermosetting PU applications.[70,71]

One of the promising alternative sources for isocyanate production precursors can be derived from the cashew fruit.[105] Large abundance, low cost, and nonintrusion into food markets make it a target for synthesizing various polymer precursors.[42,106] The dark, viscous extract is rich in anacardic acid, cardanol, cardol, and 2-methylcardol.[107] Termed cashew nutshell liquid, the phenolic compounds it contains have side chains with multiple degrees of unsaturation at the meta position.[106] Taking advantage of the chemical structures, More and Sadavarte synthesized 2,4-diisocyanate-1-pentadecylbenzene, a renewable and structural analog for toluene diisocyanate.[88,89]

One of the primary targets for producing PU precursors is vegetable oils, which contain various components with diverse unsaturated sites that can be converted to a wide range of polymer precursors.[108–112] Vegetable oils are advantageous by their nontoxic nature, biodegradability, and renewability.[91,92] Dimer acid diisocyanates were among the first fatty acid diisocyanates, synthesized in the 1970s and commercialized by the Henkel Corporation and General Mills.[113–116] Allylic bromides of some soybean oil triglycerides can be substituted using AgNCO to produce isocyanate groups.[117] Oleic acid derived from triglycerides was also used to create linear aliphatic diisocyanates by the Curtius rearrangement.[118,119] Two types of diisocyanates have been prepared by this method via a diacyl hydrazide intermediate, giving high yield and purity, including 1-isocyanato-10-[(isocyanatomethyl)thio]decane and 1,7-diisocyanatooctane.[120] However, Curtius rearrangement involving a nitrene intermediate is limited to lab operation because of safety hazards if scaled up.

Some of these vegetable oil triglycerides are also produced by algae. Recently, diisocyanates derived from algae-based oil have been synthesized using flow chemistry to complete the Curtius rearrangement, lowering the production risks associated with batch chemistry and opening new opportunities for industrial-scale production.[41] The algae oil was a waste product from algae farms producing omega-3 fatty acids. Waste biomass could be purified and converted into azelaic acid, a valuable precursor to producing algae-based polyols and diisocyanates.[33,41]

Recently, University of California San Diego researchers developed a method for producing micro-based polyols and micro-based diisocyanate, monomer units for PU polymer (Fig. 3.4). Algae are now considered one of the best renewable resources for replacing fossil fuel to limit global warming without threatening food supplies. However, the small organic contaminants such as photosynthetic pigments and other cofactors in algae oil can complicate use. Oil from the green microalgae *Nannochloropsis salina* is a common source of omega-3 fatty acids, sold as dietary supplements. The leftover oils, more than 70%, are typically thrown away or burned, but the researchers found a better use. They developed a process (Fig. 3.4) to purify and convert this waste stream into azelaic and heptanoic acid.[33] Azelaic acid is used to synthesize algae-based polyol, a building block for PU. A coproduct, heptanoic acid, can be converted to methyl heptanoate, which is used as food flavoring and fragrance. The team also developed algae-based diisocyanates through the existing Curtius rearrangement, which converts acyl azides into diisocyanates (Fig. 3.4B). Because of the explosiveness of intermediates, preparing large quantities safely is a challenge. Therefore they turned to flow chemistry, using pumps and tubing to conduct chemical reactions in a continuous flow stream, so that azelaic acid is converted to azelaic dihydrazide, a stable and easily scalable precursor. It was then reacted with nitrous acid inflow to obtain azelaic diacylazide, which was immediately isolated and went directly into a high-temperature flow reactor to be converted into heptamethylene diisocyanate.[41]

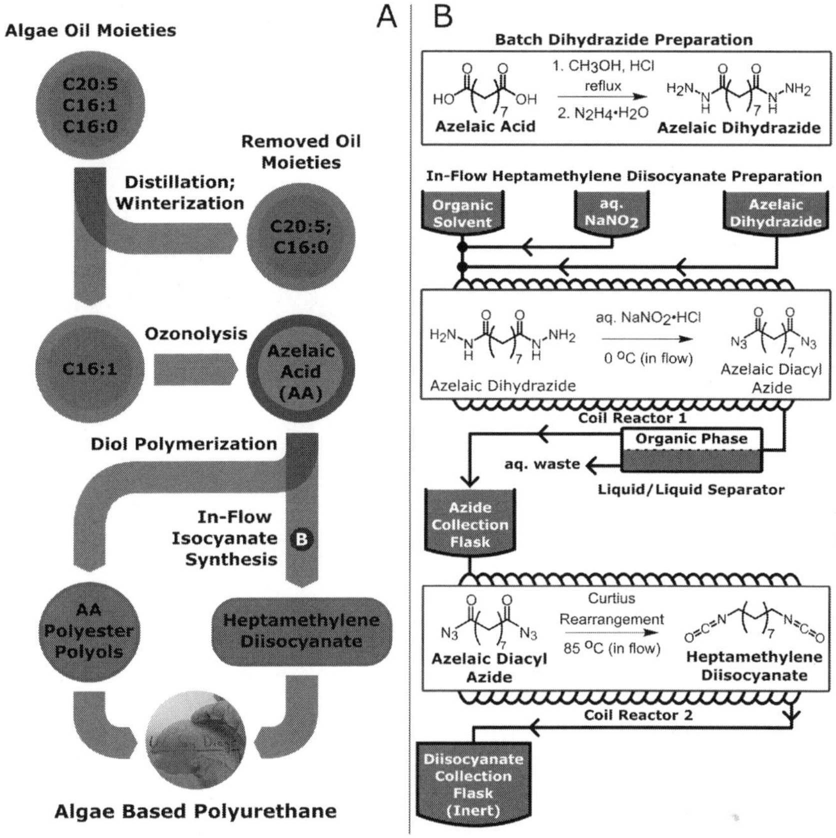

FIG. 3.4 (A) The process of separating algae oil moieties and synthesizing azelaic-based PU components. The final products are all based on azelaic acid. (B) Detailed process flow for the conversion of azelaic dihydrazide into heptamethylene diisocyanate. The azelaic dihydrazide is prepared beforehand using a batch chemistry process under reflux. Conducting the next steps in flow chemistry to prepare heptamethylene diisocyanate from azelaic dihydrazide is crucial to lowering the safety risks of the following reactions.

Conclusions

Now that we have increasingly innovative and comparable ways to capture and retain renewable feedstocks, they are becoming a more critical resource for supplementing and eventually replacing human needs for producing both plastic-based goods and combustible fuels. The expansion in renewables is now moving from lab scale to pilot before going to industrial scale. Hurdles such as quality and cost competitiveness of bio-based raw material remain, and this has slightly slowed the process. However, with the increased sustainable development support from the government and recognition of the ecological benefits of bio-based materials, the cost would be competitive if the consumer is willing to pay more for a completely renewable product.

Close-up: One company's approach to providing cost-competitive algae products

Algae bioproducts have the potential to help solve many of the world's environmental dilemmas while providing rural rejuvenation and a higher standard of living. This potential comes from the unparalleled productivity for oil and protein with algae farming compared to other crops. The overriding impediment to realizing this potential is achieving a cultivation and processing system that is cost-competitive with that of existing oil and protein commodities. Conventional algae technology is an order of magnitude too expensive, so major innovations are required. Global Algae Innovation's approach is to combine radical innovations in cultivation, harvesting, and processing with a biorefinery concept producing multiple products.

The primary products from algae are oil and protein. In the long run, algae oil will be used to replace palm oil in consumer and food products and to produce renewable jet, diesel, and gasoline fuels. The protein will be used as an animal feed and human food ingredient. The price point for these markets is approximately \$1/kg of oil or protein. The current protein markets will support about 70,000,000 acres of algae, but achieving this price point is very challenging. An alternative product spectrum is needed to accelerate deployment of algae farming, and algae oil-based biopolymers will be an important part of the mix.

The algae oil can be separated into saturated, monounsaturated, and polyunsaturated fatty acids. The saturated fatty acids are an ideal feedstock for renewable jet and diesel fuel; the monounsaturated for polyurethane production; and the polyunsaturated are needed for aquaculture feed, human nutrition, and potentially polyurethane production. Initially, a high-quality protein concentrate will be produced for human and animal food, for example in aquaculture. The average price point for these oil and protein markets is \$2/kg. While these markets are smaller than the overall protein and oil markets, this product spectrum will support the first 4,000,000 acres of algae, which is plenty to launch algae farming for commodities as a worldwide industry.

Global Algae Innovations has developed cultivation and processing technology projected to achieve a good return on investment at full scale for the initial markets described before. Currently, the process is being scaled up twentyfold, from the 8-wetted-acre R&D farm (Fig. 3.5) to a 160-acre pilot farm that will produce approximately 1500 tons of algae oil per year. Groundbreaking for this farm is planned for September 2022. After successful operation of the pilot farm, the process will be scaled up to approximately 5000 acres.

FIG. 3.5 Algae ponds operated by Global Algae Innovations in Kauai, HI. © *Reproduced with permission from The Image Group (TIG) and Andy Stenz.*

With scale-up underway, algae for commodities is just around the corner, and algae biopolymers will be at the front of the line for use of algae oil as the industry takes off.

David Hazlebeck

Chief Executive Officer and Founder, Global Algae Innovations, Inc., El Cajon, CA, United States

References

1. Hai TAP, Tessman M, Neelakantan N, et al. Renewable polyurethanes from sustainable biological precursors. *Biomacromolecules.* 2021;22:1770–1794.
2. Williams CK, Hillmyer MA. Polymers from renewable resources: a perspective for a special issue of polymer reviews. *Polym Rev.* 2008;48:1–10.

3. Papageorgiou GZ. Thinking green: sustainable polymers from renewable resources. *Polymers.* 2018;10:952.
4. Pandey S, Rajput B, Chikkali SH. Refining of plant oils and sugars to platform chemicals, monomers, and polymers. *Green Chem.* 2021;23:4255–4295.
5. Gandini A, Lacerda TM, Carvalho AJF, Trovatti E. Progress of polymers from renewable resources: furans, vegetable oils, and polysaccharides. *Chem Rev.* 2016;116:1637–1669.
6. Miao S, Wang P, Su Z, Zhang S. Vegetable-oil-based polymers as future polymeric biomaterials. *Acta Biomater.* 2014;10:1692–1704.
7. Fabris M, Abbriano RM, Pernice M, et al. Emerging technologies in algal biotechnology: toward the establishment of a sustainable, algae-based bioeconomy. *Front Plant Sci.* 2020;11.
8. Herrmann R, Jumbe C, Bruentrup M, EvansOsabuohien. Competition between biofuel feedstock and food production: empirical evidence from sugarcane outgrower settings in Malawi. *Biomass Bioenergy.* 2018;114:100–111.
9. Slade R, Bauen A. Micro-algae cultivation for biofuels: cost, energy balance, environmental impacts and future prospects. *Biomass Bioenergy.* 2013;53:29–38.
10. Blatti JL, Burkart MD. Releasing stored solar energy within pond scum: biodiesel from algal lipids. *J Chem Educ.* 2012;89:239–242.
11. Waghmare A, Patil S, LeBlanc JG, Sonawane S, Arya S, S. Comparative assessment of algal oil with other vegetable oils for deep frying. *Algal Res.* 2018;31:99–106.
12. Peyton J, Chambaretaud C, Sarbu A, Avérous L. Biobased polyurethane foams based on new polyol architectures from microalgae oil. *ACS Sustain Chem Eng.* 2020;8:12187–12196.
13. Engels H-W, Pirkl H-G, Albers R, et al. Polyurethanes: versatile materials and sustainable problem solvers for today's challenges. *Angew Chem.* 2013;52:9422–9441.
14. Mohd Noor N, Sendijarevic A, Sendijarevic V, et al. Comparison of adipic versus renewable azelaic acid polyester polyols as building blocks in soft thermoplastic polyurethanes. *J Am Oil Chem Soc.* 2016;93:1529–1540.
15. Feng Y, Liang H, Yang Z, et al. A solvent-free and scalable method to prepare soybean-oil-based polyols by thiol–ene photo-click reaction and biobased polyurethanes therefrom. *ACS Sustain Chem Eng.* 2017;5:7365–7373.
16. Furtwengler P, Avérous L. Renewable polyols for advanced polyurethane foams from diverse biomass resources. *Polym Chem.* 2018;9:4258–4287.
17. Li Y, Luo X, Shengjun H. Bio-Based Polyols and Polyurethanes. Cham: Springer; 2015.
18. Miyajima T, Nishiyama K, Satake M, Tsuji T. Synthesis and process development of polyether polyol with high primary hydroxyl content using a new propoxylation catalyst. *Polym J.* 2015;47:771–778.
19. Carothers WH. Polymerization. *Chem Rev.* 1931;8:353–426.
20. Carothers WH. Studies on polymerization and ring formation. I. An introduction to the general theory of condensation polymers. *J Am Chem Soc.* 1929;51:2548–2559.
21. Carothers WH, Dorough GL, Natta FJv. Studies of polymerization and ring formation. X. The reversible polymerization of six-membered cyclic esters. *J Am Chem Soc.* 1932;54:761–772.
22. Chen H, Lee S-Y, Lin Y-M. Synthesis and formulation of PCL-based urethane acrylates for DLP 3D printers. *Polymers.* 2020;12:1500.
23. Yoshioka M, Miyata A, Yagi T, Nishio Y. Preparation of polyols from methyl-α-d-glucoside and cyclic esters for design and fabrication of biodegradable polyurethane foams. *J Wood Sci.* 2004;50:511–518.
24. Emrani J, Benrashid R, Mohtarami S, Fini E, Abu-Lebdeh T. Synthesis and characterization of bio-based polyurethane polymers. *Am J Eng Appl Sci.* 2018;11:1298–1309.

25. Dominguez-Rosado E, Liggat JJ, Snape CE, Eling B, Prechtel J. Thermal degradation of urethane modified polyisocyanurate foams based on aliphatic and aromatic polyester polyol. *Polym Degrad Stab.* 2002;78:1–5.
26. Datta J, Kasprzyk P. Thermoplastic polyurethanes derived from petrochemical or renewable resources: a comprehensive review. *Polym Eng Sci.* 2018;58:E14–E35.
27. Omrani I, Farhadian A, Babanejad N, Shendi HK, Ahmadi A, Nabid MR. Synthesis of novel high primary hydroxyl functionality polyol from sunflower oil using thiol-yne reaction and their application in polyurethane coating. *Eur Polym J.* 2016;82:220–231.
28. Arniza MZ, Hoong SS, Idris Z, et al. Synthesis of transesterified palm olein-based polyol and rigid polyurethanes from this polyol. *J Am Oil Chem Soc.* 2015;92:243–255.
29. Pillai PKS, Floros MC, Narine SS. Elastomers from renewable metathesized palm oil polyols. *ACS Sustain Chem Eng.* 2017;5:5793–5799.
30. Pawar MS, Kadam AS, Dawane BS, Yemul OS. Synthesis and characterization of rigid polyurethane foams from algae oil using biobased chain extenders. *Polym Bull.* 2016;73:727–741.
31. Arbenz A, Perrin R, Avérous L. Elaboration and properties of innovative biobased PUIR foams from microalgae. *J Polym Environ.* 2018;26:254–262.
32. Petrović ZS, Wan X, Bilić O, et al. Polyols and polyurethanes from crude algal oil. *J Am Oil Chem Soc.* 2013;90:1073–1078.
33. Hai TAP, Neelakantan N, Tessman M, et al. Flexible polyurethanes, renewable fuels, and flavorings from a microalgae oil waste stream. *Green Chem.* 2020;22:3088–3094.
34. Patil CK, Jirimali HD, Paradeshi JS, et al. Synthesis of biobased polyols using algae oil for multifunctional polyurethane coatings. *Green Mater.* 2018;6:165–177.
35. Nghiem NP, Kleff S, Schwegmann S. Succinic acid: technology development and commercialization. *Fermentation.* 2017;3:26–40.
36. Skoog E, Shin JH, Saez-Jimenez V, Mapelli V, Olsson L. Biobased adipic acid—the challenge of developing the production host. *Biotechnol Adv.* 2018;36:2248–2263.
37. Ullmann F. Ullmann's Encyclopedia of Industrial Chemistry. vol. 24. Wiley-VCH; 2003.
38. Ulrich H. Chemistry and Technology of Isocyanates. J. Wiley & Sons; 1996.
39. Zhao J, Gimi R, Katti S, et al. Process development of a GCS inhibitor including demonstration of Lossen rearrangement on kilogram scale. *Org Process Res Dev.* 2015;19:576–581.
40. Gutmann B, Kappe CO. Continuous manufacturing in the pharma industry—an unstoppable trend? *Eur Pharm Rev.* 2015;20:37–42.
41. Hai TAP, Backer LJSD, Cosford NDP, Burkart MD. Preparation of mono- and diisocyanates inflow from renewable carboxylic acids. *Org Process Res Dev.* 2020;24:2342–2346.
42. Tawade BV, Shingle RD, Kuhire SS, et al. Bio-based di-/poly-isocyanates for polyurethanes: an overview. *Polyurethane Today.* 2017;41–46. Technical Article.
43. Goldstein IS. Chemicals and fuels from biomass review and preview. *ACS Symp Ser.* 1992;476:332–338.
44. Oyedeji O, Gitman P, Qu J, Webb E. Understanding the impact of lignocellulosic biomass variability on the size reduction process: a review. *ACS Sustain Chem Eng.* 2020;8:2327–2343.
45. Wu L, Moteki T, Gokhale AA, Flaherty DW, DeanToste F. Production of fuels and chemicals from biomass: condensation reactions and beyond. *Chem.* 2016;1:32–58.
46. Lee Y, Kwon EE, Lee J. Polymers derived from hemicellulosic parts of lignocellulosic biomass. *Rev Environ Sci Biotechnol.* 2019;18:317–334.
47. Galbis JA, García-Martín MdG, de Paz MV, Galbis E. Synthetic polymers from sugar-based monomers. *Chem Rev.* 2016;116:1600–1636.
48. Harini Bhuvaneswari G. Degradability of polymers. In: Sabu T, et al., eds. *Recycling of Polyurethane Foams.* William Andrew Publishing; 2018:29–44.

49. Gunawan NR, Tessman M, Schreiman AC, et al. Rapid biodegradation of renewable polyurethane foams with identification of associated microorganisms and decomposition products. *Bioresour Technol Rep.* 2020;11, 100513.
50. Hudson AO, Bless C, Macedo P, et al. Biosynthesis of lysine in plants: evidence for a variant of the known bacterial pathways. *BBA-Gen Subjects.* 2005;1721:27–36.
51. Xu H, Andi B, Qian J, West AH, Cook PF. The α-aminoadipate pathway for lysine biosynthesis in fungi. *Cell Biochem Biophys.* 2006;46:43–64.
52. Sanda F, Takata T, Endo T. Synthesis of a novel optically active nylon-1 polymer: anionic polymerization of L -leucine methyl ester isocyanate. *J Polym Sci A Polym Chem.* 1995;33:2353–2358.
53. Storey RF, Wiggins JS, Puckett AD. Hydrolyzable poly(ester-urethane) networks from L -lysine diisocyanate and D,L -lactide/ε-caprolactone homo- and copolyester triols. *J Polym Sci A Polym Chem.* 1994;32:2345–2363.
54. Nowick JS, Powell NA, Nguyen TM, Noronha G. An improved method for the synthesis of enantiomerically pure amino acid ester isocyanates. *J Org Chem.* 1992;57:7364–7366.
55. Goyanes SN, D'Accorso NB. Industrial Applications of Renewable Biomass Products—Past, Present and Future. Springer International Publishing; 2017:1–43. Chapter 1.
56. Hafeman AE, Li B, Yoshii T, Zienkiewicz K, Davidson JM, Guelcher SA. Injectable biodegradable polyurethane scaffolds with release of platelet-derived growth factor for tissue repair and regeneration. *Pharm Res.* 2008;25:1–28.
57. Adhikari R, Gunatillake PA, Griffiths I, et al. Biodegradable injectable polyurethanes: synthesis and evaluation for orthopaedic applications. *Biomaterials.* 2008;29:3762–3770.
58. Mathew S, Baudis S, Neffe AT, Behl M, Wischke C, Lendlein A. Effect of diisocyanate linkers on the degradation characteristics of copolyester urethanes as potential drug carrier matrices. *Eur J Pharm Biopharm.* 2015;95:18–26.
59. Guelcher SA, Srinivasan A, Dumas JE, Didier JE, McBride S, Hollinger JO. Synthesis, mechanical properties, biocompatibility, and biodegradation of polyurethane networks from lysine polyisocyanates. *Biomaterials.* 2008;29:1762–1775.
60. Wang C, Cao X, Zhang Y. A novel bioactive osteogenesis scaffold delivers ascorbic acid, β-glycerophosphate, and dexamethasone in vivo to promote bone regeneration. *Oncotarget.* 2017;8:31612–31625.
61. Joshi DC, Saxena S, Jayakannan M. Development of l-lysine based biodegradable polyurethanes and their dual-responsive amphiphilic nanocarriers for drug delivery to cancer cells. *ACS Appl Polym Mater.* 2019;7:1866–1880.
62. Wang C, Xie J, Xiao X, Chen S, Wang Y. Development of nontoxic biodegradable polyurethanes based on polyhydroxyalkanoate and L-lysine diisocyanate with improved mechanical properties as new elastomers scaffolds. *Polymers.* 1927;2019:11.
63. Han J, Chen B, Ye L, Zhang A-Y, Zhang J, Feng Z-G. Synthesis and characterization of biodegradable polyurethane based on poly(ε-caprolactone) and L-lysine ethyl ester diisocyanate. *Front Mater Sci China.* 2009;3:25–32.
64. Thakur VK, Thakur MK. Handbook of Sustainable Polymers Processing and Applications. Chapter 22, Jenny Stanford Publishing; 2015:803–856.
65. Li Y, Noordover BAJ, Benthem RATMv, Koning CE. Reactivity and regio-selectivity of renewable building blocks for the synthesis of water-dispersible polyurethane prepolymers. *ACS Sustain Chem Eng.* 2014;2:788–797.
66. Gustini L, Lavilla C, Finzel L, Noordover BAJ, Hendrix MMRM, Koning CE. Sustainable coatings from bio-based, enzymatically synthesized polyesters with enhanced functionalities. *Polym Chem.* 2016;7:6586–6597.
67. Konieczny J, Loos K. Green polyurethanes from renewable isocyanates and biobased white dextrins. *Polymers.* 2019;11(2):256.

68. Li Y, Noordover BAJ, van Benthem RATM, Koning CE. Bio-based poly(urethane urea) dispersions with low internal stabilizing agent contents and tunable thermal properties. *Prog Org Coat.* 2015;86:134–142.
69. Gebauer T, Neffe AT, Lendlein A. Influence of diisocyanate reactivity and water solubility on the formation and the mechanical properties of gelatin-based networks in water. *Mater Res Soc Symp Proc.* 2013;1569:15–20.
70. Choe H, Kim JH. Reactivity of isophorone diisocyanate in fabrications of polyurethane foams for improved acoustic and mechanical properties. *J Ind Eng Chem.* 2019;69:153–160.
71. Thiem J, Bachmann F. Synthesis and properties of polyamides derived from anhydroand dianhydroalditols. *Makromol Chem.* 1991;192:2163–2182.
72. Bachmann F, Reimer J, Ruppenstein M, Thiem J. Synthesis of novel polyurethanes and polyureas by polyaddition reactions of dianhydrohexitol configurated diisocyanates. *Macromol Chem Phys.* 2001;202:3410–3419.
73. Zenner MD, Xia Y, Chen JS, Kessler MR. Polyurethanes from isosorbide-based Diisocyanates. *ChemSusChem.* 2013;6:1182–1185.
74. Motagamwala AH, Won W, Sener C, Alonso DM, Maravelias CT, Dumesic JA. Toward biomass-derived renewable plastics: production of 2,5-furan dicarboxylic acid from fructose. *Sci Adv.* 2018;4:eaap9722.
75. Delidovich I, Hausoul PJC, Deng L, Pfützenreuter R, Rose M, Palkovits R. Alternative monomers based on lignocellulose and their use for polymer production. *Chem Rev.* 2016;116:1540–1599.
76. van Putten R-J, van der Waal JC, de Jong E, Rasrendra CB, Heeres HJ, de Vries JG. Hydroxymethylfurfural, A versatile platform chemical made from renewable resources. *Chem Rev.* 2013;113:1499–1597.
77. Rosatella AA, Simeonov SP, Frade RFM, Afonso CAM. 5-Hydroxymethylfurfural (HMF) as a building block platform: biological properties, synthesis and synthetic applications. *Green Chem.* 2011;13:754–793.
78. Zhang Z, Deng K. Recent advances in the catalytic synthesis of 2,5-furandicarboxylic acid and its derivatives. *ACS Catal.* 2015;5:6529–6544.
79. Garber JD. Furfuryl Isocyanates. United States Patent Office; 1962. US 3049552.
80. Nielek S, Lesiak T. Isocyanates of heterocyclic compounds. I. The synthesis of 2,5-Diisocyanatofuran. *J Prakt Chem.* 1988;330:825–829.
81. Fabbro C, Armani S, Carloni LE, Leo FD, Wouters J, Bonifazi D. 2,5-Diamide-substituted five-membered heterocycles: challenging molecular synthons. *Eur J Org Chem.* 2014;2014:5487–5500.
82. Cawse JL, Stanford JL, Still RH. Polymers from renewable sources, 1. Diamines and diisocyanates containing difurylalkane moieties. *Makromol Chem.* 1984;185:697–707.
83. Benigni R, Passerini L. Carcinogenicity of the aromatic amines: from structure-activity relationships to mechanisms of action and risk assessment. *Mutat Res.* 2002;511:191–206.
84. Myers RC, Ballantyne B. Comparative acute toxicity and primary irritancy of various classes of amines. *Toxic Subst Mech.* 1997;16:151–194.
85. Greim H, Bury D, Klimisch HJ, Oeben-Negele M, Ziegler-Skylakakis K. Toxicity of aliphatic amines: structure-activity relationship. *Chemosphere.* 1998;36:271–295.
86. IARC Working Group on the Evaluation of Carcinogenic Risks to Humans. General discussion of common mechanisms for aromatic amines. In: *Some Aromatic Amines, Organic Dyes, and Related Exposures.* 99. International Agency for Research on Cancer; 2010.
87. Belgacem MN, Quillerou J, Gandini A. Urethanes and polyurethanes bearing furan moieties—3. Synthesis, characterization and comparative kinetics of the formation of urethanes. *Eur Polym J.* 1993;29:1217–1224.

88. Boufi S, Belgacem MN, Quillerou J, Gandini A. Urethanes and polyurethanes bearing furan moieties. 4. Synthesis, kinetics and characterization of linear polymers. *Macromolecules.* 1993;26:6706–6717.
89. Boufi S, Gandini A, Belgacem MN. Urethanes and polyurethanes bearing furan moieties: 5. Thermoplastic elastomers based on sequenced structures. *Polymer.* 1995;36:1689–1696.
90. Gandini A. The behaviour of furan derivatives in polymerization reactions. *Adv Polym Sci.* 1977;25:47–96.
91. Sousa AF, Vilela C, Fonseca AC, et al. Biobased polyesters and other polymers from 2,5-furan dicarboxylic acid: a tribute to furan excellency. *Polym Chem.* 2015;6:5961–5983.
92. Loos K, Zhang R, Pereira I, et al. A perspective on PEF synthesis, properties, and end-life. *Front Chem.* 2020;8:1–18.
93. Wicks DA, Jr ZWW. Blocked isocyanates III: part A. Mechanisms and chemistry. *Prog Org Coat.* 1999;36:148–172.
94. Wicks DA, Jr ZWW. Blocked isocyanates III part B: uses and applications of blocked isocyanates. *Prog Org Coat.* 2001;41:1–83.
95. Rolph MS, Markowski ALJ, Warriner CN, O'Reilly RK. Blocked isocyanates: from analytical and experimental considerations to non-polyurethane applications. *Polym Chem.* 2016;7:7351–7365.
96. Lucas F, Ernst M, Zipfel HF, Panchenko A. Blocking Agents for Isocyanates. European Patent Application; 2020. EP3643733A1.
97. Neumann CND, Bulach WD, Rehahn M, Klein R. Water-free synthesis of polyurethane foams using highly reactive diisocyanates derived from 5-hydroxymethylfurfural. *Macromol Rapid Commun.* 2011;32:1373–1378.
98. Martone PT, Estevez JM, Lu F, et al. Discovery of lignin in seaweed reveals convergent evolution of cell-wall architecture. *Curr Biol.* 2009;19:169–175.
99. Laurichesse S, Avérous L. Chemical modification of lignins: towards biobased polymers. *Prog Polym Sci.* 2014;39:1266–1290.
100. Sharapova OV, Chistyakov AV, Tsodikov MV, Moiseev II. Lignin as a renewable resource of hydrocarbon products and energy carriers (a review). *Pet Chem.* 2020;60:227–243.
101. Sagues WJ, Jain A, Brown D, et al. Are lignin-derived carbon fibers graphitic enough? *Green Chem.* 2019;21:4253–4265.
102. Bajwa DS, Pourhashem G, Ullah AH, Bajwa SG. A concise review of current lignin production, applications, products and their environmental impact. *Ind Crop Prod.* 2019;139, 111526.
103. Glasser WG. About making lignin great again—some lessons from the past. *Front Chem.* 2019;7:565.
104. Demuner IF, Colodette JL, Demuner AJ, Jardim CM. Biorefinery review: wide-reaching products through Kraft lignin. *BioResources.* 2019;14:7543–7581.
105. Rwahwire S, Tomkova B, Periyasamy AP, Kale BM. Chapter 3—Green thermoset reinforced biocomposites. In: *Green Composites for Automotive Applications.* Woodhead Publishing; 2019:61–80.
106. Voirin C, Caillol S, Sadavarte NV, Tawade BV, Boutin B, Wadgaonkar PP. Functionalization of cardanol: towards biobased polymers and additives. *Polym Chem.* 2014;5:3142–3162.
107. Parambath A. Cashew Nut Shell Liquid. Cham: Springer; 2017. https://doi.org/10.1007/978-3-319-47455-7.
108. Xia Y, Larock RC. Vegetable oil-based polymeric materials: synthesis, properties, and applications. *Green Chem.* 2010;12:1893–1909.
109. Islam MR, Beg MDH, Jamari SS. Development of vegetable-oil-based polymers. *J Appl Polym Sci.* 2014;131.

110. Maisonneuve L, Chollet G, Grau E, Cramail H. Vegetable oils: a source of polyols for polyurethane materials. *Oilseeds Fats Crops Lipids.* 2016;23:D508.
111. Dubois V, Breton S, Linder M, Fanni J, Parmentier M. Fatty acid profiles of 80 vegetable oils with regard to their nutritional potential. *Eur J Lipid Sci Technol.* 2007;109:710–732.
112. Karak N. Vegetable Oil-Based Polymers Properties, Processing and Applications. Woodhead Publishing; 2012:1–336.
113. Dhahran MRK, Kuder RC. Diisocyanates. US Patent 3691225; 1972.
114. Coady CJ, Krajewski JJ, Bishop TE. Polyacrylated Oiligomers in Ultraviolet Curable Optical Fiber Coatings. US patent 4608409; 1986.
115. Bishop TE, Coady CJ, Zimmerman JM. Ultraviolet Curable Buffer Coatings for Optical Glass Fiber Based on Long-Chain Oxyalkylene Diamines. US Patent 4609718; 1986.
116. Casella AF. Interbonded Compresses Polyurethane Foam Material and Method of Making Same. US Patent 3622435; 1971.
117. Çaylı G, Küsefoğlu S. Biobased polyisocyanates from plant oil triglycerides: synthesis, polymerization, and characterization. *J Appl Polym Sci.* 2008;109:2948–2955.
118. Hojabri L, Kong X, Narine SS. Fatty acid-derived diisocyanate and biobased polyurethane produced from vegetable oil: synthesis, polymerization, and characterization. *Biomacromolecules.* 2009;10:884–891.
119. Hojabri L, Kong X, Narine SS. Novel long-chain unsaturated diisocyanate from fatty acid: synthesis, characterization, and application in bio-based polyurethane. *J Polym Sci A Polym Chem.* 2010;48:3302–3310.
120. More AS, Lebarbé T, Maisonneuve L, Gadenne B, Alves C, Cramail H. Novel fatty acid-based di-isocyanates towards the synthesis of thermoplastic polyurethanes. *Eur Polym J.* 2013;49:823–833.

PART III

Redefining the analytics

CHAPTER

4

Biodegradation: The biology

Natasha R. Gunawan[a,b], Michael T. Read[b], and Woodrow R. Brown[b]

[a]Algenesis Materials, Cardiff, CA, United States,
[b]University of California San Diego, La Jolla, CA, United States

Introduction

Among sustainable alternatives to conventional materials, those marketed as biodegradable are becoming more prevalent. Biodegradation is the process wherein materials break down into less toxic components in the environment rather than accumulating after use. Microorganisms cleave apart the chemical bonds, converting the components into cell biomass and inorganic products.[1] This process occurs naturally for raw materials in nature, such as plant matter, but it can also be used to bioremediate synthetic products persisting in the environment, such as certain polymers (plastics).

Plastic waste is of increasing concern with trends suggesting that 33,000 metric tons of total plastic waste will be produced by 2050.[2] Most conventional plastics are nondegradable, meaning they do not break down chemically into less toxic components; rather, they may fragment into smaller pieces, known as microplastics if less than 5 mm.[3] These persist in the environment, leading to a variety of ecotoxicological effects. Plastic debris has been found in all major ocean basins. Of special concern are the bioaccumulation effects of microplastics on marine organisms, which can disrupt the health of marine ecosystems.[4]

Biodegradable plastics could provide a solution to the problem. Rather than accumulating after use, they break apart into products that are incorporated into the environment with little impact. This chapter uses the term "biodegradable plastic" for those that can be wholly broken down in chemical structure and converted into nontoxic components by resident microorganisms. The term does not indicate the type of environment in

https://doi.org/10.1016/B978-0-323-99982-3.00001-8

which the product will biodegrade, so consumers must be made aware of how to dispose of the product at the end of use. The local environment is a large determining factor of biodegradability, in addition to the type of plastic and other properties. Because of their chemical structures, most conventional plastics cannot be broken apart by biological processes. Helping consumers understand the process for biodegradation can allow them to make informed decisions about products. Here we shed light on polyester polyurethanes (PUs) and explain how they can be designed to biodegrade completely.

Definitions

For consumers looking to live more sustainably, there are several alternatives to conventional plastics, though some are not biodegradable. The following characterizations offer a basis for a system that should allow consumers to make rational choices.

Plant materials in their raw form (as found in nature) eventually break down. Thus they can be used to make biodegradable products, provided that they are not reacted with a nondegradable structure or combined with a nondegradable material. Examples among plant fibers are cotton, jute, bamboo, and hemp.[5] These are polymeric in nature, which makes them useful for applications such as apparel. However, they are not always suitable replacements of nondegradable counterparts due to their limited physical specifications and tendency to break down rather quickly. Natural rubber is an example of a plant material used for its polymeric structure. It is generally thought to be biodegradable, but studies have shown the process to be slow and, as with any biodegradable product, dependent on the environment.[6] Rubber is vulcanized or blended with other plastics, creating a nonbiodegradable product. Raw plant materials differ greatly from bio-based plastics such as PUs from algae. Bio-based plastics are made by extracting compounds from a renewable biomass source and converting those compounds into starting materials, which are then reacted to make a synthetic polymer that may or may not be biodegradable. A bio-based plastic can consist of starting materials that are either entirely or partly bio-based.[7] The aim is to reduce the need for nonrenewable feedstocks such as petroleum, from which conventional plastics are made.

Although biodegradable and bio-based plastics are both environmentally friendly, they are not the same.[8] Bio-based plastics can be designed to have the same chemical structure as nondegradable plastics and have been more widely produced in recent years.[2] A petroleum-based nondegradable plastic persists in the environment for the same amount of time as a bio-based plastic with the same chemical structure. A recent example is the PlantBottle by the Coca-Cola Company; although its chemical feedstock is derived partly from sugar cane, it is a polyethylene-terephthalate (PET), common and nondegradable.[2] Biodegradable plastics are chemically designed to biodegrade, regardless of the source of the starting material.[8] Biodegradable plastics are able to break apart and have the breakdown products incorporated into the natural environment in a reasonable amount of time, according to the ASTM International definition. The biodegradation process occurs through naturally occurring microorganisms in environments such as compost, soil, and landfill.

Although not all bio-based plastics are biodegradable and not all biodegradable plastics are bio-based, products with both qualities exist, such as polyester PU with monomers sourced from a microalgae oil waste stream.[9] Another example is polyhydroxyalkanoate (PHA), made from sugar derived from corn.[10] Fossil fuel, the feedstock for synthesizing most plastics, is a finite resource that causes a myriad of environmental problems, such as in extraction (fracking, drilling), the greenhouse gas emissions from processing, and oil spills. Using bio-based feedstock can decrease the carbon footprint of a product. And making it biodegradable provides a solution to a separate issue: accumulation in the environment. This illuminates the promising technology of bio-based polyester PUs, which are renewably sourced and biodegradable. Here, we include a list of key terms useful for discussing the topic of biodegradation of polyester PUs (Table 4.1).

TABLE 4.1 Key terms in the biodegradation of polyester polyurethanes.

Term	Definition
Biodegradable plastic	A plastic designed to undergo breakdown of its chemical structure due to naturally occurring microorganisms in a specific environment, without harm to the surrounding environment.[11]
Bio-based plastic	A plastic wherein starting materials are sourced from biological renewable products rather than fossil fuels.[7]
Compostable plastic	A plastic that undergoes biodegradation in compost environments specifically and thus can be composted at the end of use.[11]
Oxo-degradable plastic	A plastic with additives that oxidize in certain environments, causing the plastic to fragment into smaller pieces without change to its chemical structure.[12]
Recyclable plastic	A plastic that may undergo physical processing back into desired products, often by melting and incorporating the product with virgin plastic. This process may reduce the need for new plastic production.[13]
Depolymerization	The process wherein a polymer decomposes into simpler, lower molecular weight compounds (such as monomers and oligomers).[1]

Continued

TABLE 4.1 Key terms in the biodegradation of polyester polyurethanes—cont'd

Term	Definition
Mineralization (in the context of biodegradation)	The process wherein polymer components (often after depolymerization) are incorporated by microorganisms into cell biomass and converted into inorganic compounds such as carbon dioxide, ammonia, nitrates, and phosphates.[14]
Hydrolysis	The process wherein chemical bonds are broken apart (cleaved) through addition of a water molecule, often catalyzed by enzymes in the context of biodegradation.[15]
Enzymes	Catalysts in biological organisms that facilitate biochemical reactions, such as breaking of particular chemical bonds.[16]
Hydrolase enzyme	Enzymes that catalyze hydrolysis reactions. These enzymes can hydrolyze a range of functional groups including esters and amides.[17]
Polyurethanase	Enzymes that catalyze degradation of polyurethanes through cleavage of particular bonds in polyurethanes.[18]
Aerobic biodegradation	Biodegradation that occurs under oxygenated conditions.[19]
Anaerobic biodegradation	Biodegradation that occurs under conditions absent of oxygen.[19]
Respirometric analysis	Analysis of gaseous components that are breakdown products of the mineralization step. It includes analysis of CO_2 production in aerobic biodegradation and CO_2 and CH_4 production in anaerobic biodegradation.[19,20]

Why don't conventional materials biodegrade?

Microorganisms can cleave various chemical bonds in nature for their own metabolic processes, using enzymes to catalyze the reaction.[21] In the case of plastics, which are synthetic polymers, the types of bonds holding the monomers together determine whether the plastic is biodegradable. In conventional, nonbiodegradable plastics, the bonds tend to be very stable and difficult to cleave.[22]

Common plastic types are polyethylene (PE, low-density and high-density), polyvinyl chloride (PVC), polypropylene (PP), and polystyrene (PS). A big contributing factor to their usefulness and wide applicability lies in their durability, which is due to their structures being chemically stable, without specific sites for hydrolytic cleavage.[22] These four are thermoplastics, which have a backbone consisting solely of a series of long carbon chains, and are resistant to the biological processes of biodegradation.

Thermoplastics differ from thermoset plastics, which have a backbone of heteroatoms such as esters and amides.[22] This allows the general structures of thermoset plastics, such as PU, polylactic acid, and polyethylene terephthalate, to have sites susceptible to degradation under certain conditions. However, biodegradability also depends on how the polymer is processed and how the structures are arranged, which will be discussed under "Effects of chemical structures" later. PUs provide another example: polyether PUs are not generally biodegradable, due to the chemical stability of ether bonds, whereas polyester PUs are susceptible to hydrolytic cleavage.[23] Several studies have demonstrated hydrolysis of urethane bonds in PUs. However, for polyester PUs, the primary site of hydrolytic cleavage occurs in the ester linkages, and it has been speculated that urethane bond hydrolysis occurs only after ester bonds are cleaved, leaving behind lower molecular weight products.

In light of this, there is an issue with a common category marketed as biodegradable and known as oxo-degradable. These consist of a biodegradable additive incorporated into a nondegradable plastic structure, typically polyethylene. They partly degrade due to the additive, but the remaining plastic material is still nondegradable, so they simply break into fragments.[12] Although visually they appear to be disintegrating, they are actually forming nondegradable microplastics, which bioaccumulate in marine organisms, causing ecotoxicological problems.[4] Also, oxo-degradable plastics may provide misleading data based on respirometric methods, such as carbon dioxide production. The consumer has been misled into thinking the entire oxo-degradable plastic is biodegradable when in fact, only the additive is degrading.[24] This emphasizes the importance of understanding the methods used for studying biodegradation, which is expanded upon under "Environments suitable for biodegradation" later, which also discusses respirometry methods in detail.

Currently, plastics are disposed of by landfilling (79%), incineration (12%), and recycling (9%), all of which have limitations.[25] Landfills use a significant amount of space and may produce toxic leachates. An increased risk of adverse health effects in residents near landfills has been reported.[26] Incineration reduces the need for space but involves many environmental and health issues, including lowered air quality from the release of toxic airborne compounds by combustion.[27] To counter negative environmental impacts, waste management facilities have turned to recycling some plastic types. For example, PE, PVC, and PP can be melted and incorporated into a new material with virgin plastic.[13] Although certain plastics can be recycled, this does not ensure that they will be. Only about 10% of all plastics have ever been recycled; most were landfilled.[28]

Biodegradability can be achieved by creating plastics that incorporate chemical bonds that are readily hydrolyzed by microorganisms through enzymes they naturally possess. For this reason, polyester PUs have great potential to be made biodegradable, as they can be designed to have

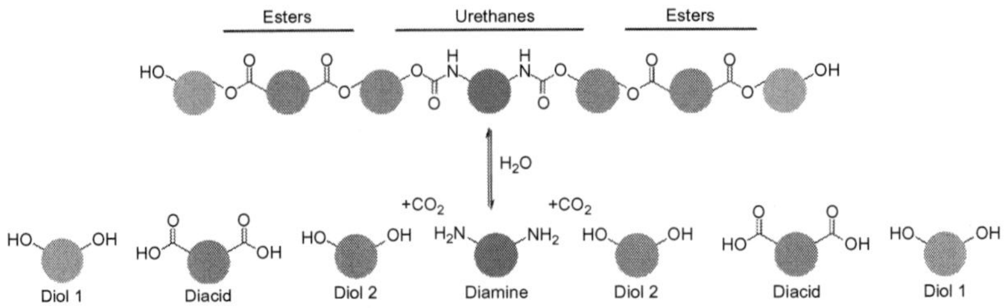

FIG. 4.1 Depolymerization of a general structure of PU. Hydrolysis of ester and urethane linkages results in lower-molecular compounds: diols, diacid, and diamine. Notably, PU depolymerization results in partly depolymerized products, oligomers. These oligomers and monomers then undergo mineralization by microorganisms if environmental conditions are suitable.

chemical bonds that can be cleaved in nature, specifically ester bonds as opposed to ether bonds in polyether PUs. Polyester PUs consist of ester and urethane linkages that can feasibly be hydrolyzed into less-persistent components, which are eventually incorporated into nature. Ester hydrolysis produces a carboxylic acid and an alcohol, while urethane hydrolysis produces an amine, an alcohol, and carbon dioxide. In PUs, complete hydrolysis of ester and urethane bonds results in regeneration of the starting products, including diols, diacids, and a diamine (Fig. 4.1). This diamine product is a result of hydrolysis of diisocyanate, one of the starting products in PU synthesis.

Microorganisms have enzymes necessary to cleave bonds prevalent in their surroundings, such as xylan and cellulose (plant polymers), which are widespread in nature.[29] Biodegradable plastics are designed with chemical bonds that can be broken by the enzymes that microorganisms naturally have. Conventional plastics are synthetic polymers with stable chemical bonds. Plastic mass production, mainly of nondegradable plastics, only began in 1950.[2] This is recent on a geological scale, so microorganisms most likely have not yet evolved to develop the appropriate metabolic processes for breaking down conventional plastics.

Biological processes

Biodegradation involves several steps. First, microorganisms must colonize the surface, leading to physical fragmentation in the surface, such as holes, discoloration, fractures, and cracks.[30] Depolymerization then occurs if the plastic has a biodegradable chemical structure, wherein bonds within the polymer are cleaved, leading to lower molecular weight components (monomers and oligomers). Microorganisms then take up these components using their metabolic processes, assimilating the organic compounds into microbial biomass, as well as into inorganic compounds that are released back into the environment (mineralization).[31]

Colonization of the surface is highly dependent on the environment. Certain environments are more suitable for biodegradation than others, and each will have a different set of microorganisms due to conditions such as temperature, pH, types of available nutrients, and oxygen levels.[32,33] Other factors such as exposure to UV and mechanical forces in the environment can also aid in physical fragmentation of the plastic at this first step. Biodegradation does not stop at the colonization step, and the presence of microorganisms does not indicate that the material is biodegrading. For nondegradable plastics, the first step can instead cause the formation of microplastics through physical fracturing of the surface.

For some microorganisms, colonization results in the formation of extracellular polymeric substances—biofilms—as a mode of attachment to the substrate surface. In marine environments, this is useful for latching onto the surface of the moving polymer. One study analyzed several microorganisms for the presence of biofilms on plastics and found that when the culture was starved for carbon, there was an increase in biofilm formation.[34,35] Another study corroborated this finding by demonstrating that bacterial growth on a biodegradable plastic was correlated with increased biofilm formation.[36]

After colonization, depolymerization is initiated by secretion of enzymes (see "Methods of measuring biodegradation" later). Enzymes are biological catalysts for various reactions and are used by microorganisms for their own metabolic purposes. In the context of biodegradation, extracellular enzymes, typically hydrolases, help catalyze the hydrolysis of the chemical bonds holding the polymer structure together. Plastics can be designed with bonds that can be feasibly hydrolyzed, though not all microorganisms will secrete those necessary for biodegradation, which highlights the significance of the environment and the microorganisms within it. For biodegradable plastics with hydrolyzable bonds, the polymer structure is cleaved at those sites, resulting in lower molecular weight breakdown products (monomers or oligomers). Microorganisms can then catabolize these lower molecular weight substrates, converting the elements into cell biomass and eventually into inorganic compounds such as carbon dioxide, water, and ammonia, which are released into the environment.[31] For a fully biodegradable plastic product, the components then would be incorporated into the natural environment. One of the standardized methods for tracking successful biodegradation is through analyzing production of carbon dioxide.[19]

When designing a plastic to achieve biodegradability, one must consider how likely it is to be disposed of in the environment intended for breakdown, and the methods for accurately determining successful biodegradation. Polyester PUs are unique in that they can be designed with the biodegradation process in mind. Several studies have indicated that the ester bonds can be feasibly cleaved by microorganisms in various environments.[23,37] Also, the urethane bonds are hydrolyzed, often after the

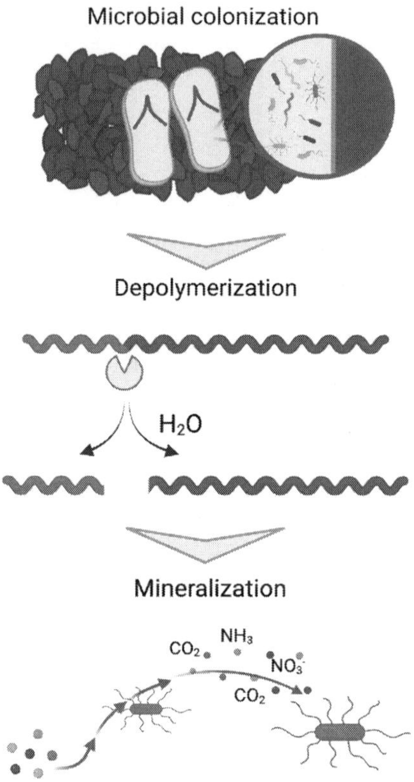

FIG. 4.2 Steps in biodegradation of synthetic polymers. (1) Microbial colonization of the surface of plastic when it is incubated in a nutrient-rich environment such as compost. Certain microorganisms then secrete enzymes (*yellow* (*light gray* in print version)) that aid in the next step. (2) Depolymerization into lower molecular weight compounds such as monomers and oligomers through hydrolase enzymes. (3) Mineralization, wherein microorganisms take up the lower molecular weight compounds and turn them into inorganic compounds such as carbon dioxide, ammonia, and nitrates that are incorporated by the microorganisms into cell biomass for growth.

ester bonds are targeted. This ability is highly dependent on the accessibility of these bonds to hydrolase enzymes. Fig. 4.2 illustrates the entire process from colonization to mineralization.

Environments suitable for biodegradation

The label "biodegradable" does not specify which environments the product will biodegrade in. Polylactic acid (PLA) is a known compostable plastic. It readily biodegrades under high-temperature compost conditions through being mineralized by microorganisms.[38] But, like nondegradable plastics such as PET, it persists in marine environments.[39]

A plastic designed to biodegrade in compost, which is aerobic, may not do so to the same degree in a landfill, which is anaerobic. Since most plastic waste is landfilled—approximately 75% in 2018—it is critical that designers, manufacturers, regulators, and consumers be informed of the environmental factor.[28]

The presence or absence of oxygen (aerobic vs. anaerobic) determines the types of organisms that are abundant, the nutrients available, and the mechanisms available for their metabolic processes. This impacts the types of reaction that must be operative for biodegradation to occur. Under aerobic conditions, oxidative reactions tend to occur, whereas under anaerobic conditions, reductive reactions tend to occur.[33] This concept is useful for considering two environments where plastics are most likely to end up: a compost bin (aerobic) and a landfill (anaerobic). Anaerobic biodegradation results in higher levels of methane, in addition to carbon dioxide.[40] Biodegradation is slower under anaerobic conditions. In one study of polyhydroxybutyrate (PHB), aerobic microorganisms showed greater polymer-degrading ability than anaerobic microorganisms.[41]

Another major environment is the ocean. Even with improvements to waste management infrastructure, an estimated 2.4–6.5 million metric tons of total plastic waste will be dumped into the ocean by 2025.[42] One solution is to use plastics that biodegrade in marine environments. Kinds that biodegrade in other environments have chemical bonds that can be hydrolyzed by a variety of microorganisms, but the types and abundance of microorganisms in the ocean vary greatly from those in compost or landfill. Marine environments are unique in having high salinity, currents, lower temperatures, and dilute nutrient levels.[43] Plastics intended for composting degrade very slowly in seawater.[43,44] As mentioned before, UV irradiation can cause fragmentation. In marine systems, plastics tend to persist near the surface, where they are exposed to UV. This may help initiate the first few steps in biodegradable plastics, but for nondegradable types it leads to the formation of microplastics, a problem of increasing concern.[4]

In designing plastics for disposal in several environments, the effect of temperature on biodegradation rate is a major consideration. Higher temperature increases the hydrolysis of chemical bonds in some plastics. PLA degrades faster in industrial composting facilities, where the temperature range is 58–65°C, than in natural soil or landfill.[45,46] The average temperature of surface seawater is 17.4°C; at greater depth, it can reach 0–4°C.[43] Temperature is a determining factor in the type and abundance of microorganisms. One study assessed biodegradation of a polyester PU in compost under three temperatures: 25°C, 45°C, and 50°C.[37] Physical degradation did not change significantly, but the type of the most abundant fungi on the plastic did. *Fusarium solani* was dominant at 25°C, while *Candida ethanolica* was dominant at 45°C and 50°C. This indicates that temperature can affect the microbiome even in similar environments but may not directly change the biodegradation rate.

The biotic and abiotic conditions such as temperature, pH, oxygen content, and nutrient composition in different environments can drastically affect the types of microorganisms present and thus the biodegradation rate.[47] Dog bone-shaped samples of polyester PU were studied under two contrasting environments: Baltic Sea water and a compost pile.[44,47] The water was more alkaline (pH 8.2 vs. 5.5). Based on weight loss and visual analyses, the PU degraded more rapidly in compost. This is not to say that pH alone affects biodegradation rate. Instead, the authors speculated, pH was a key influence on the types of microorganisms present, with fungi being more abundant in compost and bacteria more abundant in seawater.

Nutrient composition varies among environments, and this affects the microbiome. Iron content is a key factor in seawater and is known to limit biological activity.[48] Other essential elements in seawater are nitrogen and phosphorus, which cycle through the ocean in the form of nitrates and phosphates and influence the types of microorganisms present.[49] It is likely that the abundance of PU-degrading microorganisms is also influenced by nutrient composition.[50] Compost and soil environments are very biologically active, and many microorganisms known to biodegrade PUs have been isolated from there.

Methods of measuring biodegradation

Research methods for assessing PU biodegradation yield information on different aspects of the process. Typically, one of the following is assessed (Fig. 4.3):

- Physical degradation by microbial colonization
- Growth of microorganisms on PUs
- Cleaving of PU structure at the molecular level
- Uptake of PU components by microorganisms

Outside the laboratory, products are exposed to microorganisms through the environment in which they are discarded—landfill, compost bin, ocean. Inside the laboratory, biodegradation is initiated in several ways. As the closest comparison to the real world, researchers incubate the material in a model environment. To test specific biological interactions, isolated microorganisms are simply placed on the plastic. Extracted enzymes thought to hydrolyze particular bonds are tested likewise. Laboratory cultures of microorganisms or purified enzymes, though not directly comparable to the real world, provide a much less complex medium for assessment, compared to raw environmental material such as compost or landfill. Each type of experiment encompasses different methods of analysis and provides information to different extents.

In the first step of biodegradation, microbial colonization leads to physical changes in the plastic material, such as cracking, discoloration, and

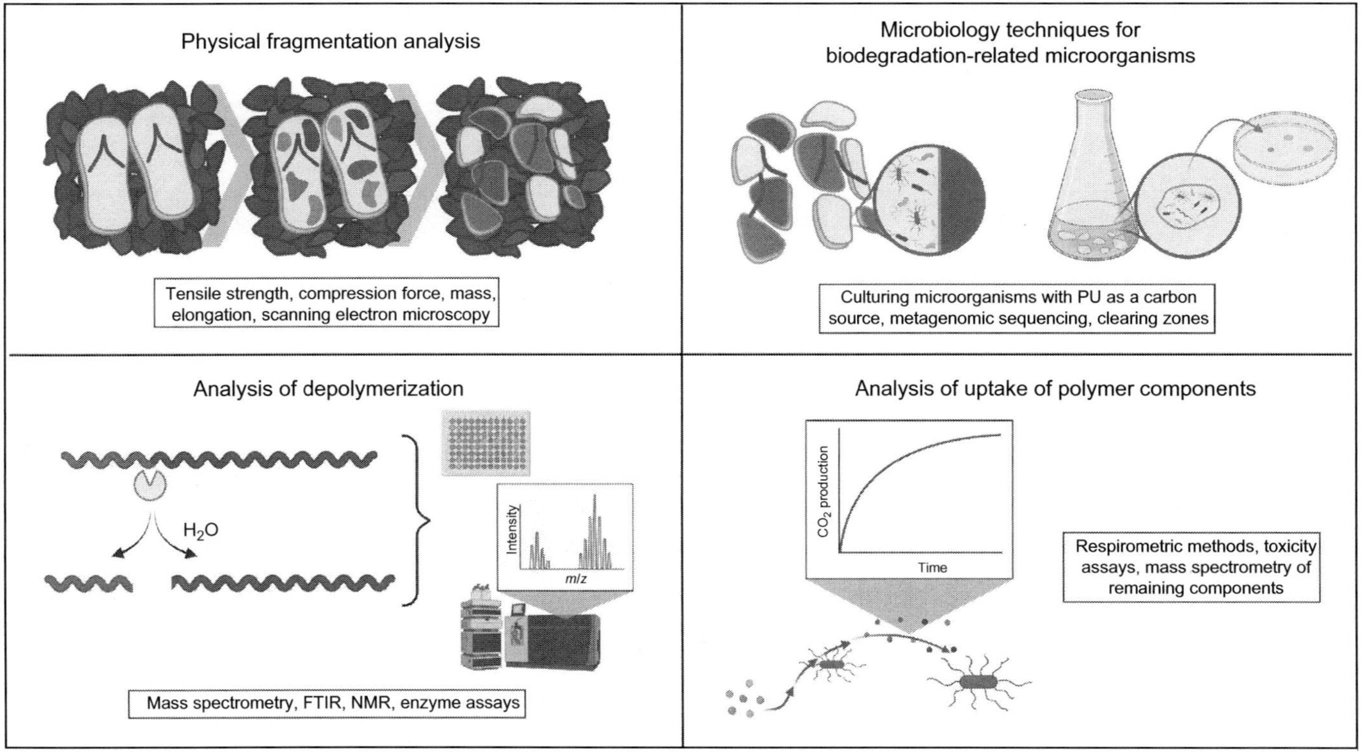

FIG. 4.3 Research processes and examples of methods for analyzing biodegradation of polyurethane plastics.

surface erosion, which are used as indicators in research studies.[30] To assess physical degradation due to microorganisms, visual analyses are used. High-resolution imaging such as scanning electron microscopy indicates how a surface changes over time during colonization, providing qualitative information. Mechanical tests assess physical changes in this first step. In a study on polyester PUs in compost and soil, mass loss and compression force loss were used as indicators that the material was physically changing after environmental incubation, in addition to scanning electron microscopy for visualizing how microorganisms interacted with the PU surface.[51] Studies have also used loss of tensile strength and loss of elongation.[37,52]

After heavy colonization of fungi on a PE surface, the structure was physically weak and readily disintegrated under mild pressure.[53] Because of its recalcitrant structure, PE is generally inert to environmental biodegradation, so surface deterioration alone does not confirm that a material is biodegrading. In fact, it is likely that microorganisms are simply physically fracturing the surface, leading to the formation of microplastics without any depolymerization. A decrease in mechanical properties indicates that the material is degrading to a certain extent, but physical tests must be paired with others.

Microbial colonization has implications other than biodegradation. Microorganisms attach themselves to many substrates, such as the nonbiodegradable plastics PE, PVC, and PS.[54] A common method for showing that degradation is in fact occurring is to grow microorganisms in a medium where the sole source for necessary elements such as carbon and nitrogen is the plastic of interest.[55] If microorganisms are growing, they must be assimilating the polymer structure to some extent (mineralization step).

Some studies have used the soluble polyester PU Impranil DLN (from Covestro AG) in media. Changes in turbidity indicate that the PU is being taken up by microorganisms. This method is extensively seen in Table 4.2, which covers the bacteria and fungi known to biodegrade polyester PUs in current literature. Because of its solubility, Impranil DLN is a useful substrate, as it can be incorporated into solid media with agar as well as in liquid cultures. Studies that use Impranil agar plates, either supplemented with carbon and nitrogen or not, look at "clearing zones" as indicators that the inoculated microorganisms are hydrolyzing the PU,[55,76] a qualitative metric of biodegradation. Other studies have attempted to use Impranil to quantitate the rate. This is done by assessing changes in optical density of the liquid culture. But the effect might not exactly represent conventional PUs, which are typically insoluble in common laboratory liquids.

Other studies used the growth of the microorganism on a culture supplemented with PU. This is assessed by changes in optical density related to an increase in the number of cells. The method can be helpful but should be paired with controls to ensure that growth is related to biodegradation of the PU. To wholly understand biodegradation requires assessment of chemical changes in the plastic substrate as well as biological processes.

Recently, more analytical chemistry techniques have been developed that are useful for assessing plastic biodegradation. Common examples are Fourier transform infrared spectroscopy (FTIR), nuclear magnetic resonance (NMR), and mass spectrometry (MS). Each provides information on how a sample is changing based on the chemical groups present in a sample, the chemical structures, and the breakdown products. One study assessed fungal biodegradation of a soluble polyester PU with analytical methods to confirm successful cleavage of particular bonds. FTIR showed that the fungus *Cladosporium pseudocladosporioides* caused a decrease of carbonyl groups (C=O) and N—H bonds.[73] Gas chromatography-mass spectrometry (GC-MS) showed a decrease in ester compounds and an increase in alcohols and hexane diisocyanate. Both analyses indicated that ester and urethane bonds were being hydrolyzed by the fungus.

Analysis of biodegrading samples can be difficult when they are in media with complex constituents; for example, MS signals from samples with a background signal from compost. Extraction methods for different biodegradation conditions had to be developed. In a study to determine the amounts of 4,4′-methylenedianiline in soil samples by ultra-performance liquid chromatography-mass spectrometry (UPLC-MS), it was necessary to develop a method for extracting the MDA from the soil.[77] This was done by solid-phase extraction (SPE), which allowed for concentration of soil extract solutions while limiting the amount of contaminants from soil. With these types of methods, there is risk of losing the compound of interest, often a breakdown product of the PU, and so controls must be used to account for any loss of product. Chapter 5 reviews current challenges with analytical methods.

Enzymatic degradation provides a less complex media to analyze. Using specific enzymes allows isolated study of the depolymerization step without any uptake of the breakdown products by microorganisms. So the level of depolymerization can be determined by pairing enzymatic degradation with analytical chemistry techniques such as MS.

The molecular weight of a polymer reflects biodegradation. It can be determined by gel permeation chromatography (GPC). As a plastic degrades, the decrease in the polymer chain indicates that the depolymerization step is occurring. The caveat with GPC is that a proper solvent must be used. Chemical crosslinking in some PUs inhibits solvation.

One concern is the toxicity of breakdown products and how long they persist prior to being taken up by microorganisms. Standard toxicity assays look at different chemical compounds and their impacts on the environment. To assess plastic leachate toxicity, larval survival and barnacle settlement are used.[78] In one study, seven types of recyclable (nondegradable) plastic were placed in seawater for 24 hours. All produced larval toxicity and inhibition of barnacle settlement. Chemical analysis revealed a complex mixture of substances released in leachates, demonstrating the importance of considering persistence of any breakdown products of conventional materials.

Truly biodegradable plastics consist of chemical compounds that can be taken up by microorganisms. In one study, the polyester PU precursors succinic acid, adipic acid, terephthalic acid, 1,4-butanediol, and ethylene glycol were tested on different isolated organisms for biodegradability. All precursors were degraded by several bacteria, with at least one bacteria able to degrade each precursor by 100%.[79] Because PU precursors are generally similar to the breakdown products, this is a suitable method for assessing whether PU products are toxic.

Respirometric methods look at the gaseous components released during biodegradation, specifically carbon dioxide for aerobic and methane for anaerobic. Since the mineralization step consists of microorganisms converting polymer components into cell biomass and inorganic compounds such as carbon dioxide, biodegradation can be tracked by the increase in carbon dioxide production compared to the baseline production of the surrounding environment. The ASTM D5338 procedure determines aerobic biodegradation rate in compost, and ASTM D7475 determines anaerobic biodegradation in landfills.[19,20] These tests do not provide information on chemical structure.

If a material cannot be shown to achieve 100% biodegradation, the residue may form microplastics. Since respirometric methods do not provide information on chemical structure, it is difficult to determine to what extent biodegradation is occurring. This is concerning for oxo-degradable plastics, which consist of a plastic with additives that accelerate breakdown.[12] For plastics with nondegradable chemical structures, these oxo-additives will simply cause the formation of nondegradable microplastics, as the structure of the remaining plastic is still recalcitrant to biodegradation. The additive portion of the product may biodegrade, producing carbon dioxide in the process, but the product as a whole will plateau past that point.[24,80]

Organisms: Bacteria and fungi

For most biodegradable plastics, bacteria and fungi that can cleave the synthetic polymer structure have been identified. The Plastics Microbial Biodegradation Database (PMBD) is a comprehensive resource for microorganisms and enzymes known to biodegrade plastics.[81] The current literature is helpful for researchers studying plastic biodegradation processes. However, methods for determining successful biodegradation vary greatly, and consumers must think critically about whether any given one is legitimate.

Several bacteria and fungi specifically target polyester PUs, and a few target polyether PUs. However, the methods for studying polyether PU biodegradation are dubious and have not produced promising results. Table 4.2 presents known microorganisms that contribute to the biodegradation of polyester PU, as well as the methods used for culturing these microorganisms, based on current literature. The methods used to identify these microorganisms vary among studies.

TABLE 4.2 Bacteria and fungi known to biodegrade polyester PUs, and methods to identify successful biodegradation.

Kingdom	Genus	Physical form of PU	Biodegradation method and results
Bacteria	*Acinetobacter calcoaceticus*[56]	Impranil DLN agar plate	Microorganism esterase activity in medium containing PU as sole carbon source
Bacteria	*Acinetobacter gerneri* (P7)[57]	Impranil DLN agar plate	Evident but slow bacteria growth on PU as sole carbon source
Bacteria	*Alicycliphilus* sp.[58]	Minimal medium with surface-coating PU (hydroform) as carbon source	Microorganism growth and esterase activity
Bacteria	*Arthrobacter globiformis*[56]	Impranil DLN agar plate	Microorganism esterase activity in enriched medium and medium containing PU as sole carbon source
Bacteria	*Bacillus subtilis*[59]	Impranil DLN agar plate	Impranil degraded in 3–4 days, demonstrating ability to use as sole carbon and energy source
Bacteria	*Comamonas acidovorans* (TB-35)[60]	PU solid in basal medium	Grown on PU as sole carbon and nitrogen source
Bacteria	*Corynebacterium turicella* sp.[61]	Dumbbell-shaped test pieces of PU in liquid medium supplemented with other carbon source	Cracking and loss of tensile strength
Bacteria	*Enterobacter agglomerans*[61]	Dumbbell-shaped test pieces of PU in liquid medium supplemented with other carbon source	Cracking and loss of tensile strength
Bacteria	*Pseudomonas aeruginosa*[62]	Medium with PU-diol as sole carbon source in agar plates	Significant reduction in PU diol, measured with high-performance thin layer chromatography

Continued

TABLE 4.2 Bacteria and fungi known to biodegrade polyester PUs, and methods to identify successful biodegradation—cont'd

Kingdom	Genus	Physical form of PU	Biodegradation method and results
Bacteria	*Pseudomonas chlororaphis*[55]	Impranil DLN agar plate with rich media	Clearing zone observed
Bacteria	*Pseudomonas fluorescens*[63]	Impranil DLN agar plate with rich media	Clearing zone observed, Impranil concentration decrease in liquid medium
Bacteria	*Pseudomonas protegens* (Pf-5, BC2-12, CHA0)[64]	Impranil DLN agar plate	Clearing zone observed
Bacteria	*Pseudomonas aeruginosa*[61]	Dumbbell-shaped test pieces of PU in liquid medium supplemented with other carbon source	Cracking observed, dry weight and tensile strength loss
Bacteria	*Pseudomonas chlororaphis*[55]	Impranil DLN in liquid medium	Complete clearing observed
Bacteria	*Pseudomonas putida*[65]	Impranil DLN in liquid medium	Nearly complete Impranil concentration loss, determined by absorbance ratios
Bacteria	*Pseudomonas* sp.[64]	Impranil DLN in liquid medium	Impranil DLN solution cleared by extracted lipase enzyme, demonstrating high esterase activity using *p*-nitrophenyl acetate assay
Bacteria	*Rhodococcus equi* TB-60[66]	Toluene-2,4-dicarbamic acid dibutyl ester(TDCB)	Significant TDCB concentration loss
Bacteria	*Serratia rubidaea*[67]	Dumbbell-shaped test pieces of PU in liquid medium supplemented with other carbon source	Dry weight (13%) and tensile strength loss (10%)
Bacteria	*Staphylococcus epidermidis*[68]	400- and 200-μm films in nutrient-free phosphate buffered saline (PBS)	In PU PBS viable cells present for 30 days vs. no more viable bacteria after 25 days in control PBS; urease activity observed in bacteria

Fungi	*Acremonium flavum* JQ966574[37]	PU coupons	Physical deterioration of PU coupons measured by loss in tensile strength and percentage elongation at break
Fungi	*Alternaria dauci*[69]	Impranil DLN agar plate	Clearing zone of Impranil observed
Fungi	*Alternaria* sp.[69]	Impranil DLN agar plate	Clearing zone of Impranil observed
Fungi	*Alternaria* sp.[70]	PU coupons; Impranil DLN agar plate	One of microorganisms recovered from surface of PU coupon buried in soil, clearing zone observed on Impranil agar plate
Fungi	*Arthrographis kalrae*[37]	PU coupons buried in compost pile	One of main microorganisms present on coupons buried in compost, 50% reduction in elongation strength and >70% reduction in tensile strength
Fungi	*Aspergillus terreus*[71]	Impranil DLN agar plate and liquid	Evident but slow growth of the bacterium on PU as sole carbon source
Fungi	*Aspergillus tubingensis*[72]	PU film	Colonized PU, surface degradation and scarring observed
Fungi	*Aspergillus fumigatus*[37]	PU coupons buried in compost pile	One of main microorganisms present on coupons buried in compost, 50% reduction in elongation strength and >70% reduction in tensile strength
Fungi	*Bionectria* sp. E2910B[69]	Impranil DLN agar plate; liquid medium with Impranil DLN as sole carbon source	Clearing zone of Impranil observed on agar plate and in liquid culture
Fungi	*Candida rugosa* (JQ966580)[37]	PU coupons buried in compost pile	One of main microorganisms present on coupons buried in compost, 50% reduction in elongation strength and >70% reduction in tensile strength

Continued

TABLE 4.2 Bacteria and fungi known to biodegrade polyester PUs, and methods to identify successful biodegradation—cont'd

Kingdom	Genus	Physical form of PU	Biodegradation method and results
Fungi	*Chaetomium globosum*[71]	Impranil DLN agar plate	Evident but slow growth of the bacterium on PU as sole carbon source
Fungi	*Cladosporium asperulatum* (BP3.I.2)[73]	Mineral medium with Impranil DLN	Impranil degraded 78% (by concentration) after 14 days incubation in mineral medium, decrease of C=O and N—H bonds shown by FTIR
Fungi	*Cladosporium pseudocladosporioides* (T1. PL.1)[73]	Mineral medium with Impranil DLN	Impranil degraded 87% (by concentration) after 14 days incubation in mineral medium, decrease of C=O and N—H bonds shown by FTIR, hydrolysis of ester and urethane bonds shown by GC-MS
Fungi	*Cladosporium tenuissimum*[73]	Mineral medium with Impranil DLN	Impranil degraded ≥80% (by concentration) in mineral medium, decrease of C=O and N—H bonds shown by FTIR
Fungi	*Cladosporium ascomycetes*[74]	Impranil DLN agar plate	Clearing zone of Impranil observed
Fungi	*Cladosporium montecillanum*[73]	Mineral medium with Impranil DLN	Impranil degraded 75% (by concentration) after 14 days incubation in mineral medium, decrease of C=O and N—H bonds shown by FTIR
Fungi	*Curvularia senegalensis*[74]	Impranil DLN agar plate	Clearing zone of Impranil observed
Fungi	*Cylindrocladiella parva* (AY793455)[70]	PU coupons	One of microorganisms recovered from surface of PU coupon buried in soil
Fungi	*Edenia gomezpompae*[69]	Impranil DLN agar plate	Clearing zone of Impranil observed

Fungi	*Emericella nidulans* JQ966573[37]	PU coupons buried in compost pile	One of main microorganisms present on coupons buried in compost, 50% reduction in elongation strength and >70% reduction in tensile strength
Fungi	*Fusarium solani*[74]	Impranil DLN agar plate	Clearing zone of Impranil observed
Fungi	*Geomyces pannorum* (S9-A3/2, S9-A4)[75]	Dumbbell-shaped test pieces of PU; Impranil DLN agar plate	One of main organisms that made up biofilm formed on PU buried in soil with tensile strength loss up to 60%, clearing zone of Impranil observed
Fungi	*Geomyces pannorum* (AF015789)[70]	PU coupons cut into strips 4.5 by 0.5 by 0.15 cm	Zones of clearance on Impranil agar very obvious and extended <1 cm outwards from the colony edge.
Fungi	*Geomyces pannorum* (S33-A1)[75]	Dumbbell-shaped test pieces of PU	One of main organisms that made up biofilm formed on PU buried in soil with tensile strength loss up to 60%, clearing zone of Impranil observed
Fungi	*Lasiodiplodia* sp. (E2611A)[69]	Impranil DLN agar plate; liquid medium with Impranil DLN as sole carbon source	Clearing zone of Impranil observed on agar plate and in liquid culture
Fungi	*Lichtheimia* sp. (JQ966582)[37]	PU coupons buried in compost pile	One of main microorganisms present on coupons buried in compost, 50% reduction in elongation strength and >70% reduction in tensile strength
Fungi	*Malbranchea cinnamomea* (JQ966583)[37]	PU coupons buried in compost pile	One of main microorganisms present on coupons buried in compost, 50% reduction in elongation strength and >70% reduction in tensile strength

Continued

TABLE 4.2 Bacteria and fungi known to biodegrade polyester PUs, and methods to identify successful biodegradation—cont'd

Kingdom	Genus	Physical form of PU	Biodegradation method and results
Fungi	*Nectria gliocladioides* (S6-A1, S6-B1, S9-A6, S9-B4, S9-C2)[75]	Dumbbell-shaped test pieces of PU in soil; Impranil DLN agar plate	One of main organisms that made up biofilm formed on PU buried in soil with tensile strength loss up to 60%, clearing zones of Impranil observed
Fungi	*Nectria* sp.[69]	Impranil DLN agar plate	Clearing zone of Impranil observed
Fungi	*Penicillium ochrochloron* (S6-A3)[75]	Dumbbell-shaped test pieces of PU in soil; Impranil DLN agar plate	One of main organisms that made up biofilm formed on PU buried in soil with tensile strength loss up to 60%, clearing zones of Impranil observed
Fungi	*Penicillium inflatum* (AY373920)[70]	PU coupons	One of microorganisms recovered from the surface of PU coupon buried in soil, zones of clearance on Impranil agar very obvious and extended <1 cm outwards from the colony edge
Fungi	*Penicillium venetum* (AY373939)[70]	PU coupons	One of microorganisms recovered from the surface of PU coupon buried in soil, zones of clearance on Impranil agar very obvious and extended <1 cm outwards from the colony edge
Fungi	*Penicillium viridicatum* (AY373935)[70]	PU coupons	One of microorganisms recovered from the surface of PU coupon buried in soil, zones of clearance on Impranil agar very obvious and extended <1 cm outwards from the colony edge
Fungi	*Penicillium chrysogenum*[73]	Mineral medium with Impranil DLN	Impranil degraded 74% in mineral medium, decrease of C=O and N—H bonds shown by FTIR

	Pestalotiopsis microspora[69]	Impranil DLN agar plate; liquid medium with Impranil DLN as sole carbon source	Clearing zone of Impranil observed, highest rates of decrease of Impranil in liquid minimal medium, loss of C=O bond shown by FTIR
Fungi	*Pestalotiopsis*[69]	Impranil DLN in liquid culture; Impranil DLN agar plate	Increased growth of fungal material on plate with PU vs. plate without, half time for PU clearance in liquid = 5 days, clearing zone of Impranil on agar observed
Fungi	*Phaeosphaeria*[69]	Impranil DLN agar plate	Clearing zone of Impranil observed
Fungi	*Plectosphaerella*[69]	Impranil DLN agar plate	Clearing zone of Impranil observed
Fungi	*Plectosphaerella*[70]	Impranil DLN agar plate	One of microorganisms recovered from the surface of PU coupon buried in soil, zones of clearance on Impranil agar extended <1 cm outwards from the colony edge
Fungi	*Pleosporales*[69]	Impranil DLN agar plate; liquid medium with Impranil DLN as sole carbon source	Clearing zone of Impranil observed, decrease of Impranil in liquid minimal medium
Fungi	*Thermomyces*[37]	PU coupons buried in soil	One of main microorganisms present on coupons buried in compost, 50% reduction in elongation strength and >70% reduction in tensile strength

Enzymes

Enzymes play a crucial role in biodegradation, specifically at the depolymerization step where chemical bonds in the polymer are cleaved by hydrolysis. Microorganisms secrete enzymes, typically hydrolase enzymes, which catalyze the hydrolysis process. Key types are esterases, ureases, urethane hydrolases, lipases, and proteases, which are useful for microorganisms in their natural metabolic processes. For PUs, ester and urethane bonds are susceptible to enzymatic hydrolysis. Most of the currently studied polyurethanases (enzymes that hydrolyze PUs) target the ester bonds, while urethane bonds are less susceptible.[82] It may be that urethane bonds are only hydrolyzed once the PU is broken into its lower molecular weight oligomers by ester cleavage.

Using enzymes in biodegradation studies is less complex than using microorganisms. It eliminates the mineralization step, which allows tracking of the initial breakdown products (diols, diacids, diamines).

Not all microorganisms have enzymes that degrade PUs. Several studies have been done to isolate types that do. Concurrently, assays to confirm the enzymatic activity must be developed. Often the substrates that do this mimic the ester or urethane bonds in PUs. For example, ethyl carbamate (urethane) was used as a substrate for quantifying polyurethanase activity in the presence of alcohol dehydrogenase.[83] When ethyl carbamate is hydrolyzed, the products are ethanol, carbon dioxide, and ammonia. Ethanol is then oxidized by alcohol dehydrogenase and NAD is reduced into NADH, causing a shift in absorbance at 340nm, which is monitored to quantify enzyme activity (Fig. 4.4). This study provides one example of an enzyme assay for targeting urethane bonds. However, ethyl carbamate does not completely mimic the structure of PUs.

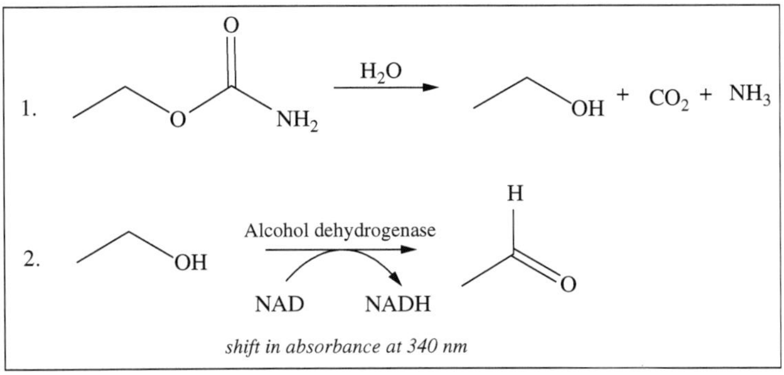

FIG. 4.4 Steps in an assay using ethyl carbamate (urethane) as the substrate to quantify polyurethane biodegradation enzyme activity.[83]

Activity is also assessed by colorimetric changes when a soluble PU substrate (Impranil DLN) is hydrolyzed, and by other assays using Impranil DLN,[83,84] *p*-nitrophenyl laurate, and *p*-nitrophenyl acetate.[72] Advances in technology have allowed genetic analysis of specific microorganisms and their hydrolase enzymes. For example, whole genome sequencing was done on a strain of *Pseudomonas* sp. using PU oligomers and monomers as their sole carbon source.[85] Considering this analysis alongside databases of other known *Pseudomonas* strains, the researchers proposed a degradation pathway for 2,4-diaminotoluene, a precursor for PU synthesis, to identify dioxygenases that might be key to the process. Table 4.3 presents enzymes found to degrade PUs, with the assays used to determine activity.

TABLE 4.3 Enzymes that degrade polyester PUs.

Enzyme type	Microorganism source	Enzyme assay
Carboxylic ester hydrolase (pudA)[86]	*Delftia acidovorans*	Solid PU
Polyurethanase (pueA)[87]	*Pseudomonas chlororaphis*	Genetic analysis on microorganism
Polyurethanase (pueB)[87]	*Pseudomonas chlororaphis*	Genetic analysis on microorganism
Esterase[72]	*Aspergillus tubingensis*	*p*-Nitrophenyl laurate and *p*-nitrophenyl acetate
Esterase[74]	*Curvularia senegalensis*	Solid PU: reduction in tensile strength
Esterase protease (pulA)[88]	*Pseudomonas fluorescens*	Soluble PU medium plates (Impranil DLN): zones of clearance; esterase activity determined against hide powder azure
Serine hydrolase[69]	*Pestalotiopsis microspora* E2712A	Soluble PU medium as sole carbon source (Impranil DLN)
Esterase[62]	*Pseudomonas aeruginosa*	PU-diol medium
Polyurethanase[84]	*Pseudomonas chlororaphis*	Protein fractions spotted on YES agar with soluble PU (Impranil DLN)
Esterase[76]	*Pseudomonas fluorescens*	Soluble PU medium (Impranil DLN)
Lipase[64]	*Pseudomonas protegens* (strains A506, Pf0-1, BC2-12, CHA0, Pf-5)	Soluble PU medium (Impranil DLN)
Lipase[76]	*Pseudomonas* sp.	Soluble PU medium (Impranil DLN)
Dioxygenase[90]	*Pseudomonas* sp.	2,4-Diaminotoluene; genetic analysis of amino acid sequence similarity

Effects of chemical structures

Polyester PUs biodegrade feasibly, and polyether PUs do not.[23] This is due to the chemical stability of ether bonds, which are not susceptible to hydrolytic cleavage, even in the presence of mild acids and bases (Fig. 4.5).[89] Breaking the C—O bond in ethers requires significant energy, so microbial scission is rare. This is also true for other ether-based compounds, such as polyethylene glycol (PEG), polypropylene glycol (PPG), and diphenyl ethers in herbicides.

In a study that tested several formulations of PUs and fungi, all polyester PUs were susceptible but polyether PUs were resistant.[24] Since then, several studies have shown biodegradation of polyester PUs in soil and compost, as well as by isolated microorganisms.[72] Few if any studies have found biodegradation of polyether PUs, with many results suggesting that while urethane bonds can be hydrolyzed, the effect may be limited by the other bonds present, such as ether bonds in polyether PUs.

In nature, ester bonds are more prevalent than ether bonds, and microorganisms have the necessary enzymes to hydrolytically cleave ester bonds for their own metabolic purposes.[90] Hydrolysis at high temperature (thermohydrolysis) depends on the stability of chemical bonds. For the three bonds of interest in PUs—ester, urethane, and ether—the thermohydrolytic stability order is ether ≫ urethane > ester.[91] This is consistent with the principle that ether bonds are more chemically stable than urethane or ester bonds. In polyester PUs, the ester bond is thought to be the primary site of cleavage. Several enzymes that hydrolyze urethane bonds have been found, but it is unclear whether urethanes can be hydrolyzed directly. Urethane bonds might only be susceptible to hydrolysis on the surface or following breakdown of the polymer into lower molecular mass products.[15,92] In some cases, polyether PUs have been shown to degrade at urethane sites, but ether bonds are fairly stable and resistant to hydrolysis.

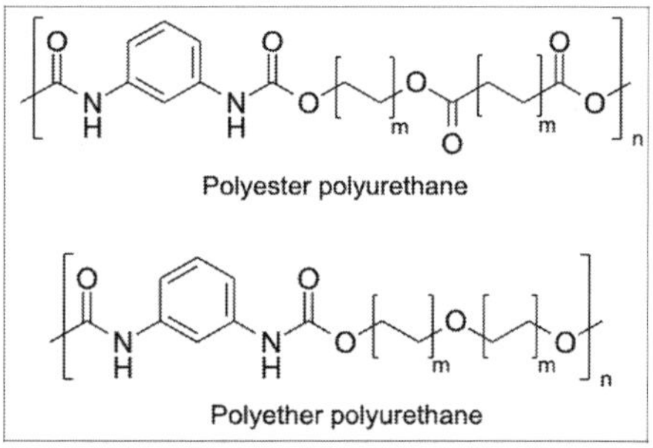

FIG. 4.5 Structures of polyester polyurethane and polyether polyurethane. The first has ester and urethane bonds, the second ether and urethane bonds.

PU is versatile, and a variety of polyester PUs can be made with different starting materials (diols, diacids, diisocyanates). The long carbon chain portion of PU is a "soft" segment, while the rigid component consisting of the isocyanate and chain extender is "hard."[93] The ratio of soft to hard affects mechanical properties, which in turn affect biodegradation rate. Increasing the length of the soft segment—that is, the percentage of the material that is soft—improves susceptibility to enzymatic cleavage,[6,94] as does incorporating functional groups that increase hydrophilicity.[30]

The structural complexity of a plastic influences biodegradation rate. Crosslinked structures are more resistant to degradation because they allow less access of enzymes to susceptible bonds.[15] However, crosslinking does not necessarily eliminate biodegradation if ester and urethane bonds can be hydrolyzed feasibly. Similarly, crystallinity slows down biodegradation; so a polymer structure in its amorphous form will biodegrade differently from its crystalline form. For example, although the ester linkages in amorphous PET are susceptible, when PET chains are organized in a crystalline structure—as in conventional plastic bottles—biodegradation is difficult.[85,95] Crystalline structures are fairly resistant to hydrolysis because of the limited accessibility of hydrophilic sites.

Lower molecular weight polymers degrade more readily than higher molecular weight polymers. This can explain why microorganisms must depolymerize a plastic into lower molecular weight products (monomers and oligomers) prior to the mineralization step. In a study of biodegradable PUs consisting of polycaprolactone segments, a sample with the lowest molecular weight had the greatest mass loss after enzymatic degradation.[96]

Hydrolysis is the key process that leads to biodegradation. The chemical and physical properties of PUs govern the feasibility of enzymatic hydrolysis by microorganisms in the natural environment, so developing a biodegradable polyester PU requires thoughtful design at the synthesis step.

Why definitions and measurements matter: Greenwashing

The hope is that consumers can be empowered to better understand plastic biodegradation. However, companies must also be held accountable for promoting products labeled as biodegradable in a misleading way. A clear set of regulations and standardization methods could decrease the risk of companies profiting from the perception of sustainability without proof to back up the claim.[97] Greenwashing is an issue of ethical harm.[98] Products that use false claims can make consumers wary of any sustainable products on the market, undermining those that are truly biodegradable and benefit the environment. Consumers need to be educated about what biodegradable plastics are, how biodegradation is measured, and the importance of context in whether the desired end result is achieved.

Polyester PU products are promising because they can be created with less reliance on nonrenewable feedstocks such as petroleum, and because they can be designed to biodegrade. Biodegradability is part of the solution to the overwhelming problem of plastic accumulation in the environment.

Close-up: Harnessing biodegradation

Plastics are fantastic: pliable, strong, and hardy, they truly are a miraculous material. But our frivolity with it, like our addiction to single-use plastics, has led to a reality where islands of plastic waste populate the oceans and all life is threatened by the scale of the pollution.

It's not just unsightly; the effects of microscopic plastic waste are being felt. Microplastics and the harmful compounds released from degradation accumulate in water, food, and air to the detriment of our health. For example, phthalates, plasticizers that increase durability and are found in many household goods, are linked to asthma, a variety of cancers, and alarming changes to male reproductive systems, including diminished sperm count and DNA damage to germ cells. Plastic waste is making us and our planet sick.

How do we deal with this kind of waste? "Reduce reuse recycle" doesn't easily apply to compounds that we can barely see, and especially the nonplastic compounds that leach into the environment. Microorganisms might have the answer. Specific strains of fungi and bacteria possess the ability to depolymerize plastic polymers into monomers, and further into the basic components, sugars and acids, that constitute the polymer. Specific ligninolytic fungi, for instance, use an array of extremely potent enzymes to break down natural polymers such as lignin. These same enzymes can have a degradative effect on synthetic polymers like plastics. Harnessing these systems could open the door for large-scale, high-throughput biodegradation, and thus an effective end-of-life solution for plastics. In this way, plastic waste could be transformed into a valuable carbon source to fuel microorganisms that can produce biofuel, bioplastics, and antibiotics, rather than simply being cycled out into the world again as new plastic products.

Tapping into these systems could allow us to break the plastic cycle, which ends in pollution no matter how many times a piece is recycled, and to provide an actual end-of-life solution for these waste streams. In addition, fungi are well-documented degraders of many of the toxic compounds leached from plastics, such as endocrine disruptors, and lend them well to the detoxification of wastes as they degrade.

Samantha G.T. Jenkins
BIOHM, Cleveland, OH, United States

References

1. Alexander M. Biodegradation and Bioremediation. 2nd ed. Academic Press; 1994:177–235.
2. Geyer R. Production, use, and fate of synthetic polymers. In: *Plastic Waste and Recycling*. Elsevier; 2020:13–32. https://doi.org/10.1016/B978-0-12-817880-5.00002-5.
3. Huerta Lwanga E, Mendoza Vega J, Ku Quej V, et al. Field evidence for transfer of plastic debris along a terrestrial food chain. *Sci Rep*. 2017;7(1):14071. https://doi.org/10.1038/s41598-017-14588-2.
4. Andrady AL. Microplastics in the marine environment. *Mar Pollut Bull*. 2011;62(8):1596–1605. https://doi.org/10.1016/j.marpolbul.2011.05.030.
5. Mwaikambo LY. Review of the history, properties and application of plant fibres. *Afr J Sci Technol*. 2006;7(2):120–133.
6. Shah AA, Hasan F, Shah Z, Kanwal N, Zeb S. Biodegradation of natural and synthetic rubbers: a review. *Int Biodeterior Biodegrad*. 2013;83:145–157. https://doi.org/10.1016/j.ibiod.2013.05.004.
7. USDA BioPreferred. BioPreferred Program. Published online; 2021.
8. Iwata T. Biodegradable and bio-based polymers: future prospects of eco-friendly plastics. *Angew Chem Int Ed*. 2015;54(11):3210–3215. https://doi.org/10.1002/anie.201410770.
9. Phung Hai TA, Neelakantan N, Tessman M, et al. Flexible polyurethanes, renewable fuels, and flavorings from a microalgae oil waste stream. *Green Chem*. 2020;22(10):3088–3094. https://doi.org/10.1039/D0GC00852D.
10. Snell KD, Peoples OP. PHA bioplastic: a value-added coproduct for biomass biorefineries. *Biofuels Bioprod Biorefin*. 2009;3(4):456–467. https://doi.org/10.1002/bbb.161.
11. ASTM International. ASTM D883-20b: Standard Terminology Relating to Plastics; 2020. https://doi.org/10.1520/D0883-20B.
12. European Bioplastics. Oxo-Degradable Plastics; 2021. https://www.european-bioplastics.org/bioplastics/standards/oxo-degradables/. Accessed 12 September 2021.
13. Merrington A, ed. Recycling of plastics. In: *Applied Plastics Engineering Handbook: Processing and Materials*. 1st ed. Elsevier/William Andrew; 2011. PDL handbook series;.
14. Knapp JS, Bromley-Challenor KCA, eds. Recalcitrant organic compounds. In: *Handbook of Water and Wastewater Microbiology*. Academic Press; 2003.
15. Howard GT. Biodegradation of polyurethane: a review. *Int Biodeterior Biodegrad*. 2002;49(4):245–252. https://doi.org/10.1016/S0964-8305(02)00051-3.
16. Patel AK, Singhania RR, Pandey A. Production, purification, and application of microbial enzymes. In: *Biotechnology of Microbial Enzymes*. Academic Press; 2017:13–41. [chapter 2].
17. Johnson AN, Barlow DE, Kelly AL, Varaljay VA, Crookes-Goodson WJ, Biffinger JC. Current progress towards understanding the biodegradation of synthetic condensation polymers with active hydrolases. *Polym Int*. 2021;70(7):977–983. https://doi.org/10.1002/pi.6131.
18. do Canto VP, Thompson CE, Netz PA. Polyurethanases: three-dimensional structures and molecular dynamics simulations of enzymes that degrade polyurethane. *J Mol Graph Model*. 2019;89:82–95. https://doi.org/10.1016/j.jmgm.2019.03.001.
19. ASTM International. ASTM D5338: Test Method for Determining Aerobic Biodegradation of Plastic Materials Under Controlled Composting Conditions, Incorporating Thermophilic Temperatures. ASTM International; 2021. https://doi.org/10.1520/D5338-15R21.
20. ASTM International. ASTM D7475: Test Method for Determining the Aerobic Degradation and Anaerobic Biodegradation of Plastic Materials under Accelerated Bioreactor Landfill Conditions. ASTM International; 2020. https://doi.org/10.1520/D7475-20.
21. Roohi BK, Kuddus M, et al. Microbial enzymatic degradation of biodegradable plastics. *Curr Pharm Biotechnol*. 2017;18(5). https://doi.org/10.2174/1389201018666170523165742.

22. Zheng Y, Yanful EK, Bassi AS. A review of plastic waste biodegradation. *Crit Rev Biotechnol.* 2005;25(4):243–250. https://doi.org/10.1080/07388550500346359.
23. Nakajima-Kambe T, Shigeno-Akutsu Y, Nomura N, Onuma F, Nakahara T. Microbial degradation of polyurethane, polyester polyurethanes and polyether polyurethanes. *Appl Microbiol Biotechnol.* 1999;51(2):134–140. https://doi.org/10.1007/s002530051373.
24. New Plastics Economy. Oxo-Degradable Plastic Packaging is not a Solution to Plastic Pollution, and does not Fit in a Circular Economy. Published online; May 2019.
25. Rhodes CJ. Plastic pollution and potential solutions. *Sci Prog.* 2018;101(3):207–260. https://doi.org/10.3184/003685018X15294876706211.
26. Vrijheid M. Health effects of residence near hazardous waste landfill sites: a review of epidemiologic literature. *Environ Health Perspect.* 2000;108(Suppl. 1):101–112. https://doi.org/10.1289/ehp.00108s1101.
27. Yang Z, Lü F, Zhang H, et al. Is incineration the terminator of plastics and microplastics? *J Hazard Mater.* 2021;401, 123429. https://doi.org/10.1016/j.jhazmat.2020.123429.
28. Environmental Protection Agency. Plastics: Material-Specific Data; 2021. https://www.epa.gov/facts-and-figures-about-materials-waste-and-recycling/plastics-material-specific-data. Accessed 12 September 2021.
29. van den Brink J, de Vries RP. Fungal enzyme sets for plant polysaccharide degradation. *Appl Microbiol Biotechnol.* 2011;91(6):1477–1492. https://doi.org/10.1007/s00253-011-3473-2.
30. Alshehrei F. Biodegradation of synthetic and natural plastic by microorganisms. *J Appl Environ Microbiol.* 2017;5(1):8–19.
31. Zumstein MT, Schintlmeister A, Nelson TF, et al. Biodegradation of synthetic polymers in soils: tracking carbon into CO_2 and microbial biomass. *Sci Adv.* 2018;4(7), eaas9024. https://doi.org/10.1126/sciadv.aas9024.
32. Leja K, Lewandowicz G. Polymer biodegradation and biodegradable polymers—a review. *Pol J Environ Stud.* 2010;19(2):255–266.
33. Reineke W. Aerobic and anaerobic biodegradation potentials of microorganisms. In: Beek B, ed. *Biodegradation and Persistance. Vol 2K. The Handbook of Environmental Chemistry.* Springer-Verlag; 2001:1–161. https://doi.org/10.1007/10508767_1.
34. Ghosh S, Qureshi A, Purohit HJ. Microbial degradation of plastics: biofilms and degradation pathways. In: *Contaminants in Agriculture and Environment: Health Risks and Remediation.* Haridwar, India: Agro Environ Media—Agriculture and Environmental Science Academy; 2019:184–199. https://doi.org/10.26832/AESA-2019-CAE-0153-014.
35. Sanin SL, Sanin FD, Bryers JD. Effect of starvation on the adhesive properties of xenobiotic degrading bacteria. *Process Biochem.* 2003;38(6):909–914. https://doi.org/10.1016/S0032-9592(02)00173-5.
36. Sivan A. New perspectives in plastic biodegradation. *Curr Opin Biotechnol.* 2011;22(3):422–426. https://doi.org/10.1016/j.copbio.2011.01.013.
37. Zafar U, Nzeram P, Langarica-Fuentes A, et al. Biodegradation of polyester polyurethane during commercial composting and analysis of associated fungal communities. *Bioresour Technol.* 2014;158:374–377. https://doi.org/10.1016/j.biortech.2014.02.077.
38. Pradhan R, Misra M, Erickson L, Mohanty A. Compostability and biodegradation study of PLA–wheat straw and PLA–soy straw based green composites in simulated composting bioreactor. *Bioresour Technol.* 2010;101(21):8489–8491. https://doi.org/10.1016/j.biortech.2010.06.053.
39. Anderson G, Shenkar N. Potential effects of biodegradable single-use items in the sea: polylactic acid (PLA) and solitary ascidians. *Environ Pollut.* 2021;268, 115364. https://doi.org/10.1016/j.envpol.2020.115364.
40. Seng B, Hirayama K, Katayama-Hirayama K, Ochiai S, Kaneko H. Scenario analysis of the benefit of municipal organic-waste composting over landfill, Cambodia. *J Environ Manage.* 2013;114:216–224. https://doi.org/10.1016/j.jenvman.2012.10.002.

41. Ishigaki T, Sugano W, Nakanishi A, Tateda M, Ike M, Fujita M. The degradability of biodegradable plastics in aerobic and anaerobic waste landfill model reactors. *Chemosphere.* 2004;54(3):225–233. https://doi.org/10.1016/S0045-6535(03)00750-1.
42. Jambeck JR, Geyer R, Wilcox C, et al. Plastic waste inputs from land into the ocean. *Science.* 2015;347(6223):768–771. https://doi.org/10.1126/science.1260352.
43. Wang G, Huang D, Ji J, Völker C, Wurm FR. Seawater-degradable polymers—fighting the marine plastic pollution. *Adv Sci.* 2021;8(1):2001121. https://doi.org/10.1002/advs.202001121.
44. Rutkowska M, Krasowska K, Heimowska A, et al. Environmental degradation of blends of atactic poly[(R,S)-3-hydroxybutyrate] with natural PHBV in Baltic Sea water and compost with activated sludge. *J Polym Environ.* 2008;16(3):183–191. https://doi.org/10.1007/s10924-008-0100-0.
45. Pranamuda H, Tokiwa Y, Tanaka H. Polylactide degradation by an *Amycolatopsis* sp. *Appl Environ Microbiol.* 1997;63(4):1637–1640. https://doi.org/10.1128/aem.63.4.1637-1640.1997.
46. Karamanlioglu M, Robson GD. The influence of biotic and abiotic factors on the rate of degradation of poly(lactic) acid (PLA) coupons buried in compost and soil. *Polym Degrad Stab.* 2013;98(10):2063–2071. https://doi.org/10.1016/j.polymdegradstab.2013.07.004.
47. Krasowska K, Janik H, Gradys A, Rutkowska M. Degradation of polyurethanes in compost under natural conditions. *J Appl Polym Sci.* 2012;125(6):4252–4260. https://doi.org/10.1002/app.36597.
48. Achterberg EP, Holland TW, Bowie AR, Mantoura RFC, Worsfold PJ. Determination of iron in seawater. *Anal Chim Acta.* 2001;442(1):1–14. https://doi.org/10.1016/S0003-2670(01)01091-1.
49. Hutchins DA, Fu F. Microorganisms and ocean global change. *Nat Microbiol.* 2017;2(6):17058. https://doi.org/10.1038/nmicrobiol.2017.58.
50. Ryckeboer J, Mergaert J, Vaes K, et al. A survey of bacteria and fungi occurring during composting and self-heating process. *Ann Microbiol.* 2003;53(4):349–410.
51. Gunawan NR, Tessman M, Schreiman AC, et al. Rapid biodegradation of renewable polyurethane foams with identification of associated microorganisms and decomposition products. *Bioresour Technol Rep.* 2020;11, 100513. https://doi.org/10.1016/j.biteb.2020.100513.
52. Kay MJ, Morton LHG, Prince EL. Chemical and physical changes occurring in polyester polyurethane during biodegradation. *Int Biodeterior Biodegrad.* 1991;27(2):205–222. https://doi.org/10.1016/0265-3036(91)90012-G.
53. Bonhomme S, Cuer A, Delort A-M, Lemaire J, Sancelme M, Scott G. Environmental biodegradation of polyethylene. *Polym Degrad Stab.* 2003;81(3):441–452. https://doi.org/10.1016/S0141-3910(03)00129-0.
54. Wright RJ, Erni-Cassola G, Zadjelovic V, Latva M, Christie-Oleza JA. Marine plastic debris: a new surface for microbial colonization. *Environ Sci Technol.* 2020;54(19):11657–11672. https://doi.org/10.1021/acs.est.0c02305.
55. Howard GT, Ruiz C, Hilliard NP. Growth of Pseudomonas chlororaphis on a polyester–polyurethane and the purification and characterization of a polyurethanase–esterase enzyme. *Int Biodeterior Biodegrad.* 1999;43(1–2):7–12. https://doi.org/10.1016/S0964-8305(98)00057-2.
56. El-Sayed AHMM, Mahmoud WM, Davis EM, Coughlin RW. Biodegradation of polyurethane coatings by hydrocarbon-degrading bacteria. *Int Biodeterior Biodegrad.* 1996;37(1–2):69–79. https://doi.org/10.1016/0964-8305(95)00091-7.
57. Howard GT, Norton WN, Burks T. Growth of *Acinetobacter gerneri* P7 on polyurethane and the purification and characterization of a polyurethanase enzyme. *Biodegradation.* 2012;23(4):561–573. https://doi.org/10.1007/s10532-011-9533-6.

58. Oceguera-Cervantes A, Carrillo-García A, López N, et al. Characterization of the polyurethanolytic activity of two *Alicycliphilus* sp. strains able to degrade polyurethane and N-methylpyrrolidone. *Appl Environ Microbiol.* 2007;73(19):6214–6223. https://doi.org/10.1128/AEM.01230-07.
59. Rowe L, Howard GT. Growth of *Bacillus subtilis* on polyurethane and the purification and characterization of a polyurethanase-lipase enzyme. *Int Biodeterior Biodegrad.* 2002;50(1):33–40. https://doi.org/10.1016/S0964-8305(02)00047-1.
60. Nakajima-Kambe T, Onuma F, Kimpara N, Nakahara T. Isolation and characterization of a bacterium which utilizes polyester polyurethane as a sole carbon and nitrogen source. *FEMS Microbiol Lett.* 1995;129(1):39–42. https://doi.org/10.1111/j.1574-6968.1995.tb07554.x.
61. Kay MJ, McCabe RW, Morton LHG. Chemical and physical changes occurring in polyester polyurethane during biodegradation. *Int Biodeterior Biodegrad.* 1993;31(3):209–225. https://doi.org/10.1016/0964-8305(93)90006-N.
62. Mukherjee K, Tribedi P, Chowdhury A, et al. Isolation of a *Pseudomonas aeruginosa* strain from soil that can degrade polyurethane diol. *Biodegradation.* 2011;22(2):377–388. https://doi.org/10.1007/s10532-010-9409-1.
63. Howard GT, Blake RC. Growth of *Pseudomonas fluorescens* on a polyester–polyurethane and the purification and characterization of a polyurethanase–protease enzyme. *Int Biodeterior Biodegrad.* 1998;42(4):213–220. https://doi.org/10.1016/S0964-8305(98)00051-1.
64. Hung C-S, Zingarelli S, Nadeau LJ, et al. Carbon catabolite repression and impranil polyurethane degradation in *Pseudomonas protegens* strain Pf-5. Parales RE, ed. *Appl Environ Microbiol.* 2016;82(20):6080–6090. https://doi.org/10.1128/AEM.01448-16.
65. Peng Y-H, Shih Y, Lai Y-C, Liu Y-Z, Liu Y-T, Lin N-C. Degradation of polyurethane by bacterium isolated from soil and assessment of polyurethanolytic activity of a *Pseudomonas putida* strain. *Environ Sci Pollut Res.* 2014;21(16):9529–9537. https://doi.org/10.1007/s11356-014-2647-8.
66. Akutsu-Shigeno Y, Adachi Y, Yamada C, et al. Isolation of a bacterium that degrades urethane compounds and characterization of its urethane hydrolase. *Appl Microbiol Biotechnol.* 2006;70(4):422–429. https://doi.org/10.1007/s00253-005-0071-1.
67. Kay MJ, Morton LHG, Prince EL. Bacterial degradation of polyester polyurethane. *Int Biodeterior Biodegrad.* 1991;27(2):205–222. https://doi.org/10.1016/0265-3036(91)90012-G.
68. Jansen B, Schumacher-Perdreau F, Peters G, Pulverer G. Evidence for degradation of synthetic polyurethanes by *Staphylococcus epidermidis. Zentralblatt Für Bakteriol.* 1991;276(1):36–45. https://doi.org/10.1016/S0934-8840(11)80216-1.
69. Russell JR, Huang J, Anand P, et al. Biodegradation of polyester polyurethane by endophytic fungi. *Appl Environ Microbiol.* 2011;77(17):6076–6084. https://doi.org/10.1128/AEM.00521-11.
70. Cosgrove L, McGeechan PL, Robson GD, Handley PS. Fungal communities associated with degradation of polyester polyurethane in soil. *Appl Environ Microbiol.* 2007;73(18):5817–5824. https://doi.org/10.1128/AEM.01083-07.
71. Howard GT. Polyurethane biodegradation. In: Singh SN, ed. *Microbial Degradation of Xenobiotics.* Berlin, Heidelberg: Springer; 2012:371–394. Environmental Science and Engineering; https://doi.org/10.1007/978-3-642-23789-8_14.
72. Khan S, Nadir S, Shah ZU, et al. Biodegradation of polyester polyurethane by *Aspergillus tubingensis. Environ Pollut.* 2017;225:469–480. https://doi.org/10.1016/j.envpol.2017.03.012.
73. Álvarez-Barragán J, Domínguez-Malfavón L, Vargas-Suárez M, González-Hernández R, Aguilar-Osorio G, Loza-Tavera H. Biodegradative activities of selected environmental fungi on a polyester polyurethane varnish and polyether polyurethane foams. Kivisaar M, ed. *Appl Environ Microbiol.* 2016;82(17):5225–5235. https://doi.org/10.1128/AEM.01344-16.

74. Crabbe JR, Campbell JR, Thompson L, Walz SL, Schultz WW. Biodegradation of a colloidal ester-based polyurethane by soil fungi. *Int Biodeterior Biodegrad.* 1994;33(2):103–113. https://doi.org/10.1016/0964-8305(94)90030-2.
75. Barratt SR, Ennos AR, Greenhalgh M, Robson GD, Handley PS. Fungi are the predominant micro-organisms responsible for degradation of soil-buried polyester polyurethane over a range of soil water holding capacities. *J Appl Microbiol.* 2003;95(1):78–85. https://doi.org/10.1046/j.1365-2672.2003.01961.x.
76. Biffinger JC, Barlow DE, Cockrell AL, et al. The applicability of Impranil®DLN for gauging the biodegradation of polyurethanes. *Polym Degrad Stab.* 2015;120:178–185. https://doi.org/10.1016/j.polymdegradstab.2015.06.020.
77. Brunet C, Aubin S, Gagné S, West R, Lesage J. Development of a method for extraction and determination of 4,4′-methylenedianiline in soils by solid-phase extraction and UPLC-MS-MS. *J Liq Chromatogr Relat Technol.* 2018;41(15–16):919–926. https://doi.org/10.1080/10826076.2018.1539673.
78. Li H-X, Getzinger GJ, Ferguson PL, Orihuela B, Zhu M, Rittschof D. Effects of toxic leachate from commercial plastics on larval survival and settlement of the barnacle *Amphibalanus amphitrite*. *Environ Sci Technol.* 2016;50(2):924–931. https://doi.org/10.1021/acs.est.5b02781.
79. Kim M-N, Lee B-Y, Lee I-M, Lee H-S, Yoon J-S. Toxicity and biodegradation of products from polyester hydrolysis. *J Environ Sci Health A.* 2001;36(4):447–463. https://doi.org/10.1081/ESE-100103475.
80. Thomas NL, Clarke J, McLauchlin AR, Patrick SG. Oxodegradable plastics: degradation, environmental impact and recycling. *Proc Inst Civ Eng.* 2012;165(3):133–140. https://doi.org/10.1680/warm.11.00014.
81. Gan Z, Zhang H. PMBD: a comprehensive plastics microbial biodegradation database. *Database.* 2019;2019:baz119. https://doi.org/10.1093/database/baz119.
82. Loredo-Treviño A, Gutiérrez-Sánchez G, Rodríguez-Herrera R, Aguilar CN. Microbial enzymes involved in polyurethane biodegradation: a review. *J Polym Environ.* 2012;20(1):258–265. https://doi.org/10.1007/s10924-011-0390-5.
83. Howard GT, Vicknair J, Mackie RI. Sensitive plate assay for screening and detection of bacterial polyurethanase activity. *Lett Appl Microbiol.* 2001;32(3):211–214. https://doi.org/10.1046/j.1472-765x.2001.00887.x.
84. Ruiz C, Main T, Hilliard NP, Howard GT. Purification and characterization of two polyurethanase enzymes from *Pseudomonas chlororaphis*. *Int Biodeterior Biodegrad.* 1999;43(1–2):43–47. https://doi.org/10.1016/S0964-8305(98)00067-5.
85. Espinosa MJC, Blanco AC, Schmidgall T, et al. Toward biorecycling: isolation of a soil bacterium that grows on a polyurethane oligomer and monomer. *Front Microbiol.* 2020;11:404. https://doi.org/10.3389/fmicb.2020.00404.
86. Nomura N, Deguchi T, Shigeno-Akutsu Y, Nakajima-Kambe T, Nakahara T. Gene structures and catalytic mechanisms of microbial enzymes able to biodegrade the synthetic solid polymers nylon and polyester polyurethane. *Biotechnol Genet Eng Rev.* 2001;18(1):125–147. https://doi.org/10.1080/02648725.2001.10648011.
87. Howard GT, Mackie RI, Cann IKO, et al. Effect of insertional mutations in the pueA and pueB genes encoding two polyurethanases in *Pseudomonas chlororaphis* contained within a gene cluster: insertional mutations in pueA and pueB genes. *J Appl Microbiol.* 2007;103(6):2074–2083. https://doi.org/10.1111/j.1365-2672.2007.03447.x.
88. Vega RE, Main T, Howard GT. Cloning and expression in *Escherichia coli* of apolyurethane-degrading enzyme from *Pseudomonas fluorescens*. *Int Biodeterior Biodegrad.* 1999;43(1–2):49–55. https://doi.org/10.1016/S0964-8305(98)00068-7.
89. White GF, Russell NJ, Tidswell EC. Bacterial scission of ether bonds. *Microbiol Rev.* 1996;60(1):216–232. https://doi.org/10.1128/mr.60.1.216-232.1996.

90. Mohanan N, Montazer Z, Sharma PK, Levin DB. Microbial and enzymatic degradation of synthetic plastics. *Front Microbiol.* 2020;11, 580709. https://doi.org/10.3389/fmicb.2020.580709.
91. Simon J, Barla F, Kelemen-Haller A, Farkas F, Kraxner M. Thermal stability of polyurethanes. *Chromatographia.* 1988;25(2):99–106. https://doi.org/10.1007/BF02259024.
92. Santerre JP, Labow RS, Duguay DG, Erfle D, Adams GA. Biodegradation evaluation of polyether and polyester-urethanes with oxidative and hydrolytic enzymes. *J Biomed Mater Res.* 1994;28(10):1187–1199. https://doi.org/10.1002/jbm.820281009.
93. Hood MA, Wang B, Sands JM, La Scala JJ, Beyer FL, Li CY. Morphology control of segmented polyurethanes by crystallization of hard and soft segments. *Polymer.* 2010;51(10):2191–2198. https://doi.org/10.1016/j.polymer.2010.03.027.
94. Carnecka M, Obruca S, Ondruska V, et al. Use of several yeast and moulds to biodegradation of modified polyurethane foams—a screening study. *J Biotechnol.* 2007;131(2):S174. https://doi.org/10.1016/j.jbiotec.2007.07.907.
95. Mecozzi M, Nisini L. The differentiation of biodegradable and non-biodegradable polyethylene terephthalate (PET) samples by FTIR spectroscopy: a potential support for the structural differentiation of PET in environmental analysis. *Infrared Phys Technol.* 2019;101:119–126. https://doi.org/10.1016/j.infrared.2019.06.008.
96. Han J, Chen B, Ye L, Zhang A, Zhang J, Feng Z. Synthesis and characterization of biodegradable polyurethane based on poly(ε-caprolactone) and L-lysine ethyl ester diisocyanate. *Front Mater Sci China.* 2009;3(1):25–32. https://doi.org/10.1007/s11706-009-0013-4.
97. Delmas MA, Burbano VC. The drivers of greenwashing. *Calif Manage Rev.* 2011;54(1):64–87. https://doi.org/10.1525/cmr.2011.54.1.64.
98. Nyilasy G, Gangadharbatla H, Paladino A. Perceived greenwashing: the interactive effects of green advertising and corporate environmental performance on consumer reactions. *J Bus Ethics.* 2014;125(4):693–707. https://doi.org/10.1007/s10551-013-1944-3.

CHAPTER

5

Polyurethane processing and degradation: The analytical chemistry

Marissa Tessman[a], *Berk Kuntasal*[b], *and Miheer Modi*[b]

[a]Algenesis Materials, Cardiff, CA, United States, [b]University of California San Diego, La Jolla, CA, United States

Introduction

Analytics are vital to each step in the life cycle of a polyurethane (PU). Key metrics must be followed from the cradle to the grave to evaluate synthesis, performance, and stability. The PU industry has developed a well-established and robust set of methods for measuring the physical and chemical properties of the significant PU precursors: polyols and diisocyanates. As with most synthesis reactions, the purity, miscibility, and component mole ratio significantly affect the end product. Polymer formation is unique to other syntheses in that the starting materials and products are not intrinsically pure but instead consist of molecular weight distribution. In the case of PU, the polyol chain length; degree of molecular weight distribution; residual water and acid content; and nitrogen, carbon, and oxygen content of the isocyanate affect the final PU product properties and quality. This chapter outlines methods for just a few fundamental properties that are crucial to PU quality and can be used to predict PU mechanical trends.

At the end of the PU lifetime, the metrics are less established. PU's chemical degradation mechanisms have been understood for decades; apparent environmental degradation has been studied to a lesser degree. The core barrier behind the true understanding of biological degradation, however, is complexity. Both PU's inhomogeneous structure and the biological matrix in which the plastic is degraded pose a barrier to scientists'

https://doi.org/10.1016/B978-0-323-99982-3.00009-2

ability to track the breakdown process fully. Most analytical methods work well under controlled environments with clean, homogeneous samples, where signal changes over time are distinguishable from background noise. These methods also have specificity: monitoring only one property or compound with great sensitivity. But PUs are not homogeneous. At their simplest, they are comprised of one or more types of polyol linked with a diisocyanate. These long polymer chains, often 10–500 KDa, then orient into a complex matrix of soft and hard regions, which help determine the material's physical and mechanical properties. The variety of polyols and diisocyanates and the diversity in processing lead to a near-infinite matrix of potential materials. Because of this, most peer-reviewed research runs the gamut of analytical techniques, tending to focus on one or two of the properties specific to the material being worked with [e.g., films, thermoplastic PU (TPU), dispersions, thermosets, foams], while most industrially accepted analytical methods for monitoring PU degradation tend to ignore material properties and focus on monitoring the biological output of CO_2. Both approaches have their advantages and disadvantages, which will be explored in this chapter.

Analytical methods of measuring polyurethane precursors

Hydroxyl number titrations

ASTM 1899: Standard test method for hydroxyl groups using reaction with p-toluenesulfonyl isocyanate (TSI) and potentiometric titration with tetrabutylammonium hydroxide[1]

In polyol analysis, hydroxyl number testing is one of the most prevalent ways to determine the extent of reaction completion and polyol stability. The test is performed as a dual equivalence point potentiometric titration of n-butylammonium hydroxide in isopropanol (0.1 M n-Bu_4NOH in IPA) against the acylated product formed by reacting the polyol of interest with p-toluenesulfonyl isocyanate in acetonitrile (0.02 M p-TSI in ACN) in minimal oxygen and water vapor conditions. Hydroxyl numbers are denoted in units of mg potassium hydroxide (KOH) per gram of sample. Hydroxyl number testing of polyol samples begins by dissolving a set quantity of the polyol in ACN using manual agitation (stirring). To avoid overloading the system and then overshooting the second equivalence point, the following formula is used to calculate the amount to be weighed out given the quantity of polyol sampled.

$$\text{Quantity of polyol required (g)} = \frac{40}{\text{Expected hydroxyl number}}$$

The titration starts by pipetting 10 mL of 0.02 M p-TSI in ACN with the sample and stirring in a sealed container for 5 min. These reagents initiate a complete acylation of the –OH groups present. After stirring the sample with the 0.02 M p-TSI in ACN, 0.5 mL of deionized water is injected into the sample to quench the reaction, i.e., neutralize the p-TSI. Postneutralization, the reaction mixture is titrated against 0.1 M n-Bu_4NOH in IPA. The reaction mixture is titrated to two equivalence points (VEQ 1 and VEQ 2), where the volumes of 0.1 M n-Bu_4NOH in IPA dispensed coincide with the sharpest drop in mixture potential. The following formula is used to calculate the hydroxyl number of the sample.

$$\text{Hydroxyl number}\left(\frac{\text{mgKOH}}{\text{g}}\right) = \frac{VEQ1 - VEQ2 \times 56.1 \times M_{\text{n-Bu4NOH}}}{m_{sample}}$$

Acid number titrations

ASTM D664: Standard test method for acid number of petroleum products by potentiometric titration[2]

Acid number testing is a complementary technique performed alongside hydroxyl number testing to determine the extent of reaction completion and overall polyol stability. Acid numbers are a relatively more straightforward potentiometric titration to perform, as they involve a single equivalence point titration of potassium hydroxide in IPA (0.05 M KOH in IPA) against a sample dissolved in a solvent comprised of 50% toluene, 48% propane-2-ol, and 2% deionized water. The solvent is an ideal mixture of polar and nonpolar compounds to aid in the complete solvation of the sample. Acid numbers are denoted in units of mg KOH/g sample.

After the sample is completely dissolved in the titration solvent, the mixture is titrated against 0.05 M KOH in IPA until it reaches VEQ1, which is shown by the volume of 0.05 M KOH in IPA that coincides with the sharpest increase in mixture potential. The following formula is used to calculate the acid number of the sample.

$$\text{Acid number}\left(\frac{\text{mg KOH}}{\text{grams of the sample}}\right) = \frac{(VEQ1 - 0.1)\text{mL} \times 56.1\,\frac{\text{g}}{\text{mol}} \times \left(\text{Molarity of KOH in IPA}\right)\frac{\text{mol}}{\text{L}}}{\text{Mass of the sample (g)}}$$

A sizeable variety of polyols are synthesized using acidic species, and the acid number test can help determine reaction progress and be used to test polyol stability. The acid number is a valuable indicator of polyol stability, as they can hydrolyze to diols and diacids in the presence of water over time.

Water content

ASTM D4672: Determination of water content of polyols[3]

Coulometric Karl Fisher titration is the industry standard for measuring water content in organics, oils, and polymers. Samples are injected quantitatively into a moisture-free titration cell containing a generator electrode, a sensing electrode, and coulometric solvent of an alcohol, amine, sulfur dioxide, and iodine. The alcohol reacts with sulfur dioxide and the amine to produce an alkyl sulfite salt. The salt is then oxidized by iodine in the presence of water to generate hydroiodic salt, which is solvated and increases the solution potential, measured at the sensing electrode. Water and iodine are consumed in a 1:1 stoichiometric ratio.

$$ROH + SO_2 + RN \rightarrow (RNH)SO_3R + H_2O + I_2 \rightarrow (RNH)I + (RNH)SO_4R$$

where ROH is an alcohol, RN is an amine, and RNH is a tertiary amine. Throughout the process, the generator electrode oxidizes iodide to iodine, and thus the titration endpoint is reached when the iodine is restored and the potential drops. Volumetric titration can be used as an alternative, which involves titrating the sample dissolved in a similar solvent with an iodine solution.

Isocyanate content

This is the primary value of interest for isocyanates. It refers to the mass percentage of isocyanate groups in the bulk diisocyanate material and is used in conjunction with hydroxyl number to determine the mass ratio of polyol and diisocyanate to react together. A standard amount of di-n-butylamine reacts with isocyanate groups to produce urea. Excess di-n-butylamine is then back titrated with hydrochloric acid and compared to a blank containing only di-n-butylamine.

Analytical methods of monitoring biodegradation

Because of the complex nature of PU construction and the conditions in which they biodegrade, a plethora of analytical techniques have been developed to characterize degradation in the most realistic and informative manner possible. Evolving technologies and discoveries have also enhanced PU analysis and rendered some older methods flawed or obsolete. While the techniques described in this section vary in application and effectiveness, the most useful PU degradation literature includes multiple techniques performed on the same sample type that highlight physical and mechanical property changes, molecular weight decrease, mineralization rate, and primary breakdown identification products.

Visual observations

The most common technique has been to monitor the observable decay of the material. Often it begins to show weaknesses in the form of surface roughening. Cracks, holes, and flakes can be seen on the surface, as well as discoloration and, in some cases, biological growth. Microscopy is used to distinguish bacteria or fungi from the polymer matrix or to image polymer flaws in greater detail. Researchers with access to imaging techniques such as scanning electron microscopy (SEM) can also observe high-resolution surface morphology such as PU foam cell opening as the lamellae crack and decay. SEM involves sputter coating a thin layer of material with a conductive surface such as iridium or colloidal gold to improve contrast, placing the sample into a high vacuum, and sending a beam of electrons across the surface in a raster pattern.[4,5] Secondary electron emissions are detected and converted to a high-resolution image. SEM is particularly adept at showing detailed images of biological growth. Fungal spores, hyphae, and bacterial biofilm formation present in holes and pockets are indicators of biodegradation.[6–8]

Visual analysis is a fast and often cheap method to see if major changes have occurred at a surface level and to draw a comparison between materials and conditions (Fig. 5.1). Multiple papers have compared images of polyester PU foam inoculated in an environment compared to a control. In each case, the foams deformed, changed color, and shrank. Similarly, the effect of physical properties can sometimes be compared. For instance, open-cell PU foams with high surface areas show more visual changes than closed-cell foams or films with no observable cells. An advantage of SEM is its speed and versatility; sample preparation is easy, and it can generate high quality 3-D images in real time.[9]

But visual observation alone cannot prove biodegradability, as physical deformation can be due to several other factors such as mechanical or chemical degradation. Mechanical degradation includes erosion, crushing, and grinding. The bulk material may disintegrate, but the polymer chain is not chemically altered. Instead, the PU is broken into tiny microplastics ranging from millimeter to micron scale. Chemical degradation includes hydrolysis and oxidation.

In many cases, chemical degradation preferentially breaks ester bonds, leaving partly degraded microplastics as well. When paired with additional measurements to ensure biodegradation, imaging is convincing evidence and easy to convey. A considerable barrier to SEM imaging is its cost of operation and maintenance and the possibility of introducing unwanted artifacts during sample preparation. SEM is especially problematic in comparison with biodegradation analysis, as artifacts can lead to false positives. Another important consideration is that SEM samples must be solid and inorganic and capable of fitting into the vacuum chamber. Size considerations require retroactive planning to ensure that samples are small enough.

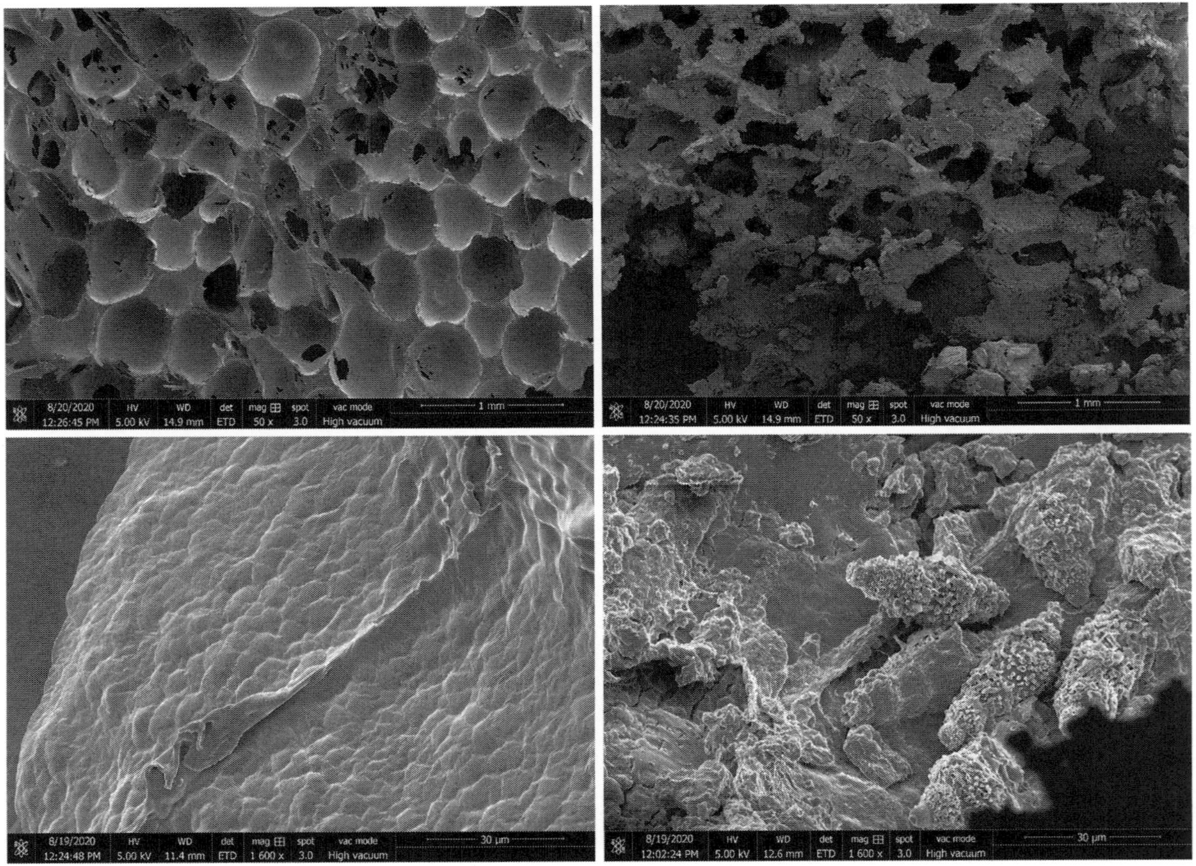

FIG. 5.1 SEM images of a shoe foam control (top left) and degraded in compost (top right) show the deterioration of the foam cell structure. Higher resolution captures microorganism growth on the surface of the foam (bottom right) compared to control (bottom left).

Mass loss

Like visual observation, mass loss has historically been used as a standard method of monitoring biodegradation. When organisms are ingesting and using PU as a food source, they uptake the carbon and release it as CO_2 gas, thereby removing carbon mass from the PU material. Mass loss measurements involve preweighing the material, placing it into a degradative environment, and reweighing it after a designated period. Before mass measurements, the material should be cleaned with a solvent such as water or ethanol to remove external matter and dried overnight in a low-temperature oven. The results are then compared to uninoculated control samples and reported as percent mass loss. The mass of the material should be on the gram scale so that significant loss can be precisely measured, and replicates are measured to account for differences between individual samples.[6,7] Mass loss is an excellent preliminary test to quantify significant changes, as definitive mass loss can indicate biodegradation. It is quick and inexpensive to perform; analytical balances are relatively inexpensive, and sample prep is simple, requiring only rinsing and drying.

There are several significant sources of error in this method. Samples are cleaned to remove as much surface contamination material as possible, but residual dirt, organic matter, and salts can add mass, particularly if the material is porous and can easily trap matter. The process of cleaning degraded samples can also cause small pieces to break off, resulting in falsely high mass loss. If the sample masses or mass changes are too small, these sources of error cause any results to be statistically unreliable, so the method is only meant for significant mass changes. Traditional mass loss measurements also do not account for microplastic formation, which has been a research topic for only the past decade, thus casting some doubt on older literature reporting a mass loss.

Mechanical properties

A PU polymer comprises long chains of polyols and diisocyanates linked by urethane bonds and short diol chain extenders, which phase-separate into hard and soft domains. As described in previous chapters, the constituent identities and processing methods define the PU's chemical and physical structure. Mechanical properties of the PU are thus correlated to hard and soft segment segregation, crystallinity, and molecular weight. Because biodegradation involves organisms clipping polymer chains into digestible monomers, it is hypothesized that the PU's mechanical properties will also be affected.[6,10] There are a wide variety of possible mechanical property measurements that are dependent on the type of sample. Foams, for example, can be tested for compressive strength, compression set, and porosity, whereas common TPU measurements include tensile and tear strength, to name a few (Table 5.1). Thus mechanical

TABLE 5.1 Common mechanical property tests.

Mechanical property	Standard
Density	DIN EN ISO 845
Resilience	DIN 53512 ASTM D1054
Compression set	ASTM D395
Split tear strength	ASTM 3574
Tear strength	DIN 53329-A ASTM D624
Breaking elongation	DIN 53504
Tensile strength	DIN ISO 34-1 ASTM D412
Breaking elongation after hydrolysis	DIN 35304
Tensile strength after hydrolysis	DIN ISO 34-1
Fatigue bending	ASTM D7774
Ross Flex (cut growth)	ASTM D1052
Abrasion	ASTM D1630 ASTM D3389
Shore hardness	ISO 7619-1 ASTM D2240

property biodegradation tests are selected based on PU type. Fig. 5.2 shows four common apparatuses for testing mechanical properties.

Jiang et al. measured the change in tensile properties (tensile strength, modulus, and elongation at break) of waterborne PUs doped with polyethylene glycol segments as a function of time when subjected to hydrolytic and enzymatic conditions. Both elongation at break and tensile strength dropped significantly over 30h in lipase AK compared to hydrolytic conditions, during which the tensile properties were stable. Increasing polyethylene glycol segments resulted in slower enzymatic degradation.[11]

One of the major disadvantages of mechanical property testing for degraded material is that it requires test samples with particular dimensions and no material flaws. Biodegradation that alters the dimensions or generates flaws produces poor test results. For example, tensile strength requires a dog bone-shaped material clamped at both ends and pulled apart while the force and displacement are recorded. Deformations in the dog bone shape will cause the sample to break prematurely, so changes due to biodegradation can only be measured until significant deformation occurs. As a result, mechanical properties cannot be used

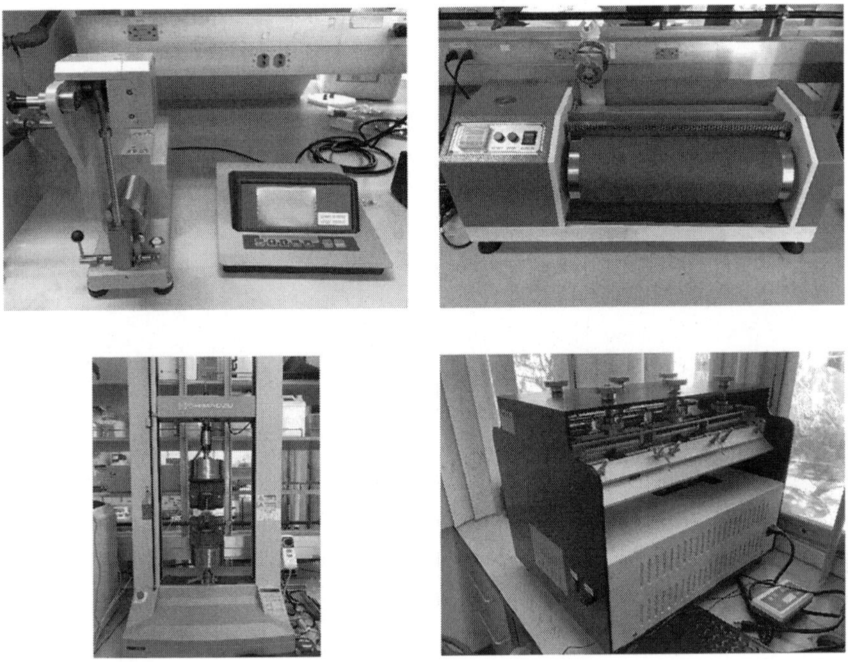

FIG. 5.2 Resilience tester (top left); abrasion tester (top right); universal testing machine for measuring tensile, compression, elongation, and tear strength (bottom left); Ross Flexing Tester (bottom right).

to measure later stages of biodegradation. There can also be significant variability in the previously mentioned mechanical properties, as no two samples will biodegrade in the same way, so accounting for that can be a challenge. Mechanical properties can also change depending on the formulation, for example, tensile strength can change because of poor formulation compared to biodegradation over time.

Gas evolution and consumption

The respirometric test (O_2 uptake); Sturm test (CO_2 release); and methane test in compost, aquatic, and soil environments are good indicators of polymer degradation for laboratory settings. Respirometry is the most widely accepted technique of proving biodegradation. ASTM (American Society of Testing and Materials), CEN (European Committee for Standardization), and ISO (International Organization for Standardization) have generated universal biodegradation test methods under a variety of conditions. ASTM, in particular, designed the standards to mimic real-life scenarios in controlled conditions as closely as possible. The complete set of standards attempts to differentiate between the extent of chemical, photo-, and biodegradation; track the environmental impact

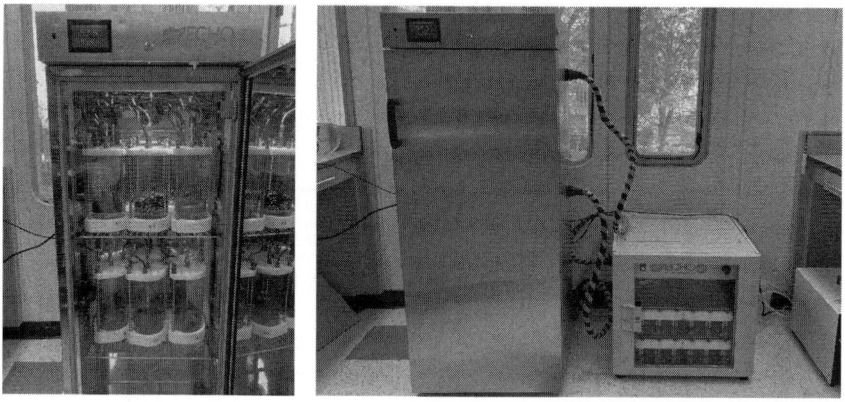

FIG. 5.3 Respirometer open to show temperature-controlled incubator with samples in compost (left). Closed incubator with 12 individually controlled flow meters for monitoring CO_2 release (right).

of degradation products; and most recently establish the likelihood of plastics to form microplastics. Carbon dioxide and methane respirometric methods monitor biogas production against the sample material's carbon content, with 100% degradation corresponding to CO_2 or methane mass balance. Fig. 5.3 shows an apparatus for measuring gas flux.

ASTM D6400: Standard specification for labeling of plastics designed to be aerobically composted in municipal or industrial facilities[12]

The method is to prevent interference during composting in commercial and municipal aerobic composting facilities. A product that meets the requirements is classified as compostable under the guidelines of the US Federal Trade Commission. The material needs to show that it will disintegrate during composting and biodegrade, and that the compost end product will not negatively affect plant growth or introduce regulated metals or other hazardous substances in large proportions. The plastic residuals must be indistinguishable from other organic materials and should not be in significant quantities at the final distribution of the compost.

The criterion for disintegration is having no more than 10% of its original weight after passing through a 2-mm sieve at 12 weeks. The criterion for biodegradation is having 90% of the organic carbon in the sample or 90% of each organic constituent converted to CO_2 in 180 days as measured by ASTM D5338. Organic constituents in the range of 1%–10% should be tested individually to meet the standards. If the concentration of an organic constituent is less than 1%, it does not need to demonstrate biodegradability.

The plastic product should not be introduced to conditions that would increase biodegradation before performing the test. In solid form, its concentrations of regulated metals should be less than 50% of those in the sludges or composts that are found in the country. The product should also not negatively affect the germination rate and biomass of plants found in the compost. Both things mentioned should be no less than 90% of the corresponding blank composts for the two different plant species.

ASTM D5338: Standard test method for determining aerobic biodegradation of plastic materials under controlled composting conditions, incorporating thermophilic temperatures[13]

Gaining more knowledge on plastic biodegradation in a composting unit can be beneficial, especially decomposition's effects on the materials enclosed in the plastic. Biologically degrading waste plastic is a safe way to dispose of it in composting facilities and residential areas. The method presented here is used to determine the degree and rate of aerobic biodegradation of all plastic materials by exposing them to a controlled and simulated composting environment at thermophilic temperatures. The materials are exposed to an inoculum that is obtained from composts from municipal solid waste. Aerobic composting is in an environment where the temperature, aeration, and humidity can be closely monitored and controlled. The biodegradation is calculated by percent biodegradability.

This value is derived by determining the percentage of carbon in the plastic sample that is converted to CO_2, and does not include the amount of carbon converted to cell biomass, which is later metabolized to CO_2 during the test. However, the method is not meant to simulate a specific composting environment, because composting facilities have varying constructions, operations, and regulatory requirements. The inoculated compost must be aerated and between 2 and 4 months old, contain organic portions of municipal solid waste, and be sieved on a screen that is less than 10 mm. If such a compost is not available, then other forms can be used. These other forms can be composted from plants, yard waste, green waste, or a mixture of green and municipal solid waste. For the first 10 days, the compost inoculum should produce between 50 and 150 mg of CO_2 per gram of volatile solids, have an ash content less than 70%, and attain a pH between 7.0 and 8.2. The inoculum should not contain large inert materials such as glass or stones, and there should be sufficient porosity to enable the best possible conditions for aerobic biodegradation. The test specimen can consist of films, dog bone shapes, powder, formed articles, and granules. It should contain enough carbon to produce CO_2 that can be measured by a trapping apparatus.

One study compared the biodegradability of two plastic materials, Mater-Bi produced by Novamont and another produced by Environmental Products, Inc., under aerobic and anaerobic conditions.[14] Both conditions use a positive control of cellulose filter paper. The aerobic system was prepared by composting. Biodegradation of the two plastics and the control were analyzed by mass loss. Temperature, pH, moisture content, and volatile solids content were closely monitored to ensure that the composting environment was under the right conditions. Tests showed that the decomposition process was occurring and the microbes were active. In 72 days, Mater-Bi lost 26.9% of its weight, while the Environmental Products material showed barely any loss. The results indicate that these two samples require more time to biodegrade fully.

ASTM D6691: Standard test method for determining aerobic biodegradation of plastic materials in the marine environment by a defined microbial consortium or natural sea water inoculum[15]

Oceanic plastic pollution has long been a major issue. This ASTM method was developed to determine the rate and degree of biodegradation when a plastic is exposed to known marine microbes under laboratory conditions. Rate is calculated as the proportion of polymer carbon converted to biogas carbon, that is, CO_2. A uniform inoculum containing the various marine microbes is prepared. A respirometer is used to measure the total CO_2 produced as a function of time. The plastic materials need to contain a minimum of 20% carbon.

The inoculum needs to consist of two things: a minimum of 10 test organisms, and seawater collected from the local area. The seawater should not contain any hydrocarbon residues, and the user needs to add inorganic nutrients: 0.5 g/L of NH_4Cl and 0.1 g/L of monopotassium phosphate (KH_2PO_4). To check the activity of the inoculum, a biodegradable material such as cellulose must be included in the test run. If the data on the biogas gathered is less than 70% of the theoretical biogas calculated, then the test is invalid. When the amount of CO_2 generated reaches its limit with the positive control, it shows the degree of biodegradability of the plastic. The generated CO_2 can be converted to percent mineralization to compare to other materials. If the readout is lower than expected, the toxicity of the plastic material should be checked. The degree and rate of biodegradation found in this model can be used to estimate the material's persistence in the marine environment.

ASTM 5511-18: Standard test method for determining anaerobic biodegradation of plastic materials under high-solids anaerobic-digestion conditions[16]

This method was created in a laboratory-controlled environment to simulate anaerobic conditions found in microbe digesters of plastic. It determines the rate and degree of anaerobic biodegradation of a particular material. Solid materials are processed through methanogenic inoculum to start biodegradation. The mixture is exposed in an anaerobic-static-batch digester containing more than 20% plastic solid waste. The total carbon content in the biogas (CO_2 and methane) is measured as a function of time to determine the degree of biodegradation, which is expressed as a percentage of solid carbon converted to gaseous carbon (95% confidence interval). Results are rapid and reproducible.

The inoculum must be derived from an anaerobic digester that contains household waste. The best way to do this is to have the digester operate in dry conditions. The inoculum will undergo postfermentation for 7 days at the temperature from which it was derived. It must have a pH between 7.5 and 8.5, a volatile fatty acid content below 1 g/kg wet weight, and ammonium between 0.5 and 2 g/kg wet weight. Inoculum analyses are done after diluting (5 parts to 1 by weight) with distilled water. The test specimen should contain enough carbon to generate enough CO_2 and methane. The test specimen can be powder, films, pellets, or the form of a dog bone. Percent biodegradation is calculated by dividing the average net carbon gas produced by the plastic material by the average amount of total carbon from the plastic material and multiplying that value by 100. Having information on the plastic material's toxicity allows determination of whether the plastic is within the range of this ASTM. Test results are validated by thin-layer chromatography showing at least 70% biodegradation for cellulose.

ASTM 5526-18: Standard test method for determining anaerobic biodegradation of plastic materials under accelerated landfill conditions[17]

This method has seven major principles of use: select and analyze the material chosen for use, obtain pretreated municipal solid waste and anaerobic inoculum, expose the material to the anaerobic starter, measure total carbon content, remove treated specimens for testing, assess the degree of biodegradability, and assess the degree of biodegradability for less than optimum conditions. Biodegradability is measured by percent solid carbon converted to biogas carbon (methane and CO_2).

Inoculum can be obtained from microbe digesters that operate at about 35°C. When the inoculum has undergone mesophilic postfermentation for 7 days, it is ready for use. The accepted pH value range, volatile fatty acid amounts, and ammonium amounts are the same as those of ASTM 5511-18. Like the other anaerobic ASTM, the plastic material needs to have enough carbon to produce enough methane and CO_2. The material can be in pellet, film, powder, and dog bone form. Percent biodegradation calculation is similar to that of ASTM 5511-18.

The test resembles a landfill where the generated gas is recovered, actively promoted, or both. Obtaining fast degradation allows an increase in economic feasibility for gas recovery and minimizes the care of the landfill needed after degradation. The test method might not consider all the possible conditions for landfills. There is a chance that, after degradation has been completed, the completely or partially degraded plastic materials or extracts will need to be submitted to ecotoxicity testing for environmental hazard.

Clear zone formation

Clear zone refers to the change in a material's physical appearance from opaque to transparent when consumed by microorganisms. PU dispersants can be added to a liquid medium or agar plate. TPUs can be cast onto plates as thin films or powders. The PUs are then inoculated with an organism or enzyme and observed over time for visual changes. In one instance, a methylene diphenyl diisocyanate (MDI) based polyester TPU was cast into a thin film, buried in soil taken from a plastic waste disposal site, and checked for clear zones after 6 months. The microorganisms near the clear zones were then characterized.[18] In another, ground castor oil-based PU was filed into powder, mixed with minimal media, and incubated for 216 h with previously isolated *Aspergillus* and *Chryseobacterium* to demonstrate biodegradability.[19] Impranil, a PU dispersant that can be added to a liquid medium or formed into agar, has been the material of choice in several studies. In an agar medium, the polymer can be dissolved to make a "turbid" plate: it is punctured, and different strains of microorganisms are injected into the punctures. The opaque polymer over the gel forms a clear halo if there is successful depolymerization. In liquid media, enzymes or organisms metabolizing Impranil cause a decrease in turbidity, which is measured spectroscopically.[6] The Impranil test is a quick and inexpensive means for screening degradation by specific organisms or enzymes within controlled parameters. At best, each strain's biodegradation level is qualitatively assessed; visual assessment is not the most accurate. Impranil is a specific PU formulation, so the method does not compare PUs across the formulation and processing spectrum.

Nuclear magnetic resonance spectroscopy (NMR)

This is a direct evaluative technique that can identify the structure of a compound through the functional group-specific relaxation of certain isotopes in response to a magnetic field. ^{1}H and ^{13}C are the most common isotopes measured, as they have natural abundances of over 99.9% and 0.01%, respectively, meaning that they will be present in the sample of interest.[20–22] A big advantage of NMR is that it is noninvasive, preserving the sample for further analysis. Classic NMR has the same disadvantage as gel permeation chromatography (GPC), in which the scope of possible analytes is limited to those soluble in the available solvents. In contrast to GPC, many deuterated solvents can be used, the most common being DMSO-d_6, $CDCl_3$, and acetone-d_6, although care must be taken to avoid a solvent that has interfering chemical shifts to that of the polymer.

NMR can determine the reaction rates of biodegradation. Copolyesters with different stoichiometric compositions were studied using high-resolution ^{13}C NMR. By evaluating the copolymer's composition and block length, the ratio of aliphatic and aromatic dicarboxylic acids was calculated, which helped determine the overall biodegradability of the compound. This method is also valuable for experimentally calculating the average sequence lengths of the fragments.

There are few resources for NMR biodegradation techniques; it is used mostly in the synthesis or controlled chemical, thermal, or ultraviolet (UV) degradation of polymers. NMR works best to analyze pure materials, as peak identification of a complex analyte can be challenging. The required purity makes the analysis of PU complex in two ways. First, PUs are copolymers of varying chain lengths, which inherently complicates the NMR spectrum with many unique chemical shifts. Second, biodegradation introduces further inhomogeneity through nonuniform bond cleavage or even through difficulty in adequately removing organic matter from the sample to be analyzed. Also, it is not as sensitive as other methods such as mass spectrometry (MS), which means greater sample quantities are needed.

Isotopic labeling

Though relatively uncommon, PUs have been subjected to isotopic labeling to trace degradation. This can be accomplished in two ways: stable isotope analysis and radiolabeling. Most PUs consist of nitrogen, carbon, and hydrogen only. The available stable isotopes include ^{17}O, ^{13}C, and ^{15}N, measurable by NMR. Applicable radioisotopes include ^{14}C and are measured by a liquid scintillation counter. This technique is sensitive and versatile in many different conditions, as radiation can be clearly distinguished from complex organic and inorganic mixtures such as enzymatic solutions or compost. Much work in radiolabeling was spearheaded by Santerre and Labow over two decades ago in enzyme degradation studies.[23,24] In collaboration, Tang et al. doped polycarbonate PUs of increasing hard segment concentration with ^{14}C-labelled hexamethylene diisocyanate or 1,4-butanediol. Postenzymatic degradation aliquots of media were analyzed for critical path method (CPM) per mL to test for migration of the radiolabeled atoms from the polymer. The assumption here was that chain cleavage by the enzyme would generate water-soluble, radiolabeled molecules. It was discovered that increasing hard segment content reduced the enzyme's effectiveness, a foundational result that has been replicated many times.[25] In a separate study, Harris et al. monitored thermal degradation in hydroxyl-terminated polybutadiene (HTPB) per isophorone diisocyanate (IPDI) PU by sealing the material in heated conditions with $^{17}O_2$ and measuring the migration of the oxygen isotope into alcohol, carboxylic acid, and ester functional groups in the solid PU via NMR.[21] Despite its role in establishing PU biodegradative trends, most isotope degradation tests have been performed on other types of plastic or organics, likely because of the lack of availability of appropriate equipment to handle radioisotopes and the prohibitive cost of custom isotopically labeled compounds.[26,27]

Fourier transform infrared spectroscopy (FTIR)

FTIR is very commonly used to determine functionality changes with biodegradation. An infrared laser beam passes through a sample of constant width. Atomic bonds absorb infrared energy to induce vibration. The absorbed wavelength is specific to the type of bond vibration that can occur, thus identifying functional groups. Signal intensity is proportional to path length and functional group concentration within the sample according to Beer's Law:

$$A = ebC$$

PU samples undergoing biodegradation are often large solid plastic pieces, so attenuated total reflectance (ATR) cells are commonly used, which require very little sample preparation. A solid sample is pressed against a zinc selenide or diamond surface. Infrared energy is reflected off the sample, thus giving a reproducible path length that allows sample concentrations to be decently compared. The most common biodegradation methods involve oxidation of the ester, urethane, or aromatic functional groups, so the appearance of carboxylic acid, alcohol, and primary amine groups and the disappearance of aromatic secondary amines, amides, and ester groups denote evident oxidative degradation. Gaytan used this reasoning to identify functional group changes in a polyether-PU-acrylate copolymer degrading in landfill inoculum over 20 days (Fig. 5.4). The carbonyl stretch increase and then decrease was interpreted to mean that hydrolysis of the urethane and acrylates was

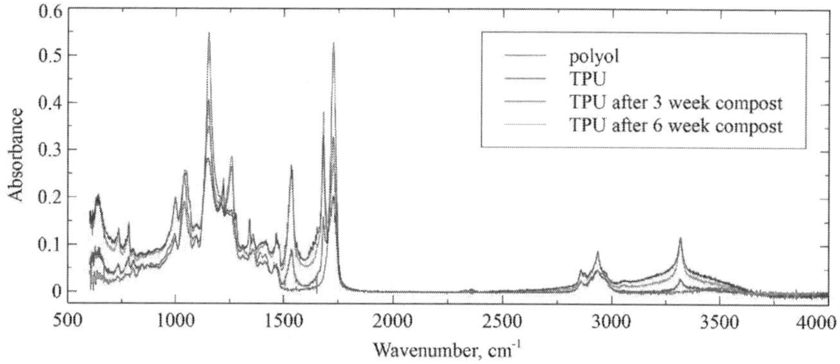

FIG. 5.4 Sample FTIR spectrum of a thermoplastic polyurethane (TPU) degraded in compost for 6 weeks. Functional group changes can be seen with respect to the ester and urethane carbonyl peaks around 1750 cm^{-1} and the hydroxyl and amine peaks around 3400 cm^{-1}.

followed by microbial uptake of the hydrolysis products. The C-N stretch from the urethane was retained, indicating that the hydrolyzed urethane moiety was not ingested after the 20 days.[28] Others have used FTIR similarly.[6,10,29,30]

FTIR offers three specific advantages: Fellgett, Jacquinot, and Connes. Fellgett is also known as the multiplex advantage because the compound's constituents are being measured simultaneously, which reduces the amount of noise (or at least maintains a high signal-to-noise ratio). The Jacquinot advantage states that because there are no slits in FTIR, the total optical output is significantly higher, increasing the signal-to-noise ratio (no loss of IR intensity). The Connes advantage states that because the frequency of the FTIR is known, many spectra can be overlapped. This property means that spectra can also be removed as needed.

FTIR peak identification, especially in the fingerprint region between 1500 and 500 cm^{-1}, can be challenging, so the best use of FTIR is to observe prominent peak changes in the well-defined regions noted before. In inhomogeneous samples such as PU subjected to biodegradation, different surface regions can undergo varying levels of attack, so measurement of one area may not give a complete picture. While FTIR is useful for identifying functional group changes, it can be used only semiquantitatively to identify the extent of degradation.

Gel permeation chromatography (GPC)

A specific kind of size exclusion chromatography, this method is used to separate polymers by their molecular weights before and after biodegradation to evaluate chain length decrease. Polymers are dissolved in an organic solvent such as tetrahydrofuran (THF) or dimethylformamide (DMF), which causes them to fold and cluster from chains to spheroid shapes. The volume of each cluster depends on the chain length, chemical makeup, and solvent used. The solution is then passed through a column packed with a porous material, usually divinylbenzene (DVB) or styrene, which traps smaller molecular volumes and allows larger volumes to pass unhindered. The resulting chromatogram of a typical polymer appears as Gaussian mono- or multimodal peaks, depending on the polymer and extent of degradation. Refractive index detectors are commonly used for polymer GPC, which monitors the change in the angle of light refraction as it passes through the sample compared to a reference solvent. Retention volume is compared to a set of standards [polymethyl methacrylate (PMMA) or polystyrene] to determine the average molecular weight, number average molecular weight, and polydispersity. Recently, evaporative light scattering detectors (ELSDs) have been used, in which the eluent is evaporated from the polymer via nebulization, converting its form into small particulates, which are measured by the Mie scattering light effect. ESLD is about ten times more sensitive than refractive index (RI) and able to detect smaller molecular weight polymers, making it ideal for degradation measurements over time.[31–33]

GPC is the only method that directly correlates degradation with molecular weight decrease and is specifically designed to accommodate inhomogeneous PU chain lengths. Trends in how molecular weights change over time can thus be observed.[34] For example, oxidative degradation of MDI polyester PUR in one case and bacterial degradation of PU film in another resulted in a decrease in molecular weight and increase in polydispersity, indicating nonspecific chain cleavage in both cases.[33,35] RI detectors are limited by their lack of sensitivity, as detection can vary dramatically because of polymer composition and temperature. There are currently no known PU standards compared to experimental PU, so any molecular weight distribution is not indicative of the absolute value but is instead relative to available standards. Relative changes in molecular weights can be detected, but because of the lack of sensitivity and daily variance, appropriate replicates and controls should be run in parallel, which increases the total number of required samples per batch. GPC is also limited by polymer solubility and is thus not ideal for large molecular weight or crosslinked PU, instead favoring thermoplastic PU.

High-pressure liquid chromatography (HPLC) mass spectrometry (MS) and variants

HPLC is a common technique used to separate and quantitate known compounds in a complex mixture based on polarity differences. Dilute samples are injected into continuously flowing eluent of a specific polarity, known as the mobile phase. The eluent then flows at high pressure through a column packed with non- or semipolar material called the stationary phase, usually silica modified with ^{18}C alkanes or proprietary substituents. Analytes will partition between the mobile and stationary phases relative to their respective polarities and elute out of the column at specific retention times. HPLC instrumentation generally relies on ultraviolet and visual (UV-vis) detection, limiting the compound range to those detectable by spectroscopic means. This also means that the sensitivity is dependent on the extinction coefficient. HPLC is best used for aromatic degradation products such as those originating from TSI- and MDI-based PUs that generally have high extinction coefficients in the UV-vis range. For example, Tang et al. compared enzymatic degradation of hexamethylene diisocyanate (HDI) and MDI-derived PU samples via HPLC analysis of the enzymatic solution after several weeks. It was quickly determined that traditional HPLC-UV could not be used for the

HDI samples, as the expected breakdown product was a UV-inactive aliphatic diamine. In comparison, breakdown products for the MDI PU were detected using HPLC-UV.[36] HPLC-UV-vis also requires the use of standards to identify when the analytes of interest elute, limiting its use to quantify known and expected breakdown compounds.

Recent degradation detection by HPLC has shifted to HPLC-MS, in which a mass spectrometer is connected to the back end as a detector instead of UV-vis. Samples are introduced from the HPLC to the MS through an ionization source called atmospheric pressure ionization (API). The eluent is passed through a capillary with an applied voltage that introduces a charge to the analyte. Protons or other charged adducts are associated with the analyte at this time. Aerosolized droplets of neutral eluent and charged analyte are then passed through a series of charged filters with an applied vacuum, during which the eluent is removed, leaving the charged analytes to be detected by conventional MS or MS^n (tandem mass spectrometry). MS increases compound detection range and sensitivity significantly. Whereas UV-vis can detect mid to high parts per million (ppm) concentrations, MS can detect several orders of magnitude lower in the ppb (parts per billion) to ppm range. Because MS detects the mass-to-charge ratio of compounds rather than their UV-vis absorbance, a broader range of compounds can not only be detected, but their masses can also be used for identification. MS is a powerful tool in identifying trace amounts of known and unknown compounds. In the previously mentioned paper by Tang et al., further analysis of the HDI and MDI PU degradation by LC-MS/MS revealed multiple PU fragments, highlighting site-specific enzymatic cleavage. Other papers have shown similar techniques and findings, demonstrating LCMS's superiority in evaluating degradation products, mechanisms, and time frame.[37–39]

Gas chromatography-mass spectrometry (GC-MS)

This technique is particularly effective at measuring and identifying small volatile compounds in complex microscale mixtures, with high sensitivity and resolving power. At optimized parameters, MS detectors can measure ppm and ppb quantities. Electrospray ionization of the separated compounds generates unique fragmentation patterns that can be used in identification. GC-MS results are also easily reproducible when using electron ionization, due to having a standardized 70-eV filament voltage. There are also large transferable electron ionization libraries shared globally that eliminate the need to make a new database for each lab. Compounds extractable in volatile solvents such as chloroform, ethyl acetate, or acetonitrile can be easily separated from larger biological molecules or salt mixtures, making them suitable for GC-MS analysis. In some instances, derivatization is necessary to increase signal or affect volatility. Polyester hydrolysis, in particular, results in the formation of diacids and diols, which tend to require silylation to achieve decently reproducible chromatographic peaks of sufficient signal-to-noise ratio.

Shah et al. used GC-MS to detect silylated 1,4-butanediol and adipic acid monomers because of polyester polyol cleavage by *Pseudomonas aeruginosa* (23). Perez-Lara analyzed PU degradation products of *Alicycliphilus* sp. BQ8 bacteria by extracting with acid or base adjusted ethyl acetate and methylating with diazomethane before injection. Dipropylene glycol, adipic acid, and adipic acid methyl ester were identified through the Wiley spectrum database matches. A tentative degradation product was identified as dipropylene glycol adipate through analysis of the MS fragmentation pattern because of the molecule not being present in the spectral database.[40] Alvarez analyzed Impranil degradation products over 14 days, including diols, diisocyanates, and ester-related compounds. Gaytan identified and tracked more than 20 degradation products and additives from a polyether-PU-acrylate copolymer called PolyLack incubated in landfill inoculum over 20 days. After incubation, his samples were extracted with a chloroform:methanol solution and solvent swapped with THF before running on the GC-MS.[28] GC-MS can also be used to monitor volatile organic compounds from degradation. One of the drawbacks of GC-MS is that it can only measure primary degradation products with sufficient volatility, so it cannot monitor partial degradation or nonspecific cleavage. To achieve a more thorough understanding of degradation, pyrolysis-GC-MS has been used. A 2017 Sandia National Laboratories report and a paper by Kim discussed a GC-MS analysis of thermogravimetric off-gassing from PU elastomers to simulate how the elastomers would respond to high temperatures. The gas was collected at intervals throughout the experiment and injected directly into the GC-MS. The results were identified by comparing the MS spectra to an existing National Institute of Standards and Technology spectral database, although some of the compounds could not be identified in this manner.[41] Similarly, Kim coupled GC-MS and pyrolysis by trapping and concentrating pyrolysis off-gassing. Compounds identified at increasing temperature intervals corresponded to soft and hard segment thermal degradative products.[42]

A primary disadvantage of GC-MS is that it cannot process and analyze samples that are not volatile, polar, or thermally labile (when compounds break down upon exposure to high temperatures). This problem can be mitigated by derivatization, which offers increased sensitivity, selectivity, and specificity, but at the disadvantage of lengthened sample preparation time and analysis costs.

Differential scanning calorimetry (DSC)

This method involves subjecting a PU elastomer and reference sample to controlled temperature change and comparing the resulting heat flow. Following classical thermodynamics, heat transfer characteristics for polymers are different when undergoing phase transitions. Thus DSC can be used to identify PU thermochemical properties such as glass transition, melt, and crystallization temperatures, noticeable by a baseline shift, negative peak (endotherm), or positive peak (exotherm), respectively.[43–45] Intermolecular forces and crystal structure dictate phase transitions, so in theory, bond cleavage or functional group chemistry due to biodegradation would alter the transition temperature ranges and enthalpy. Specifically, for PU, DSC can differentiate between hard and soft segment length and microphase segregation due to the unique temperatures at which each domain undergoes glass transition. Prisacariu states that these properties are proportional to the hard segment percentage and extent of homogeneity between the two structures.[46] Thus the extent of morphology-specific PU degradation can be observed. For example, Zafar et al. measured glass transition temperatures of elastomeric PU coupons buried in compost for 6 months. The soft and hard segment glass transition temperature (T_g) values were identified initially as −22.5°C and 68.5°C, respectively. T_g trends on the surface and center of the coupons were then identified after 14 and 28 days. Because of the decrease in soft and hard segment T_g at the surface (−24.5°C and −36.1°C) and increase in hard segment T_g in the center (97.2°C), it was concluded that as the PU degraded, the crystalline surface was becoming more amorphous while the center was becoming more crystalline.[47] Similarly, Bialkowska synthesized nonisocyanate PUs, inoculated them with the bacteria *Achromobacter xylosoxidans* G21, and demonstrated the disappearance of the crystalline melting point range around 100°C, corresponding to the organism's attack of the rigid segments.[48]

Close-up: Some insights into the role of high-end instrumentation in environmental discovery

Environmental discovery hinges on unraveling the chemical composition within complex environmental matrices. As such, a major goal of analytical chemists is to utilize instrumentation and methodologies that will provide the highest level of confidence in identifying and quantifying individual chemical species in environmental samples. With such information in hand, scientists gain insight into the history and impacts of environmental materials.

Environmental researchers continue to rely on innovations in analytical instrumentation to address the molecular-level world of each complex environmental sample. The more common techniques include a wide range of chromatography, spectroscopy, and mass spectrometry. While a plethora of advances in these techniques have recently been made, a special few have been crucial to advancing knowledge. First and foremost is the rapidly increasing resolution power of commercialized mass spectrometers. Recent advances in Orbitrap mass spectrometry now harness the amazing power of ultra-high mass resolution at the laboratory benchtop. These systems are pushing the boundaries in providing ultra-high mass accuracy that facilitates confident compound identification. Furthermore, new cutting-edge software packages automate accurate assignment of unknown compounds within mass spectrometry data. These platforms use intelligent workflows that bring together many resources, including global mass spectral libraries, chemical databases, statistical analysis packages, and visualization tools, to provide an unprecedented level of confidence in compound identification.

Ambient ionization techniques have been key technological advances for a convenient approach to generating ions from sample materials with little to no sample preparation, thus limiting contamination and increasing confidence in identifying unknown compounds. While such techniques are typically born in academic research labs, many have recently been developed into commercial platforms that are available to scientists around the world.

Richard E. Cochran

Thermo Fisher Scientific, Waltham, MA, United States

References

1. ASTM E1899-16. Standard Test Method for Hydroxyl Groups Using Reaction with p-Toluenesulfonyl Isocyanate (TSI) and Potentiometric Titration with Tetrabutylammonium Hydroxide; 2016. www.astm.org.
2. ASTM D664-18e2. Standard Test Method for Acid Number of Petroleum Products by Potentiometric Titration; 2021. www.astm.org.
3. ASTM D4672-18. Standard Test Method for Polyurethane Raw Materials: Determination of Water Content of Polyols; 2018. www.astm.org.
4. Bootz A, Vogel V, Schubert D, Kreuter J. Comparison of scanning electron microscopy, dynamic light scattering and analytical ultracentrifugation for the sizing of poly(butyl cyanoacrylate) nanoparticles. *Eur J Pharm Biopharm.* 2004;57(2):369–375. https://doi.org/10.1016/S0939-6411(03)00193-0.
5. Horisberger M, Rosset J. Colloidal gold, a useful marker for transmission and scanning electron microscopy. *J Histochem Cytochem.* 1977;25(4):295–305. https://doi.org/10.1177/25.4.323352.
6. Alvarez-Barragan J, Dominguez-Malfavon L, Vargas-Suarez M, Gonzalez-Hernandez R, Aguilar-Osorio G, Loza-Tavera H. Biodegradative activities of selected environmental fungi on a polyester polyurethane varnish and polyether polyurethane foams. *Appl Environ Microbiol.* 2016;82(17):5225–5235. https://doi.org/10.1128/AEM.01344-16.
7. Gaytan I, Sanchez-Reyes A, Burelo M, et al. Degradation of recalcitrant polyurethane and xenobiotic additives by a selected landfill microbial community and its biodegradative potential revealed by proximity ligation-based metagenomic analysis. *Front Microbiol.* 2019;10:2986. https://doi.org/10.3389/fmicb.2019.02986.

8. Mumtaz T, Khan MR, Hassan MA. Study of environmental biodegradation of LDPE films in soil using optical and scanning electron microscopy. *Micron.* 2010;41(5):430–438. https://doi.org/10.1016/j.micron.2010.02.008.
9. Reimer L. Scanning electron microscopy: physics of image formation and microanalysis, second edition. *Meas Sci Technol.* 2000;11(12):1826. https://doi.org/10.1088/0957-0233/11/12/703.
10. Barrioni BR, de Carvalho SM, Orefice RL, de Oliveira AA, Pereira Mde M. Synthesis and characterization of biodegradable polyurethane films based on HDI with hydrolyzable cross-linked bonds and a homogeneous structure for biomedical applications. *Mater Sci Eng C Mater Biol Appl.* 2015;52:22–30. https://doi.org/10.1016/j.msec.2015.03.027.
11. Jiang X, Li J, Ding M, et al. Synthesis and degradation of nontoxic biodegradable waterborne polyurethanes elastomer with poly(ε-caprolactone) and poly(ethylene glycol) as soft segment. *Eur Polym J.* 2007;43(5):1838–1846. https://doi.org/10.1016/j.eurpolymj.2007.02.029.
12. ASTM D6400-19. Standard Specification for Labeling of Plastics Designed to be Aerobically Composted in Municipal or Industrial Facilities; 2019. www.astm.org.
13. ASTM D5338-15. Standard Test Method for Determining Aerobic Biodegradation of Plastic Materials Under Controlled Composting Conditions, Incorporating Thermophilic Temperatures; 2021. www.astm.org.
14. Mohee R, Unmar GD, Mudhoo A, Khadoo P. Biodegradability of biodegradable/degradable plastic materials under aerobic and anaerobic conditions. *Waste Manag.* 2008;28(9):1624–1629. https://doi.org/10.1016/j.wasman.2007.07.003.
15. ASTM D6691-17. Standard Test Method for Determining Aerobic Biodegradation of Plastic Materials in the Marine Environment by a Defined Microbial Consortium or Natural Sea Water Inoculum; 2017. www.astm.org.
16. ASTM D5511-18. Standard Test Method for Determining Anaerobic Biodegradation of Plastic Materials Under High-Solids Anaerobic-Digestion Conditions; 2018. www.astm.org.
17. ASTM D5526-18. Standard Test Method for Determining Anaerobic Biodegradation of Plastic Materials Under Accelerated Landfill Conditions; 2018. www.astm.org.
18. Shah AA, Hasan F, Akhter JI, Hameed A, Ahmed S. Degradation of polyurethane by novel bacterial consortium isolated from soil. *Ann Microbiol.* 2008;58(3):381. https://doi.org/10.1007/BF03175532.
19. Cangemi JM, et al. Biodegradation of polyurethane derived from castor oil. *Polímeros.* 2008;18(3):5. https://doi.org/10.1590/S0104-14282008000300004.
20. Evilia RF. Quantitative NMR Spectroscopy. *Anal Lett.* 2001;34(13):2227–2236. https://doi.org/10.1081/AL-100107290.
21. Harris DJ, Assink RA, Celina M. NMR analysis of oxidatively aged HTPB/IPDI polyurethane rubber: degradation products, dynamics, and heterogeneity. *Macromolecules.* 2001;34(19):6695–6700. https://doi.org/10.1021/ma0108766.
22. Witt U, Müller R-J, Deckwer W-D. Studies on sequence distribution of aliphatic/aromatic copolyesters by high-resolution 13C nuclear magnetic resonance spectroscopy for evaluation of biodegradability. *Macromol Chem Phys.* 1996;197(4):1525–1535. https://doi.org/10.1002/macp.1996.021970428.
23. Labow RS, Duguay DG, Santerre JP. The enzymatic hydrolysis of a synthetic biomembrane: a new substrate for cholesterol and carboxylesterases. *J Biomater Sci Polym Ed.* 1995;6(2):169–179. https://doi.org/10.1163/156856294X00293.
24. Santerre JP, Labow RS. The effect of hard segment size on the hydrolytic stability of polyether-urea-urethanes when exposed to cholesterol esterase. *J Biomed Mater Res.* 1997;36(2):223–232. https://doi.org/10.1002/(SICI)1097-4636(199708)36:2<223::AID-JBM11>3.0.CO;2-H.
25. Tang YW, Labow RS, Santerre JP. Enzyme induced biodegradation of polycarbonate-polyurethanes: dose dependence effect of cholesterol esterase. *Biomaterials.* 2003;24(12):2003–2011. https://doi.org/10.1016/S0142-9612(02)00563-X.

26. Pfaender FK, Bartholomew GW. Measurement of aquatic biodegradation rates by determining heterotrophic uptake of radiolabeled pollutants. *Appl Environ Microbiol.* 1982;44 (1):159–164. https://doi.org/10.1128/aem.44.1.159-164.1982.
27. Shimp RJ, Larson RJ. Estimating the removal and biodegradation potential of radiolabeled organic chemicals in activated sludge. *Ecotoxicol Environ Saf.* 1996;34(1):85–93. https://doi.org/10.1006/eesa.1996.0048.
28. Gaytán I, Sánchez-Reyes A, Burelo M, et al. Degradation of recalcitrant polyurethane and xenobiotic additives by a selected landfill microbial community and its biodegradative potential revealed by proximity ligation-based metagenomic analysis. *Front Microbiol.* 2020;10(2986). https://doi.org/10.3389/fmicb.2019.02986.
29. Chan-Chan LH, Solis-Correa R, Vargas-Coronado RF, et al. Degradation studies on segmented polyurethanes prepared with HMDI, PCL and different chain extenders. *Acta Biomater.* 2010;6(6):2035–2044. https://doi.org/10.1016/j.actbio.2009.12.010.
30. Shah Z, Gulzar M, Hasan F, Shah AA. Degradation of polyester polyurethane by an indigenously developed consortium of Pseudomonas and Bacillus species isolated from soil. *Polym Degrad Stab.* 2016;134:349–356. https://doi.org/10.1016/j.polymdegradstab.2016.11.003.
31. Holding S, Rapra S. Characterisation of Polymers Using Gel Permeation Chromatography and Associated Techniques. Chromatography Today; 2011.
32. Knol WC, Pirok BWJ, Peters RAH. Detection challenges in quantitative polymer analysis by liquid chromatography. *J Sep Sci.* 2021;44(1):63–87. https://doi.org/10.1002/jssc.202000768.
33. Rek V, Mencer HJ, Bravar M. GPC and structural analysis of polyurethane degradation. *Polym Photochem.* 1986;7(4):273–283. https://doi.org/10.1016/0144-2880(86)90055-2.
34. Skleničková K, Abbrent S, Halecký M, Kočí V, Beneš H. Biodegradability and ecotoxicity of polyurethane foams: A review. *Crit Rev Environ Sci Technol.* 2020;1–46. https://doi.org/10.1080/10643389.2020.1818496.
35. Shah Z, Hasan F, Krumholz L, Aktas DF, Shah AA. Degradation of polyester polyurethane by newly isolated Pseudomonas aeruginosa strain MZA-85 and analysis of degradation products by GC–MS. *Int Biodeter Biodegr.* 2013;77:114–122. https://doi.org/10.1016/j.ibiod.2012.11.009.
36. Tang YW, Labow RS, Santerre JP. Isolation of methylene dianiline and aqueous-soluble biodegradation products from polycarbonate-polyurethanes. *Biomaterials.* 2003;24 (17):2805–2819. https://doi.org/10.1016/S0142-9612(03)00081-4.
37. Borrowman CK, Johnston P, Adhikari R, Saito K, Patti AF. Environmental degradation and efficacy of a sprayable, biodegradable polymeric mulch. *Polym Degrad Stab.* 2020;175, 109126.
38. Gunawan NR, Tessman M, Schreiman AC, et al. Rapid biodegradation of renewable polyurethane foams with identification of associated microorganisms and decomposition products. *Biores Technol Rep.* 2020;11, 100513. https://doi.org/10.1016/j.biteb.2020.100513.
39. Magnin A, Pollet E, Perrin R, et al. Enzymatic recycling of thermoplastic polyurethanes: Synergistic effect of an esterase and an amidase and recovery of building blocks. *Waste Manag.* 2019;85:141–150. https://doi.org/10.1016/j.wasman.2018.12.024.
40. Pérez-Lara LF, Vargas-Suárez M, López-Castillo NN, Cruz-Gómez MJ, Loza-Tavera H. Preliminary study on the biodegradation of adipate/phthalate polyester polyurethanes of commercial-type by Alicycliphilus sp. BQ8. *J Appl Polym Sci.* 2016;133(6):42992. https://doi.org/10.1002/app.42992.
41. Harrison KW, Murtagh D, Silva H, Cordaro JG. Thermal Degradation Investigation of Polyurethane Elastomers using Thermal Gravimetric Analysis - Gas Chromatography/Mass Spectrometry. United States: United States; 2017.

42. Kim B-H, Yoon K, Moon DC. Thermal degradation behavior of rigid and soft polyurethanes based on methylene diphenyl diisocyanate using evolved gas analysis-(gas chromatography)–mass spectrometry. *J Anal Appl Pyrolysis.* 2012;98:236–241. https://doi.org/10.1016/j.jaap.2012.09.010.
43. Capitain C, Ross-Jones J, Möhring S, Tippkötter N. Differential scanning calorimetry for quantification of polymer biodegradability in compost. *Int Biodeter Biodegr.* 2020;149, 104914. https://doi.org/10.1016/j.ibiod.2020.104914.
44. Han J, Chen B, Ye L, Zhang A-Y, Zhang J, Feng Z-G. Synthesis and characterization of biodegradable polyurethane based on poly(ε-caprolactone) and L-lysine ethyl ester diisocyanate. *Front Mater Sci China.* 2009;3(1):25–32. https://doi.org/10.1007/s11706-009-0013-4.
45. Reh U, Kraepelin G, Lamprecht I. Differential scanning calorimetry as a complementary tool in wood biodegradation studies. *Thermochim Acta.* 1987;119(1):143–150. https://doi.org/10.1016/0040-6031(87)88015-2.
46. Prisacariu C. Thermal behaviour of polyurethane elastomers. In: Prisacariu C, ed. *Polyurethane Elastomers: From Morphology to Mechanical Aspects.* Vienna: Springer Vienna; 2011:61–101.
47. Zafar U, Nzerem P, Langarica-Fuentes A, et al. Biodegradation of polyester polyurethane during commercial composting and analysis of associated fungal communities. *Bioresour Technol.* 2014;158:374–377. https://doi.org/10.1016/j.biortech.2014.02.077.
48. Białkowska A, Bakar M, Marchut-Mikołajczyk O. Biodegradation of linear and branched non isocyanate condensation polyurethanes based on 2-hydroxy-naphthalene-6-sulfonic acid and phenol sulfonic acid. *Polym Degrad Stab.* 2019;159:98–106. https://doi.org/10.1016/j.polymdegradstab.2018.11.011.

CHAPTER

6

TEA and LCA of bio-based polyurethanes

Matthew Wiatrowski, Eric C.D. Tan, and Ryan Davis

National Renewable Energy Laboratory, Golden, CO, United States

Introduction

What is a TEA?

Background

Techno-economic analysis (TEA) is an instrumental tool used to guide and prioritize research directions across the full range of technology maturity, from early conceptualization through commercial deployment and optimization of existing processes. As the name implies, it incorporates *technical* inputs, typically run through process modeling, to generate essential information coupled with engineering *economic* analysis to inform the process's financial feasibility. For decades, TEA modeling has been used as a principal means of setting production cost targets; benchmarking progress toward achieving such targets; and highlighting key drivers, gaps, and sensitivities related to technical or process factors, e.g., for production of fuels or products.[1–7]

The general approach to TEA modeling involves the establishment of the capital and operating costs for the process scope boundaries and input and output flow rates and yields. Modeling is best and most accurately achieved through first establishing a process simulation to track mass and energy balances across all pertinent unit operations within the process. The model can be used to capture feedstock, chemical, and waste operational costs; equipment sizing; associated kinetics; residence times; operating conditions (temperature, pressure, chemical loadings, and associated metallurgy considerations); and other parameters. Depending on

Rethinking Polyester Polyurethanes
https://doi.org/10.1016/B978-0-323-99982-3.00004-3

the level of required granularity, such process models may be conducted through "back-of-the-envelope" spreadsheet calculations or more thermodynamically rigorous process simulation software packages. Either way, once the capital and operating expenses for constructing and operating a processing facility are established, this information may be utilized to perform a financial assessment for the overall process.

Several approaches to financial assessment are possible, as documented elsewhere,[8–10] but two strategies commonly applied are levelized cost of production and discounted cash flow rate of return (DCFROR) analysis. Levelized cost of production typically allows for more straightforward means of estimating the costs per unit of product output (e.g., dollar-per-pound, gallon, kWh, or BTU) based on a single equation as a function of capital charge rates (a means of amortizing the cost of capital based on associated discount rates, financing costs, and depreciation), fixed and variable operating costs, and product output yields.[11] It is often valuable for comparing technologies across different scales, investment times, construction times, and costs.[8] However, DCFROR analysis provides a higher-fidelity approach to financial analysis. It offers more comprehensive opportunities for evaluating technology and financial details, based on an explicit calculation of a facility's cash flows for each year of construction and operation, including profits and losses, depreciation and financing charges, annual revenues, and taxes.[12] As such, it is generally the authors' preferred approach once appropriate time can be expended to establish the DCFROR framework methodology upon which the pertinent capital and operating costs and product output rates can be applied (published TEA models inclusive of DCFROR analysis framework examples are freely available at https://www.nrel.gov/extranet/biorefinery/aspen-models/).

After conducting a financial analysis based on the underlying capital and operating cost and yield information, several parameters can be solved for depending on the metric of interest. Typically, a standard metric is to solve for the minimum selling price (MSP) of the product, which represents the price the product must be sold for to satisfy the other financial specifications after considering revenues from by-products and coproducts or externalities (policy credits, for example). In this case, a particular rate of return (discount factor) and net present value (NPV) must first be specified to solve the MSP. In the example discussed later, the TEA is conducted at a target 10% internal rate of return (IRR), defined as the rate at which the NPV is zero. Alternatively, all products may be fixed at set prices, and the TEA may instead solve for the NPV at a fixed rate of return or vice versa (e.g., solve for an IRR where NPV is by definition fixed at zero). Or further yet, all product prices and rate of return and NPV may be set, and instead solve for the maximum allowable purchase price of the key feedstock to satisfy all constraints. The latter approach may be appropriate, for example, when economics for the production process is established. However, various options for feedstock sourcing may exist such that the TEA may place a value on the feedstock that the conversion facility would be willing to pay, given that feedstock is often the single largest contributor to final product costs.[13–16]

Benefits and outcomes: Why should we care?

TEA serves four primary functions in the context of precommercial R&D technology development:

- Enables economic viability screening at all stages of commercialization. TEA modeling may (and should) be done very early during R&D conceptualization stages as a means of screening technologies or ideas that will not prove economically viable or to compare technology options or approaches against each other. The economic potential for an alternative process configuration may be readily compared against an incumbent baseline. As technology maturity evolves, the level of fidelity in the models may be refined and improved.
- Supports defining a future cost goal. TEA may be leveraged to establish future cost goals for a technology and to highlight the technical performance requirements that must be achieved to meet this goal (for example, feedstock cost, yield improvements, reduced chemical loadings, improved reaction kinetics).
- Quantifies the economic implications of R&D improvements. Once future targets are established, TEA modeling may also be conducted to incorporate the latest experimental data or other process improvements to benchmark progress toward meeting the targets.
- Demonstrates key drivers and sensitivities. TEA models for an established base case may be exercised to highlight the greatest economic risks or cost drivers to prioritize future research focus through sensitivity analysis. Such sensitivity analyses could involve single-point tornado plots or multivariate studies, as well as Monte Carlo analysis. This approach may also be used as a factor in conducting a risk assessment for technology scale-up.

As an example to illustrate the first three points, Fig. 6.1 depicts a TEA cost breakdown plot reflecting historical progression on algae conversion R&D work at the National Renewable Energy Laboratory (NREL) over recent years. It also shows cost projections for further envisioned improvements moving into future years.[4,17] In this plot, "state-of-technology" (SOT) performance is benchmarked on an annual basis attributed to experimental data collected between 2018 and 2020 spanning operations for both algal biomass cultivation and harvesting (rolled into Feedstock cost in the figure) and conversion through a low-temperature pathway focused on fractionation of algae into its major lipid, carbohydrate, and

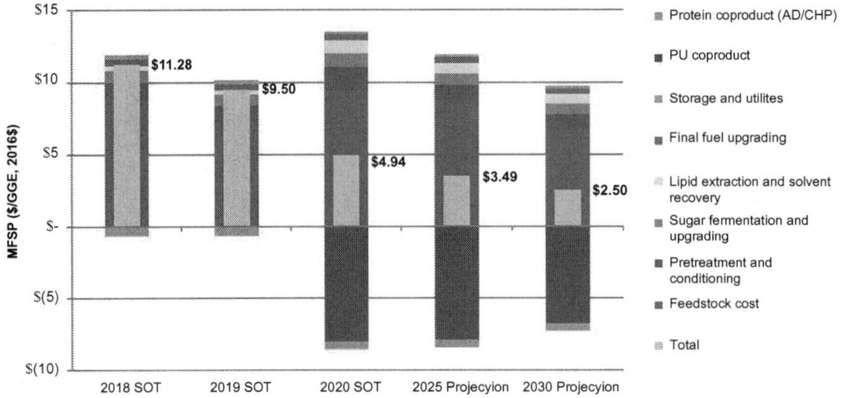

FIG. 6.1 TEA progression plot tracking state-of-technology (SOT) progression for integrated algal biorefinery fuel production versus future out-year projections.[4,17] *MFSP*, minimum fuel selling price; *$/GGE*, dollars per gallon gasoline equivalent.

protein constituents followed by conversion and utilization of each respective fraction. Experimental data is inputted to process models, with resulting mass and energy balance information used to estimate system costs following the methods described before, culminating in TEA model outputs reporting the resultant minimum fuel selling price (MFSP) in $/gallon gasoline equivalent (GGE).

In this example, MFSP was demonstrated through TEA modeling to be reduced by nearly $2/GGE from 2018 to 2019, attributed to underlying process performance improvements observed experimentally, e.g., through improved cultivation productivity and other conversion developments. A more dramatic MFSP reduction was subsequently demonstrated from 2019 to 2020, enabled by the inclusion of algal lipid-derived polyurethanes (PUs) as a value-added coproduct garnering substantial revenue for the modeled biorefinery. Although this was done at the sacrifice of a fraction of lipids that had previously been used for fuel production (reducing the total fuel yields and thus the $/GGE denominator, accordingly increasing the apparent contributions of processing steps to MFSP), the coproduct revenue generated from the introduction of algal PUs ultimately enabled a reduction in the net MFSP by over $5.50/GGE—*thus demonstrating both the important utility of TEA modeling in such a context and the significant merits of algal-derived PU as may be produced in a fuel and product algal biorefinery setting to promote economic viability*. Moving forward, the TEA models also highlight further opportunities to reduce system costs by improving performance spanning algal biomass production and biorefinery conversion to this slate of fuels and products, ultimately demonstrating a path to achieve final MFSP targets of $2.50/GGE.

What is an LCA?

Background

Just as TEA is an essential component of developing economically viable precommercial R&D technology, life cycle assessment (LCA) is its counterpart for evaluating the environmental sustainability of the technology or product. LCA uses a defined methodology to quantify the environmental impacts of a given product or process (Fig. 6.2). The first step is to define the goal and scope of the LCA, identifying a system boundary, and the environmental impacts which will be tracked for the system. An inventory of all material and energy flows into and out of the process is accounted for within the system boundary. A functional unit is also defined, and all process inputs and outputs are normalized to that functional unit (e.g., 1 kg of PU).[18]

With the process inputs and outputs or life cycle inventory (LCI) defined, the next step is to evaluate the potential environmental impacts of each input and output (i.e., perform an impact assessment). Several impact categories have been defined to measure how a process input/output can potentially affect various aspects of the environment and human health.[19,20] Any given process input may have a strongly negative impact

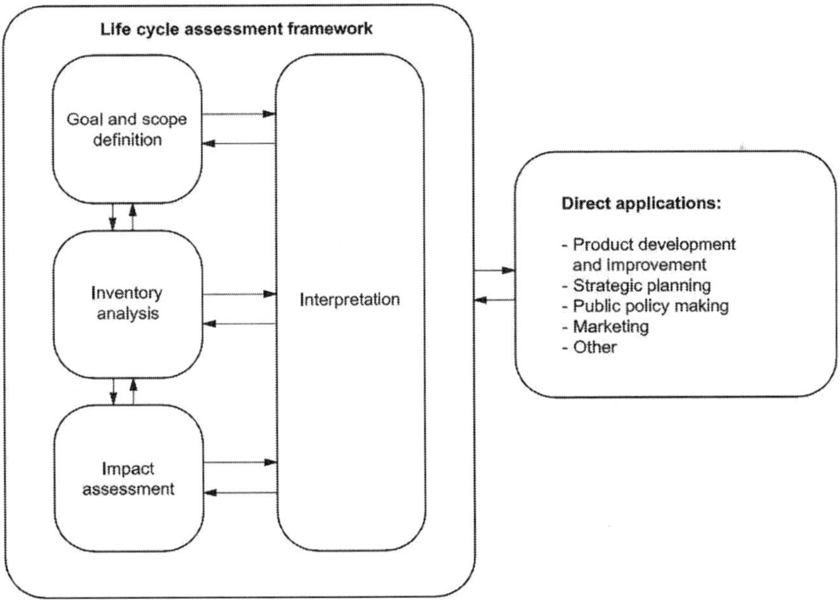

FIG. 6.2 Stages of an LCA. *Reproduced with permission from ISO 14040:2006 (Environmental management — Life cycle assessment — Principles and framework). This standard can be obtained from any ISO member and from the website of the ISO Central Secretariat at the following address: www.iso.org. Copyright remains with ISO.*

for one category but minimal impact for others. For this reason, it is important to consider a range of impact categories rather than only the primary one. This transparent and holistic approach is one of the primary benefits of LCA. Once an LCA is performed, the results can be used to guide product development and improvement, inform public policy, or be leveraged for strategic planning and marketing.[18]

Benefits and outcomes: Why should we care?

Environmental sustainability is critical in protecting our planet, specifically, to minimize the impact of human activities on areas of protection: natural environment, human health, and limited natural resources.[21] LCA is an analytical framework for quantifying and comparing the resources used and the impact to the environment and human health based on a functional unit of a product or activity.[22] In a backdrop of increasing global concern for rising global temperatures and environmental destruction, LCA addresses the core questions that must be answered to quantify the long-term sustainability of a product or process. While TEA answers the question of economic feasibility, it does not consider the environmental implications that must be included to ensure a sustainable product life cycle. By standardizing the quantification of environmental impacts, LCA allows comparison between different product options (e.g., petroleum versus bio-based pathways). It also allows for process optimization toward specific impact categories (e.g., greenhouse gas emissions) while simultaneously identifying and quantifying any environmental shortcomings of a given product or process. Therefore it identifies opportunities to improve the environmental performance of products or services at various life cycle stages. The identification of hot spots throughout the life cycle is a core aspect of LCA, with the intention to improve the overall potential environmental impacts. Identifying and understanding the hot spots and trade-offs helps to enable sustainable product development.

The foundations

TEA and LCA: Determining the total impacts of materials

To demonstrate the approach of performing TEA and LCA on bio-based PU, we will walk through examples of each. For simplicity and brevity, the scope of the process will consider the production of a bio-based flexible PU from a clean oil feedstock (in this case, linseed oil is assumed as a proxy for vegetable and algae oils more generally). The benefit of this approach is that it uses an oil feedstock that is commonly known and used commercially, allowing for simple quantification of upstream cost and sustainability implications. However, this approach could equally be applied to upgrading other plant- or algae-based oils, provided that factors such as the unique lipid profile of the feedstock are taken into account. Further TEA must be performed when considering oils for which the upstream process is more novel, such as algae oil. This TEA must include design considerations for the cultivation of algal biomass and the extraction of the algal lipids. Any other coproducts produced from the biomass must also be considered. The scope and complexity of the process is too great to be described in detail here; however, examples considering the TEA[4] and LCA[23] of algae-based PU can be found in the literature.

Example TEA of bio-based polyurethane

The PU production process used for this case study example draws from a published TEA.[24] All steps required for defining the TEA and key assumptions will be discussed here at a sufficient level of detail; additional details can be found in the literature,[24,25] which will be referenced throughout the chapter. For the reasons discussed previously, a DCFROR analysis was chosen for the TEA methodology. The primary metrics considered were minimum selling price (MSP) of the product (allowing for comparison to the price point of conventional petroleum-based PUs) and maximum oil purchase price (MOPP), providing a feedstock oil price target which can be applied to other plant- or algae-based oils with similar lipid profiles.

The scope of the process was defined as a PU production facility using linseed oil as a feedstock and producing flexible slabstock PU foam. A process flow diagram is shown in Fig. 6.3, generally consisting of an epoxidation and ring-opening step (using formic acid and hydrogen peroxide to form hydroxyl groups on the unsaturated bonds of the oil), purification of the polyol intermediate, and reaction with toluene diisocyanate

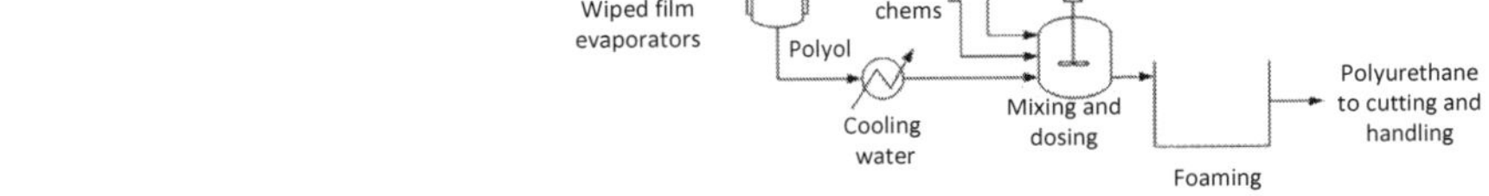

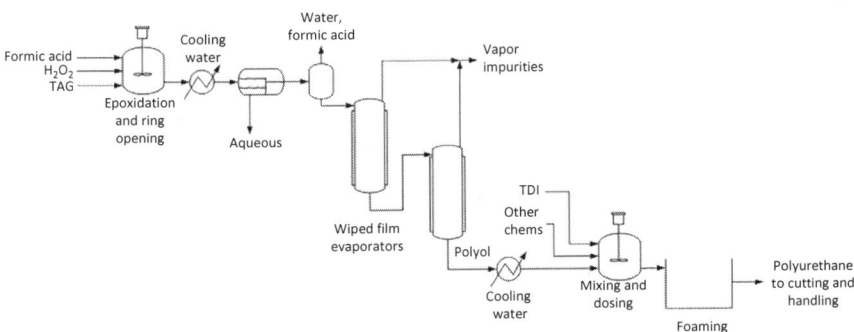

FIG. 6.3 Process flow diagram of the modeled polyurethane production facility. *From Wiatrowski M, Davis R.* Algal Biomass Conversion to Fuels Via Combined Algae Processing (CAP): 2020 State of Technology and Future Research. *National Renewable Energy Lab. (NREL), Golden, CO; 2021:36.*

(TDI) to form PU foam. Note that the process and its associated assumptions described here are meant to serve as an example of a single feasible method for producing bio-based PU, and that other approaches and PU product materials are also possible.

The design rate for the facility was specified as 2800 tons per year (TPY) of PU foam product, the average rate for a mid-sized conventional PU facility.[24] While world-class PU facilities can reach capacities on the order of 100,000 tons per year, the smaller size is maintained to conservatively reflect economy of scale implications for the facility, which can be significant in the results of a TEA. In a complete study, it would be prudent to perform a sensitivity analysis on the scale of the facility in order to understand how the economic viability of the process changes at different sizes.

Another important consideration is the characterization of the lipid profile of a feedstock. The relative amounts of the various saturated and unsaturated fatty acids can vary greatly depending on the source of lipids. These variations affect the amount of each reactant needed to form the intermediate polyol and subsequent PU product, which can significantly influence any TEA and LCA results tied to the process model. For example, one raw material that has been shown to strongly influence both TEA and LCA results is TDI, which is fed in specific proportions to the hydroxyl functionality of the polyol (which is directly proportional to the number of double bonds in the initial lipid feedstock). Also, the lipid profile will ultimately determine the final product properties; not all potential lipid feedstocks will produce a PU foam of sufficient quality, as some are better suited for rigid foams and others for coatings or other applications. While this relationship between feedstock and product quality is beyond the scope of the TEA, it is a vital screening consideration that should be reviewed. The lipid profile for linseed oil feedstock was assumed to consist of triglycerides with an average of six double bonds.[24] This value can vary substantially across various plant and algae oils; for example, the average number of double bonds for *Scenedesmus acutus* is estimated at 2.9, which results in significantly lower use of TDI compared to a linseed oil-based PU product.[25]

Once a facility scale, feedstock, and high-level process design have been identified, the next step for the TEA is to define the critical process conditions and assumptions used in each unit operation. These assumptions can be based on experimental data or results from the literature and are used in conjunction with fundamental engineering assumptions and equations (e.g., thermodynamic equations of state) to model the mass and energy flows of the process. Table 6.1 provides a summary of assumptions used in the process model for this example study.

In the epoxidation and ring-opening reaction, hydrogen peroxide and formic acid form performic acid, which subsequently reacts with the double bonds of the fatty acids to form an epoxide ring. Epoxides then react

TABLE 6.1 Key process assumptions used in the TEA model.

Parameter	Specification	Basis
Plant design		
Facility design rate	2800 tons polyurethane/year	Dong 2021[24]
Feedstock	Linseed oil (double bond # = 6)	Dong 2021[24]
Epoxidation and ring opening		
Temperature	75°C	Tessman 2016[26]
Pressure	1 atm	Tessman 2016[26]
Residence time	6 h	Tessman 2016[26]
H_2O_2	1.5× double bonds	Dong 2021[24]
Formic acid	Equimolar with double bonds	Tessman 2016[26]
Formic acid:H_2O epoxide rxn ratio	1:1	Wiatrowski 2020[4]
Power	0.54 MWh/ton polyol	Davis 2019[25]
Cooling water	220 w/w polyol	Davis 2019[25]
Low-pressure steam	0.02 w/w polyol	Davis 2019[25]
Nitrogen	0.02 w/w polyol	Davis 2019[25]
Polyol purification		
Flash evaporation pressure	5–15 mmHg	Davis 2019[25]
Wiped film evaporator temperature	(1) 140°C (2) 260°C	Davis 2019[25]
Polyurethane foam production		
TDI	0.5 mol/mol hydroxyl group	Wiatrowski 2020[4]
Water	0.0281 w/w polyol	Davis 2019[25]
DEA	0.0026 w/w polyol	Davis 2019[25]
Stannous octoate	0.0023 w/w polyol	Davis 2019[25]
DABCO	1 mol/mol polyol	Sardon 2015[27]
Surfactant	0.0049 w/w polyol	Davis 2019[25]
Power	0.004 w/w polyol	Davis 2019[25]

References are provided for additional details on the basis of each assumption.

with either formic acid or water (present in the reactor from the peroxide, fed as an aqueous 70% solution) to form one or two hydroxyl groups per double bond, respectively. Epoxides are assumed to react with water and formic acid in a 1:1 ratio, and the reaction is assumed to proceed to completion. Resultant polyols are separated from the aqueous phase and then sent to a flash evaporator to remove residual water and formic acid. Two sequential wiped film evaporators are used to separate any high-boiling impurities that may cause odors in the final product, yielding a high-purity polyol intermediate for further reaction.

Polyols then react with the crosslinking component (TDI) to produce the PU foam. Several other reagents and catalysts are added to aid the reaction and provide ideal product qualities, including stannous octoate, 1,4-diazabicyclo[2.2.2]octane (DABCO), diethanolamine (DEA), and surfactant. The PU foam is sent to the cutting and handling section after the foaming reaction and an initial curing period. Utility needs for the process, including low pressure steam, cooling water, electricity, and wastewater treatment, are also accounted for.

Finally, the mass and energy balances generated from the process model are used to size the equipment and account for the raw material needs of the process. Equipment and raw material costs are consistent with those reported in the literature.[25] With these costs established, the DCFROR analysis can be performed. Financial assumptions used in the DCFROR are presented in Table 6.2. Direct and indirect costs are also accounted for and are presented with the summary of other capital costs in Table 6.3. Operating costs are presented in Table 6.4.

TABLE 6.2 Financial assumptions used in the TEA, based on a mature *n*th plant.[12]

Financial assumptions	Value
Plant life	30 years
Cost year dollar	2016$
Capacity factor	90%
Discount rate	10%
General plant depreciation	MACR
General plant recovery period	7 years
Federal tax rate	21%
Financing	40% equity
Loan terms	10-year loan at 8% APR
Construction period	3 years

TABLE 6.2 Financial assumptions used in the TEA, based on a mature *n*th plant—cont'd

Financial assumptions	Value
First 12 months' expenditures	8%
Next 12 months' expenditures	60%
Last 12 months' expenditures	32%
Working capital	5% of fixed capital investment
Start-up time	6 months
Revenues during start-up	50%
Variable costs during start-up	75%

APR, annual percentage rate; *MACR*, modified accelerated cost recovery.

TABLE 6.3 Summary of capital expenditures for the modeled polyurethane production facility producing 2800 TPY of product.

Process area		Purchased cost	Installed cost
Epoxidation and ring opening		$6,904,166	$6,904,166
Foam production		$4,869,156	$4,869,156
Total installed cost		**$11,773,322**	**$11,773,322**
Warehouse	Based on external calculation		$2,107,643
Site development	9.0%	of total installed cost	$1,059,599
Additional piping	4.5%	of total installed cost	$529,799
Total direct costs (TDC)			**$15,470,364**
Prorateable expenses	10.0%	of TDC	$1,547,036
Field expenses	10.0%	of TDC	$1,547,036
Home office and construction fee	20.0%	of TDC	$3,094,073
Project contingency	10.0%	of TDC	$1,547,036
Other costs (start-up, permits, etc.)	10.0%	of TDC	$1,547,036
Total indirect costs			**$9,282,218**
Fixed capital investment (FCI) = TDC + Total indirect costs			**$24,752,582**
Land			$1,848,000
Working capital	5.0%	of FCI	$1,237,629
Total capital investment (TCI)			**$27,838,211**

Direct and indirect cost calculations are consistent with those reported in the literature.[25]

TABLE 6.4 Summary of operating expenditures for the modeled polyurethane production facility.

Process input	Rate	Unit	Cost $	Unit	$/year ×1000
Linseed oil (feedstock)	144.1	kg/h	0.97	$/kg	1102
Formic acid	45.3	kg/h	0.73	$/kg	260
H_2O_2	71.7	kg/h	0.99	$/kg	560
Toluene diisocyanate	128.4	kg/h	3.05	$/kg	3085
Stannous octoate	0.4	kg/h	14.02	$/kg	49
DEA (diethanolamine)	0.8	kg/h	2.27	$/kg	15
Surfactant	1.6	kg/h	2.75	$/kg	34
DABCO	0.2	kg/h	14.02	$/kg	23
Nitrogen	3.8	kg/h	0.06	$/kg	2
Low pressure steam	0.4	MMBtu/h	3.90	$/MMBtu	13
Cooling water	1.5	MMBtu/h	0.30	$/MMBtu	3
Electricity	116.1	kW	0.07	$/kWh	62
Wastewater treatment	39.3	kg organics/h	0.39	$/kg organics	122

Example LCA of bio-based polyurethane

To assess the environmental sustainability of the example renewable PU bioproduct requires a "cradle-to-gate" attributional LCA to quantify and interpret the environmental impacts resulting from the consumption of energy and raw materials and releases to the environment (air, water, and soil). The study follows the international standards (ISO 14040 and 14044) for life cycle assessment, encompassing the goal and scope definition, inventory analysis, impact assessment, and interpretation of results phases.[18]

This LCA case study aims to quantify the environmental impacts associated with the production of a bio-based PU via the process example described before. The study is done to determine the environmental impacts associated with the production of the linseed oil-derived PU. It will identify the hot spots, the areas with the most significant contribution to each environmental impact. While the results will be compared with the conventional petroleum-based PU counterpart, they are not used in

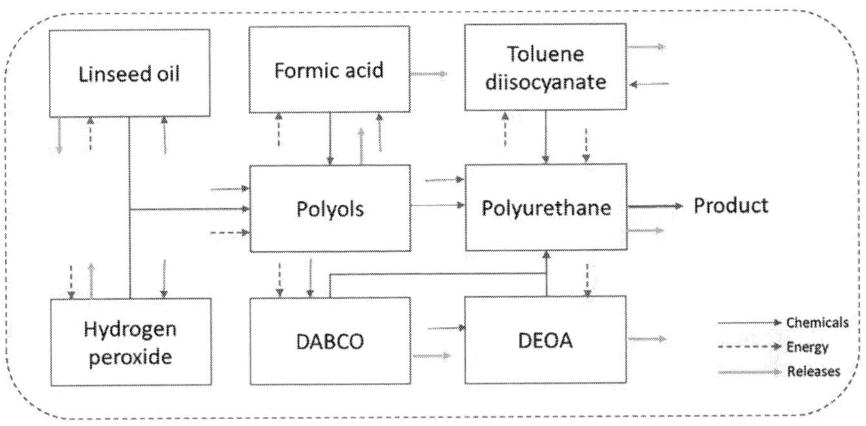

FIG. 6.4 Life cycle assessment system boundary for the example case study producing polyurethane from linseed oil.

comparative assertions, i.e., claiming the environmental superiority or equivalence of bio-based versus the benchmark PU production.

Fig. 6.4 shows the unit processes within the system boundary for production of renewable PU in the LCA. It consists of linseed oil feedstock production, raw materials, chemicals production, a collection of unit processes with elementary and product flows, and the boundaries between the technological system and nature. This is a cradle-to-gate LCA, as it starts from the linseed oil production and ends with the PU production. Such a study may benefit by expanding the system boundary to include the use stage and end-of-life stage (beyond the scope of this exercise). The full LCA assessment results would depend on the final fate of PU, in which it can be landfilled, recycled, composted, or used for other purposes. The PU production stage includes the epoxidation and ring-opening reaction of lipid in the presence of formic acid and hydrogen peroxide to polyols, the precursor for PU production through polymerization with toluene diisocyanate, consistent with the process operations discussed previously.

The functional unit is 1 kg of PU produced at the plant. The normalized material and energy inputs and outputs associated with the renewable PU process are summarized in Table 6.5. These data are derived from the material and energy flows obtained from this design concept and rigorous mass balances from the process simulation using Aspen Plus. The corresponding background data or secondary data sources are mainly based on the DATASMART life cycle inventory (LCI) package using a dataset for North American SimaPro customers[28] and the ecoinvent database, a common source of LCI data and one of the most consistent and transparent databases that supports environmental assessments of products and

TABLE 6.5 Required material and energy inputs and system outputs to produce 1 kg of renewable polyurethane.

Product	Amount	Unit
Renewable polyurethane	1.00	kg
Inputs		
Linseed oil (feedstock)	4.47×10^{-1}	kg
Formic acid	1.40×10^{-1}	kg
Hydrogen peroxide	2.22×10^{-1}	kg
Tin catalyst (stannous octoate)	1.37×10^{-3}	kg
DABCO (1,4-diazabicyclo[2.2.2]octane)	6.41×10^{-4}	kg
DEA (diethanolamine)	2.58×10^{-3}	kg
Surfactant	4.86×10^{-3}	kg
Toluene diisocyanate	3.99×10^{-1}	kg
Low-pressure steam	1.38	MJ
Cooling duty	4.88	MJ
Water	1.40×10^{-1}	m^3
Electricity	3.60×10^{-1}	kWh
Waste streams		
Solid waste (incineration)	–	kg
Wastewater	1.22×10^{-1}	kg

processes.[29] The data gaps are filled with the US LCI database.[30] The impact assessment method applied here is TRACI 2.1,[19] a midpoint oriented methodology. It evaluates the potential effects from the standpoints of ozone depletion, acidification, eutrophication, ecotoxicity, human health, global warming potential (GWP), carcinogenicity, noncarcinogenicity, smog, respiratory function, and fossil fuel depletion. The LCA model is developed in SimaPro, a commercial LCA software tool used for life cycle assessments.[31] Using different LCA tools almost inevitably yields different results. The differences originate primarily from discrepancies and uncertainties in the software databases for both inventory and impact assessment.[32] Also, choosing different life cycle impact assessment (LCIA) methods can influence the uncertainty of the results.[33] Consequently, to minimize uncertainty and avoid the necessity of harmonization, the environmental impacts for our example are based on the same databases and the same LCA tool and life cycle impact assessment method as those for the conventional PU benchmark.[34]

Linseed production at the farm is obtained from the Ecoinvent database. Seed yield is 3113 kg/ha at a moisture content during storage of 6%.[35] Feedstock production activity starts after the harvest of the previous crop, and the inputs of seed, mineral fertilizers, and pesticides are considered. The dataset includes all relevant machine operations and corresponding machine infrastructure and sheds. Machine operations include soil cultivation, sowing, fertilization, weed control, pest and pathogen control, combine-harvest, transport from field to a regional processing center (10 km), and seed drying. Direct field emissions are included. This activity ends after the drying of seeds at the regional processing center. As linseed oil production inventory data is not available in the database, this study assumes that the process is similar to that of rapeseed oil.[36]

DABCO, also known as triethylenediamine (TEDA), is currently not available in the databases. It can be procured from several sources, including ethylenediamine or ethanolamine, diethanolamine, or diethylenetriamine, with a variety of catalysts.[37] The life cycle inventory of the DABCO manufacturing process could be quantified to support the application of life cycle assessment via the bottom-up approach but would require detailed knowledge of the process and substantial resources, i.e., process design and simulation.[38] Alternatively, it could be derived based on the idealized conversion from ethanolamine,[39] i.e., $3\ H_2NCH_2CH_2OH \rightarrow N(CH_2CH_2)_3N + NH_3 + 3\ H_2O$, in which every kilogram of DABCO produced will require 1.66 kg ethanolamine (applied for this work).

Bio-based polyurethane TEA and LCA results and comparison to petroleum polyurethanes

TEA results and discussion

The results from this example study indicate that the modeled bio-based PU foam would require an MSP of \$2.13/lb (\$4.70/kg) to achieve a specified IRR of 10%. This value is close to the average market selling price of flexible petroleum-based PU foam, which is estimated at \$2.04/lb (\$4.50/kg) as a multiyear average of recent market prices,[25] indicating that the economic prospect for replacing conventional PU with its bio-based counterpart is economically feasible. Of course, this result is tied specifically to linseed oil, assumed to be purchased at \$0.44/lb (\$0.97/kg), and that other feedstocks such as algae oil may have a price premium. To make these results more generally applicable to other feedstocks, an MOPP is also determined, setting the PU equivalent to its market value. This strategy yields a MOPP of \$0.25/lb (\$0.55/kg) to achieve an IRR of 10%. This value may be useful for extrapolating to alternative oils; however, the TEA performed here is specific to the lipid profile and resultant chemistries, reactant demands, and yields for linseed oil, and an oil feedstock with significant deviations from that profile cannot be directly compared to this MOPP value without adjustment.

This TEA is intended to serve as an engineering feasibility assessment to quantify the economic potential of bio-based PUs, and to serve as an example of how a TEA may be performed in this context. The assumptions should be refined and tailored to a specific oil feedstock and/or PU product according to the information available. For example, similar technology may be employed to produce a rigid PU foam, or a coating, adhesive, sealant, or elastomer. Production processes, product values, and feedstocks can vary greatly, so the method should be considered independently for each type of product.

While these results are useful for screening and benchmarking, the TEA model is also valuable for performing sensitivity analysis. A single-point sensitivity analysis was done to determine key cost drivers for this example (Fig. 6.5). Plant scale had the greatest impact on process economics, with the MSP varying from \$1.59/lb to \$3.19/lb for plant scales of 10,000 and 1000 tons per year, respectively. This demonstrates the importance of maximizing plant scale for optimal economics and suggests that there would be value in choosing a feedstock based on its regional availability in addition to its cost and performance. Total installed capital cost, which varied by ±25%, also had a significant impact. Sensitivity to capital costs highlights the importance of refining capital estimates as the product development evolves. Other variables with significant economic impact are the prices of various raw materials, especially TDI and linseed oil. Although these material prices are likely independent of any progress that can be made in R&D, high sensitivity to a specific raw material may indicate that research into alternative reactants is warranted.

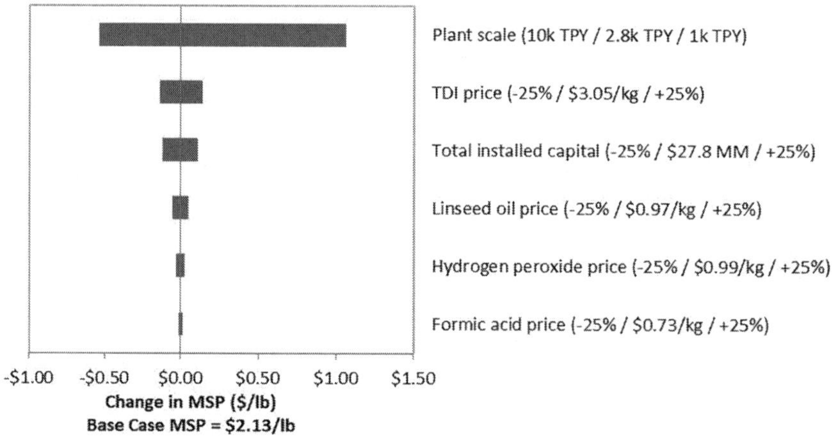

FIG. 6.5 Results of a single-point sensitivity analysis for the bio-based polyurethane example study. Descriptions in parentheses refer to the values corresponding to the (low / base case / high) MSP outcome, respectively.

LCA results and discussion

Fig. 6.6 and Table 6.6 present results for a wide range of potential environmental impacts associated with the production of renewable PU. The inclusion of multiple impact categories allows a holistic view of the production options. Contribution analysis (Fig. 6.6) was performed to identify environmental hot spots in the cradle-to-gate life cycle. Feedstock production (linseed oil) and TDI are the two major contributors to all impact categories except ozone depletion. The potential ozone depletion impact, i.e., the reduction in the total ozone volume in the Earth's stratosphere, is

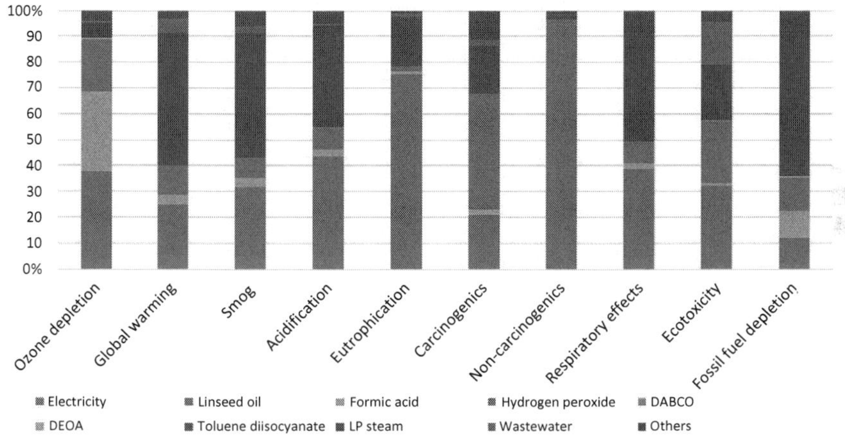

FIG. 6.6 Contribution analysis of the cradle-to-gate life cycle impacts for the production of 1 kg of renewable polyurethane (characterized midpoint results using TRACI 2.1).

TABLE 6.6 Potential total environmental impacts for producing 1 kg of renewable and conventional polyurethane.

Impact category	Unit[a]	Renewable PU	Conventional PU	Comparison[b] %
Ozone depletion	kg CFC-11 eq	3.23×10^{-7}	2.69×10^{-8}	1101
Global warming	kg CO_2 eq	5.20	4.88	7
Smog	kg O_3 eq	2.65×10^{-1}	2.48×10^{-1}	7
Acidification	kg SO_2 eq	3.04×10^{-2}	2.13×10^{-2}	43
Eutrophication	kg N eq	1.49×10^{-2}	7.98×10^{-3}	86
Carcinogenic	CTUh	3.47×10^{-7}	2.79×10^{-7}	2

Continued

TABLE 6.6 Potential total environmental impacts for producing 1 kg of renewable and conventional polyurethane—cont'd

Impact category	Unit[a]	Renewable PU	Conventional PU	Comparison[b] %
Noncarcinogenic	CTUh	3.91×10^{-6}	3.04×10^{-7}	1187
Respiratory effects	kg PM2.5 eq	2.53×10^{-3}	3.19×10^{-3}	−21
Ecotoxicity	CTUe	21.9	16.0	37
Fossil fuel depletion	MJ surplus	8.41	11.4	−26%

[a] CFC-11, *trichlorofluoromethane;* CTUe, *comparative toxic unit—ecotoxicity;* CTUh, *comparative toxic unit—human.*
[b] *Increase (+) or decrease (−) relative to conventional PU.*

evenly attributed to linseed oil (34%) and formic acid (31%). Fertilizers and chemicals used during linseed production at the farm are responsible for 94% of the potential noncarcinogenic impact and 75% of the eutrophication potential. The GWP for renewable PU production is estimated to be 5.2 kg CO_2 eq per kg PU. TDI, the chemical used in the polyols polymerization step, is the most significant global warming contributor (49%), followed by linseed oil production (20%).

Similarly, the fossil fuel depletion impact (8.4 MJ/kg PU) is also predominantly (59%) attributed to toluene diisocyanate. Hydrogen peroxide and linseed oil are the two major contributors to the total carcinogenic impact (0.35 μCTUh/kg PU), 45% and 20%, respectively. Thus improving the polyols polymerization step by having lower toluene diisocyanate usage, a renewable (nonfossil) means of sourcing TDI, or a more eco-friendly replacement could improve the environmental sustainability associated with renewable PU production.

Products derived from renewable biological resources are not intrinsically sustainable.[40] As a reference, the environmental impacts of our example are also compared with those from conventional PU production in the US LCI database.[34] As shown in Fig. 6.7, in which impact categories are normalized by the highest value per category, renewable PU production in the example exhibits higher environmental impacts in eight of the 10 categories. Respiratory effects (−21%) and fossil fuel depletion (−26%) are the two favorable to the renewable PU production system. The percentage difference relative to the conventional PU basis for each impact category is presented in Table 6.6. Agricultural activities (linseed growing) require the inputs of seed, mineral fertilizers, and pesticides. While these inputs are not pertinent to nonrenewable (i.e., conventional) PU production, they directly contribute to the environmental impacts for renewable

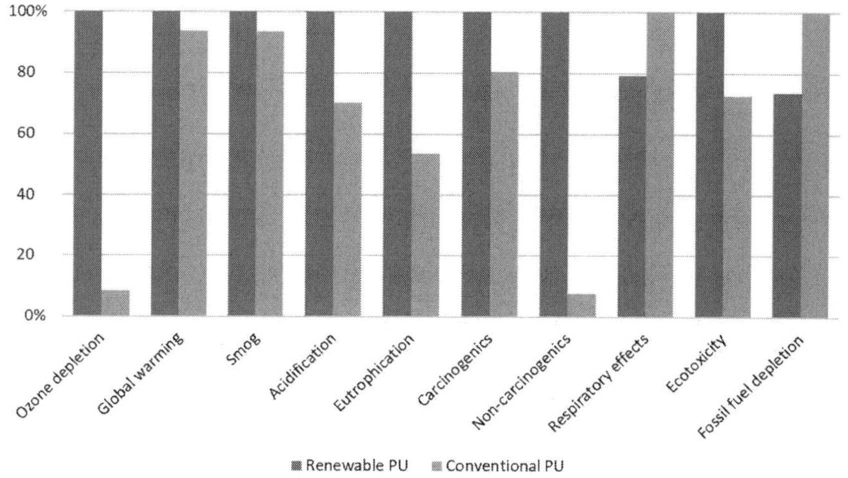

FIG. 6.7 Impact categories normalized by the highest value per category for the production of 1kg of renewable and conventional polyurethane based on characterized midpoint results using TRACI 2.1.

production. This results in higher impacts in ozone depletion (1101%), acidification (43%), eutrophication (86%), carcinogenic (24%), noncarcinogenic (1187%), and ecotoxicity (37%).

TDI is a key reactive raw material required to produce PU and is the most significant greenhouse gas contributor (49%) of the renewable PU production process (Fig. 6.6). However, TDI consumption in the example is 0.4kg TDI/kg PU (Table 6.5), about 40% higher than that for conventional PU.[34] As noted before, TDI usage is directly proportional to the number of double bonds in the lipid feedstock; alternative feedstocks with fewer double bonds would result in lower TDI usage. For example, TEA in the literature considering the conversion of algae oil to PU reflected a TDI consumption of 0.27kg TDI/kg PU, resulting from the lower number of double bonds in the algae oil feedstock.[4,25] While the lipid used is ultimately dictated by product performance, the lipid profile of the feedstock may be tailored to minimize TDI usage while still meeting product quality specifications. With a carbon intensity of 6.36kg CO_2 eq/kg TDI, a decrease in the TDI consumption to that of the conventional PU benchmark would improve the GWP from 5.20 to 4.47kg CO_2 eq/kg (a 14% improvement), lower than the conventional product's GWP (4.88kg CO_2 eq/kg).

An uncertainty analysis was done to assess the data quality's uncertainty and the comparative significance for the two PU production systems. The approach uses a data quality pedigree matrix[41] and Monte Carlo analysis in SimaPro.[31] Fig. 6.8 summarizes the results, which show high uncertainty for GWP and smog formation potential. This is because

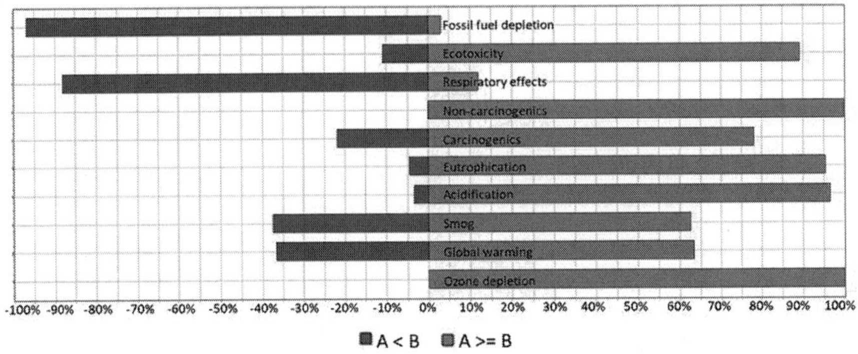

FIG. 6.8 Monte Carlo analysis of the comparison of the life cycle impacts of renewable polyurethane (A) and conventional polyurethane (B) systems for 1 kg production (characterized midpoint results, TRACI 2.1).

the LCIA scores for the two systems were similar (within ±3% of the average) for these two categories; and given variations in other data inventory, these results can readily shift in favor of either system when the data are varied. Moreover, the analysis showed that over 1000 runs, the LCIA indicated a noticeable advantage to the renewable PU system for respiratory effects and fossil fuel depletion, outperforming the conventional system over 88% and 97% of the time, respectively. The results for other environmental impact categories were also found to be consistent throughout the uncertainty analysis.

The objectives of this LCA case study were to provide an example assessment for the environmental impacts and identify environmental hot spots associated with the renewable PU production system. The comparisons estimated here against petroleum-based PU are specific to the linseed oil feedstock choice, regarding both the feedstock-specific LCA metrics and reaction-specific values (e.g., the quantity of TDI consumed based on reaction stoichiometries reflecting the linseed oil's fatty acid profile), and therefore are not equally applicable to "any" bio-derived oil feedstock, which may fare better than in this example. The study also demonstrates the importance of including multiple impact categories for a holistic view and to ensure the veracity of renewable PU production assessment.

Summary

The importance of this approach: Economic and environmental

Using TEA and LCA at all stages of product development is paramount to the success of a product to be scaled up and deployed. In the context of bio-based PU, still in nascent states of product development, TEA and LCA will play an essential role. These complementary assessment strategies can serve as screening tools for selection of economically feasible products and processes that meet specific environmental impact criteria. Throughout technology development, TEA and LCA can drive research targets and prioritize research directions to maximize impact. Here, we have provided an example of one process for producing a bio-based flexible PU foam product from plant-based lipids (in this case, linseed oil). However, similar methodologies could be applied to other bio-derived oils, including algae oil, with appropriate adjustments to accommodate different feedstock production attributes and conversion/reaction details specific to the feedstock oil profile. This study illustrated that the combination of TEA and LCA analyses could be a practical approach to assess the technical and economic viability of new processes and technologies and the environmental impacts. TEA of this example case confirmed its potential to be cost-competitive with petroleum-based PU and identified key cost drivers, including raw material requirements (TDI and the lipid feedstock) and facility capital expenditures. The LCA, however, revealed more challenging environmental impacts for this particular bio-based PU system, indicating potential improvements in several categories but penalties in a larger number of categories when compared to petroleum-based PU, highlighting the importance of considering both types of analyses together through an iterative process to reach viability across feedstock selection, reactant sourcing, conversion chemistries and conditions, etc. (beyond the scope of this example). The use of TDI was shown to be a key driver in many of the impact categories, highlighting the importance of reducing or replacing this reactant in the foam production process, as is being investigated experimentally through numerous alternative PU production pathways.

Close-Up: The importance of metrics

Techno-economic analysis (TEA) and life cycle assessment (LCA) underpin applied research and development for bio-based product technology. The reading and interpretation of these tools must be done with care. The configuration of the models and the data used for parameter development are often obscured. Good TEA and LCA illuminates uncertainty and areas for further development while clearly delineating the known from the assumed from the best guesses and allows smart decision making. Poor use of them has the potential to distort the identification of priorities, contribute to the misallocation of resources, or justify government policies that miss their intended mark. When they are used to guide policy or regulatory decisions about bio-based products and markets, as they are with biofuels through the US Renewable Fuel Standard or California's Low-Carbon Fuel Standard, the burden on analysts and policymakers to continually question the assumptions, improve methods, and engage in open discourse on the results and conclusions becomes paramount.

As the developing bio-based polyurethane industry expands in scope and impact, TEA and LCA will be used at many stages, from the assessment of individual technologies and products to the quantification of the impact of the sector on a national and global scale. The greater the claims from the developers of these exciting new technologies, the greater the expectations should be for rigorous and transparent TEA and LCA. This chapter showcases some leading examples. Consumers of TEA and LCA results should be inquisitive and question which research results are used and what is and is not known about the process. Key questions to keep in mind when interpreting the results are: What technical progress, integration, and scale-up beyond the experimental results is assumed in the analysis? Are the assumptions for processes outside the scope of the research reasonable? What is inside and outside the system boundaries, and how does the delineation affect the results? How is uncertainty handled?

With questions like these in mind, we can begin to understand the potential of algae-derived polyurethane products to contribute to addressing the climate emergency through affordable products that offer a life cycle advantage compared to petroleum-based alternatives.

Daniel B. Fishman
US Department of Energy, Golden, CO, United States

Disclaimer

This work was authored by the National Renewable Energy Laboratory, operated by Alliance for Sustainable Energy, LLC, for the U.S. Department of Energy (DOE) under Contract No. DE-AC36-08GO28308. Funding provided by U.S. Department of Energy Office of Energy Efficiency and Renewable Energy, Bioenergy Technologies Office. The views expressed in the article do not necessarily represent the views of the DOE or the U.S. Government. The U.S. Government retains and the publisher, by accepting the article for publication, acknowledges that the U.S. Government retains a nonexclusive, paid-up, irrevocable, worldwide license to publish or reproduce the published form of this work, or allow others to do so, for U.S. Government purposes.

References

1. Wooley RJ, Putsche V. Development of an ASPEN PLUS Physical Property Database for Biofuels Components. Golden, CO: National Renewable Energy Lab. (NREL); 1996. https://doi.org/10.2172/257362.
2. Schell D, Torget R, Power A, Walter P, Grohmann K, Hinman N. A technical and economic analysis of acid-catalyzed steam explosion and dilute sulfuric acid pretreatments using wheat straw or aspen wood chips. *Appl Biochem Biotechnol.* 1991;28(1):87–97.
3. Davis RE, Grundl NJ, Tao L, et al. Process Design and Economics for the Conversion of Lignocellulosic Biomass to Hydrocarbon Fuels and Coproducts: 2018 Biochemical Design Case Update; Biochemical Deconstruction and Conversion of Biomass to Fuels and Products Via Integrated Biorefinery Pathways. Golden, CO: National Renewable Energy Lab. (NREL); 2018. https://www.osti.gov/biblio/1483234.

4. Wiatrowski M, Davis R. Algal Biomass Conversion to Fuels Via Combined Algae Processing (CAP): 2020 State of Technology and Future Research. Golden, CO: National Renewable Energy Lab. (NREL); 2021:36.
5. Biddy MJ, Davis R, Humbird D, et al. The techno-economic basis for coproduct manufacturing to enable hydrocarbon fuel production from lignocellulosic biomass. *ACS Sustain Chem Eng*. 2016;4(6):3196–3211.
6. Laser M, Larson E, Dale B, Wang M, Greene N, Lynd LR. Comparative analysis of efficiency, environmental impact, and process economics for mature biomass refining scenarios. *Biofuels Bioprod Biorefin*. 2009;3(2):247–270.
7. Singh A, Rorrer NA, Nicholson SR, et al. Techno-economic, life-cycle, and socioeconomic impact analysis of enzymatic recycling of poly (ethylene terephthalate). *Joule*. 2021;5 (9):2479–2503.
8. Short W, Packey DJ, Holt T. A Manual for the Economic Evaluation of Energy Efficiency and Renewable Energy Technologies. Golden, CO: National Renewable Energy Lab; 1995.
9. Seider WD, Lewin DR, Seader JD, Widagdo S, Gani R, Ng KM. Product and Process Design Principles: Synthesis, Analysis and Evaluation. Wiley Global Education; 2016.
10. Liu G, Li M, Zhou B, Chen Y, Liao S. General indicator for techno-economic assessment of renewable energy resources. *Energy Convers Manag*. 2018;156:416–426. https://doi.org/10.1016/j.enconman.2017.11.054.
11. Drennan T, Williams R, Baker A. Alternative Liquid Transportation Fuels Simulation Model: Technical Documentation. National Energy Technology Laboratory; 2010.
12. Humbird D, Davis R, Tao L, et al. Process Design and Economics for Biochemical Conversion of Lignocellulosic Biomass to Ethanol: Dilute-Acid Pretreatment and Enzymatic Hydrolysis of Corn Stover. Golden, CO: National Renewable Energy Lab. (NREL); 2011. https://www.osti.gov/biblio/1013269-process-design-economics-biochemical-conversion-lignocellulosic-biomass-ethanol-dilute-acid-pretreatment-enzymatic-hydrolysis-corn-stover.
13. Davis R, Kinchin C, Markham J, et al. Process Design and Economics for the Conversion of Algal Biomass to Biofuels: Algal Biomass Fractionation to Lipid- and Carbohydrate-Derived Fuel Products. Golden, CO: National Renewable Energy Lab. (NREL); 2014. https://www.osti.gov/biblio/1159351.
14. Quinn JC, Davis R. The potentials and challenges of algae based biofuels: a review of the techno-economic, life cycle, and resource assessment modeling. *Bioresour Technol*. 2015;184:444–452. https://doi.org/10.1016/j.biortech.2014.10.075.
15. Jones SB, Zhu Y, Anderson DB, et al. Process Design and Economics for the Conversion of Algal Biomass to Hydrocarbons: Whole Algae Hydrothermal Liquefaction and Upgrading. Richland, WA: Pacific Northwest National Lab. (PNNL); 2014.
16. Huq NA, Hafenstine GR, Huo X, et al. Toward net-zero sustainable aviation fuel with wet waste–derived volatile fatty acids. *Proc Natl Acad Sci U S A*. 2021;118(13). https://doi.org/10.1073/pnas.2023008118.
17. Bioenergy Technologies Office 2019 R&D State of Technology. Bioenergy Technologies Office, U.S. Department of Energy; 2019:150.
18. Finkbeiner M, Inaba A, Tan R, Christiansen K, Klüppel HJ. The new international standards for life cycle assessment: ISO 14040 and ISO 14044. *Int J Life Cycle Assess*. 2006;11 (2):80–85. https://doi.org/10.1065/lca2006.02.002.
19. Bare J. TRACI 2.0: the tool for the reduction and assessment of chemical and other environmental impacts 2.0. *Clean Technol Environ Policy*. 2011;13(5):687–696.
20. Owens JW. LCA impact assessment categories. *Int J Life Cycle Assess*. 1996;1(3):151–158. https://doi.org/10.1007/BF02978944.
21. Tan ECD. Sustainable biomass conversion process assessment. In: *Process Intensification and Integration for Sustainable Design*. John Wiley & Sons, Ltd; 2021:301–318. https://doi.org/10.1002/9783527818730.ch14.
22. Smith RL, Tan ECD, Ruiz-Mercado GJ. Applying environmental release inventories and indicators to the evaluation of chemical manufacturing processes in early stage development.

ACS Sustain Chem Eng. 2019;7(12):10937–10950. https://doi.org/10.1021/acssuschemeng.9b01961.
23. Cai H, Ou L, Wang M, et al. Supply Chain Sustainability Analysis of Renewable Hydrocarbon Fuels Via Indirect Liquefaction, Ex Situ Catalytic Fast Pyrolysis, Hydrothermal Liquefaction, Combined Algal Processing, and Biochemical Conversion: Update of the 2020 State-of-Technology Cases. Argonne, IL: Argonne National Lab. (ANL); 2021. National Renewable ….
24. Dong T, Dheressa E, Wiatrowski M, et al. Assessment of plant and microalgal oil-derived non-isocyanate polyurethane products for potential commercialization. *ACS Sustain Chem Eng.* 2021. https://doi.org/10.1021/acssuschemeng.1c03653. Published online September 15.
25. Davis R, Wiatrowski M, Kinchin C, Humbird D. Conceptual Basis and Techno-Economic Modeling for Integrated Algal Biorefinery Conversion of Microalgae to Fuels and Products. 2019 NREL TEA Update: Highlighting Paths to Future Cost Goals Via a New Pathway for Combined Algal Processing. Golden, CO: National Renewable Energy Lab. (NREL); 2020. https://doi.org/10.2172/1665822.
26. Tessman MA. Synthesis, Analysis, and Applications of Renewable Polyols. Published online; 2016.
27. Sardon H, Pascual A, Mecerreyes D, Taton D, Cramail H, Hedrick JL. Synthesis of polyurethanes using organocatalysis: a perspective. *Macromolecules.* 2015;48(10):3153–3165. https://doi.org/10.1021/acs.macromol.5b00384.
28. LTS. DATASMART LCI Package; 2020. https://ltsexperts.com/services/software/datasmart-life-cycle-inventory/.
29. Wernet G, Bauer C, Steubing B, Reinhard J, Moreno-Ruiz E, Weidema B. The ecoinvent database version 3 (part I): overview and methodology. *Int J Life Cycle Assess.* 2016;21(9):1218–1230. https://doi.org/10.1007/s11367-016-1087-8.
30. US LCI. U.S. Life-Cycle Inventory, v.1.6.0. Golden, CO: National Renewable Energy Laboratory; 2008.
31. PRé Consultants. SimaPro. PRé Sustainability; 2019. https://simapro.com/.
32. Herrmann IT, Moltesen A. Does it matter which Life Cycle Assessment (LCA) tool you choose?—a comparative assessment of SimaPro and GaBi. *J Clean Prod.* 2015;86:163–169. https://doi.org/10.1016/j.jclepro.2014.08.004.
33. Chen X, Matthews HS, Griffin WM. Uncertainty caused by life cycle impact assessment methods: case studies in process-based LCI databases. *Resour Conserv Recycl.* 2021;172:105678. https://doi.org/10.1016/j.resconrec.2021.105678.
34. Hischier R. Polyurethane, Flexible Foam, at Plant/RER With US Electricity U. Published online December 4; 2012.
35. Ecoinvent. Linseed Seed, at Farm {CH}| Linseed Seed Production, at Farm | Alloc Def, U. Published online October 28; 2016.
36. Gnansounou E. Rape Oil, at Oil Mill/RER With US Electricity U. Published online August 15; 2006.
37. Roose P, Eller K, Henkes E, Rossbacher R, Höke H. Amines, aliphatic. In: *Ullmann's Encyclopedia of Industrial Chemistry.* American Cancer Society; 2015:1–55. https://doi.org/10.1002/14356007.a02_001.pub2.
38. Smith RL, Ruiz-Mercado GJ, Meyer DE, et al. Coupling computer-aided process simulation and estimations of emissions and land use for rapid life cycle inventory modeling. *ACS Sustain Chem Eng.* 2017;5(5):3786–3794. https://doi.org/10.1021/acssuschemeng.6b02724.
39. DABCO. Wikipedia; 2021. Accessed 26 October 2021 https://en.wikipedia.org/w/index.php?title=DABCO&oldid=1044698532.
40. Tan ECD, Lamers P. Circular bioeconomy concepts—a perspective. *Front Sustain.* 2021;2:53. https://doi.org/10.3389/frsus.2021.701509.
41. Weidema BP, Wesnæs MS. Data quality management for life cycle inventories—an example of using data quality indicators. *J Clean Prod.* 1996;4(3):167–174. https://doi.org/10.1016/S0959-6526(96)00043-1.

PART IV

Reformulating polyester polyurethanes

CHAPTER

7

Polyurethanes: Foams and thermoplastics

Nitin Neelakantan
Algenesis Materials, Cardiff, CA, United States

Polyurethanes: The basics

Many of the unique properties of urethanes are derived from their chemical structures and the fact that the term is used informally for an entire class of polymers. "Urethane" itself refers to a molecule otherwise known as ethyl carbamate (Fig. 7.1A), which is neither a type of polyurethane (PU) nor a polymer at all. Urethane is not even used to make PUs. However, it is useful in illustrating the simplest case of the urethane bond, which describes the moiety of NH—C=O—O, highlighted in red. This bond, repeated in a polymeric structure as shown in Fig. 7.1B, is universal to all PUs. PUs are widely used polymers encountered daily and often in ways one has not considered. The chapter will attempt to explain how such a dizzying array of products can be made simply by altering the chemical structure around the urethane bond. This will be done for two ubiquitous PU applications, foams and thermoplastics.

History and chemistry

The first PU dates back to 1937, when Dr. Otto Bayer first synthesized it.[1] His objective was to develop a material to compete with nylon fibers. IG Farbenindustrie, for whom he worked, successfully applied for the first patent on producing PUs that year.[2] A year later, the first US patent was awarded to Rinke et al.[3] This patent derived from a polymerization that Rinke's group had done by reacting 1,4-butanediol and 1,8-octanediisocyanate.[4]

Rethinking Polyester Polyurethanes
https://doi.org/10.1016/B978-0-323-99982-3.00013-4

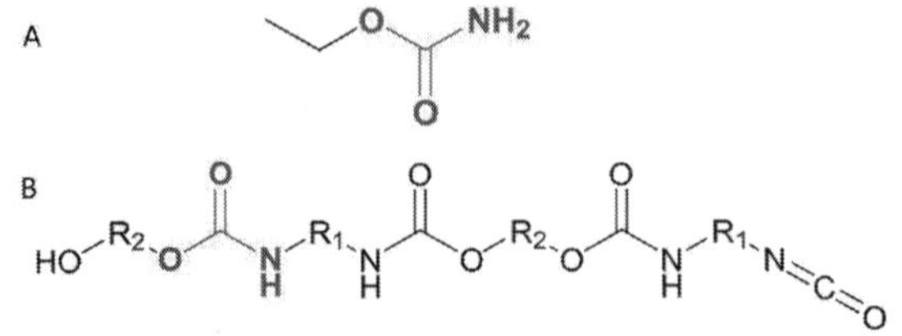

FIG. 7.1 (A) Ethyl carbamate, with urethane group in *red*. (B) Polyurethane, with urethane group in *red*.

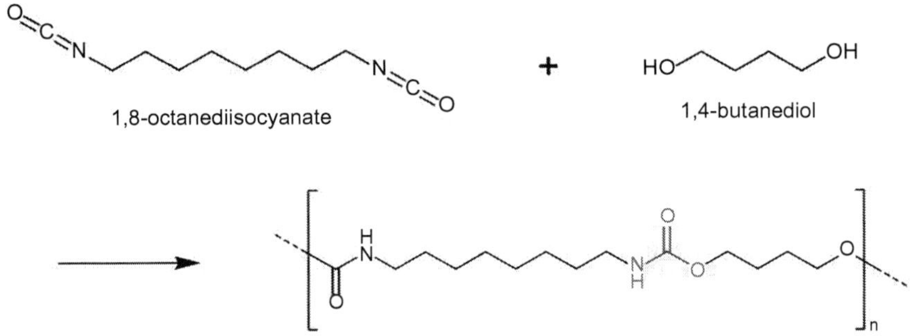

FIG. 7.2 Simplified reaction scheme of 1,4-butanediol and 1,8-octanediisocyanate.

Fig. 7.2 shows the scheme for this reaction, which exemplifies a standard reaction of the two necessary ingredients needed to make a PU: alcohol and an isocyanate. This is one of the more straightforward cases, as each component has the lowest amount of functionality (i.e., reactive groups per molecule) necessary to sustain a polymerization reaction, which is 2. The reactive group of an alcohol is the —OH, and for the isocyanate, —NCO. Each molecule of 1,4-butanediol can react with two isocyanate groups, and the polymer chain can be extended or polymerized.

Starting materials to make polyurethanes

The starting reactive alcohol for PUs is typically much more complex than a 4-carbon difunctional alcohol or diol, such as BDO. Before reacting with isocyanates, such diols are polymerized into polyols, i.e., polymeric alcohols. The polyol constitutes the soft segment, contrasted with the isocyanate, or hard segment. The performance attributes of PUs are derived in part from having phase separation between the two segments.[5] For this reason, polyols are a near ubiquitous component of PUs, often constituting 50% or more by weight of a typical PU formulation. By contrast, isocyanates are often used as is, though there are examples of polymeric isocyanates in use. The vast majority of the isocyanates produced are methylene diphenyl diisocyanate and toluene diisocyanate. Both have aromatic ring structures, which are planar and do not undergo free rotation, so they provide rigidity in the bulk polymer.

It is essential to understand where the raw materials for polyols and isocyanates come from and why these materials were chosen, to see the connection between those choices and their consequences. Starting materials for polyols will be considered first, since their structures are far more diverse than those of their isocyanate counterparts. The first polyols used to synthesize PUs were made from polyesters (Fig. 7.3A). Because this section is about foams, the focus will be on using polyols in PU foams. These polyester-based products displayed poor weathering characteristics; specifically, their resistance to heat and humidity was subpar. Consequently, the use of polyesters was displaced by polyether polyols,[6] the structure of which is shown in Fig. 7.3B. Polyether polyols, unlike polyester polyols, are resistant to hydrolysis, that is, breakdown by water.[7] Product durability is an important design criterion, so polyether foams were preferred. As will be shown subsequently, what is not so easy to see is this fateful decision's environmental impact.

Except for the carbon-oxygen double bond (C=O, carbonyl) in the polyester, the molecules are structurally very similar. Despite this, they have very different chemical properties and impart very different mechanical properties on PUs that are otherwise made by the same formulation. To understand why, we must first look at the intermolecular forces in each structure. This can be done by comparing an ether such as diethyl ether and an ester such as ethyl acetate. They have similar chemical compositions but differ in the carbon-oxygen bonding. The ether has an oxygen-carbon-oxygen single bond arrangement, whereas the ester has the characteristic carbonyl. Table 7.1 presents the structure, the difference in properties, and the difference in polarity of each molecule. The key takeaway is that ethyl acetate molecules have stronger interactions with each other than do diethyl ether molecules, and those interactions are harder to break. The oxygen atoms bonded to the carbonyl carbon pull electron density from the carbon, inducing a partial positive charge on the carbon and a partial negative charge on themselves. This difference in electronegativity

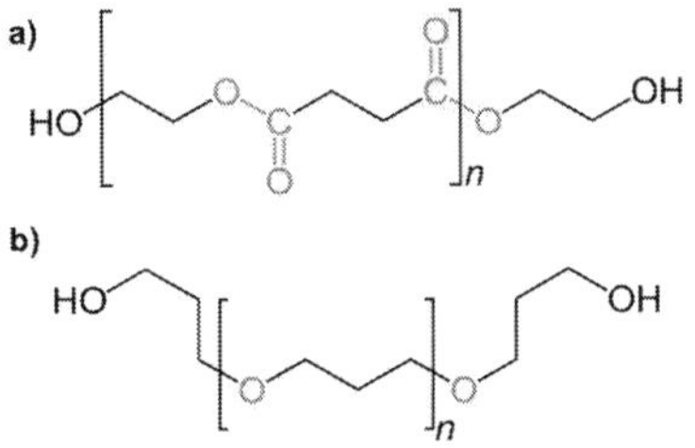

FIG. 7.3 Structure of (A) polyester polyol as compared to (B) polyether polyol. The differences are highlighted in *blue*.

TABLE 7.1 Comparative properties of diethyl ether and ethyl acetate.

Molecule	Structure	Dipole moment (D)	Viscosity (cP)	Water solubility (g/L)	Melting point (°C)
Diethyl ether	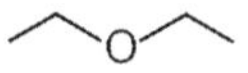	1.15	0.224	61	−116.3
Ethyl acetate	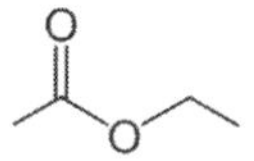	1.78	0.426	83	−83.6

creates a dipole moment, a key metric of polarity. Although the oxygen atom in diethyl ether also induces this effect, it is less potent, as measured by the dipole moment. The chemist's way of saying this is that ethyl acetate is more polar than diethyl ether. The consequence is that ethyl acetate has a higher melting point, higher viscosity, and higher water solubility, among other things.

Similarly, a polyester polyol is more polar than a comparable polyether polyol. Therefore polyester polyols will have (1) higher affinity for water, (2) higher viscosity, and (3) higher melting temperature than comparable polyether polyols. These characteristics create the respective complications in manufacturing lines: (1) more care must be exercised to keep polyesters dry to prevent hydrolysis, (2) blending polyester polyols with other ingredients is more complex, and (3) heating and insulation must be used to keep polyester polyols from freezing or becoming too thick to use.

For those reasons, polyether polyols take up the lion's share of the global market today, but polyester polyols are still used in foam. They impart to PUs good mechanical properties and solvent resistance. The starting monomers for either class of polyol come from petroleum (though polyesters can be made from plant and microbial sources), which provide the raw hydrocarbons necessary for their production. For isocyanates, the primary players are methylene diphenyl diisocyanate, followed by toluene diisocyanate. These two molecules have dominated the world's consumption since the beginning of the mass production of PUs, providing the rigid segment of the PU. Like polyols, their precursors are readily available from petroleum.

Foams

Applications

PU foam products are encountered almost daily. The diversity of the physical characteristics masks how common these materials are (Fig. 7.4). Houses are kept warm by rigid PU foam insulation, which

FIG. 7.4 Diverse PU foam products clockwise from top left: Spray insulation, steering wheel, kitchen sponges, shoe soles.

remains unseen except on those rare forays into the attic. Beneath the carpet will be a PU material known as an underlay. Shoes have soles of PU foam. Cars contain several PU materials: the door panel is rigid PU foam, the driver's seat is cushioned by high resilience PU soft foam, and the steering wheel is likely to be made with a type of PU foam known as integral skin. Bedroom mattresses and furniture cushions have PU "memory foam," which conforms to the shape of the user. These items exploit the diverse properties of PU, such as wear resistance, resilience, comfort, and toughness.

Structure-property relationships

The diversity in PU foams can be derived from the nearly limitless parameters that can be modified at the formulation and processing levels to dictate the outcome. Table 7.2 presents the effects of changing some of these major parameters, just in the two basic ingredients of polyols and isocyanates, and how they influence the final foam properties. The polyol length correlates to its molecular weight and the number of hydroxyl groups. Since one hydroxyl reacts with one isocyanate, polyols are measured by their hydroxyl number, an empirical measure of how many

TABLE 7.2 Effects of polyol and isocyanate parameters on final foam properties.

Parameter	Description	Effect on foam properties
Polyol molecular weight	Size of each polyol molecule	Higher = softer foam, slower reaction Lower = harder foam, faster reaction
Polyol functionality	Number of reactive OH groups per molecule	More = faster reaction, higher crosslinking, higher resilience Less = softer, better elongation
Isocyanate functionality	Number of reactive NCO groups per molecule	More = faster gelling, improved mechanical properties Less = slower gelling, less likely to shrink
Isocyanate index	Ratio of NCO groups to OH groups	Lower = softer foam, but weaker mechanical properties Higher = harder foam, but too high can make it brittle

reactive groups there are in a polyol relative to their molecular weight. Simply changing the hydroxyl number, but maintaining the chemical makeup of a polyol, can change how flexible or rigid the foam is.

Conversely, at the same polyol length, increasing the number of reactive —OH groups will increase crosslinking and reaction rate, which increases resilience in the final product.[8] The index is another way to modify the same formulation and create a different outcome: The amount of isocyanate is quantified by the index, a numerical value representing the ratio of isocyanate to hydroxyl groups. For instance, a PU reacted at an index of 98 implies that the calculated amount of isocyanate added was 98% of the hydroxyl equivalent in the polyol, i.e., there is a slight excess of polyol in the reaction. At 102, the converse is true, and a slight excess of isocyanate was added in the reaction. All other things being equal, the 102-indexed formulation may be harder and more brittle and take longer to cure than its 98-indexed counterpart, despite being otherwise the same.

While polyols and isocyanates can make up >90% of a PU foam, low percentage additives can often make or break the product. These additives are common and some are even necessary for the foam to react correctly. Table 7.3 presents their effect on final foam properties. Foams require catalysts to react at a rate consistent with the desired characteristics. Catalysts facilitate either the gelling or the blowing reaction. Gelling catalysts promote the reaction of isocyanate with a polyol, which causes the reactive resin to turn from a liquid to a solid, or to gel. Blowing catalysts facilitate the reaction of isocyanate with water, which produces carbon dioxide. Blowing catalysts cause the foam to blow up like a balloon and expand

TABLE 7.3 Effects of additives on foam properties.

Additive type	Description	Effect on foam
Catalyst: gelling/ blowing	A chemical that facilitates the reaction of a polyol with isocyanate (gelling) or water with isocyanate (blowing)	Faster cure time with higher gel catalysts, higher rising foam (lower density), and/or larger cells with higher blowing catalyst
Blowing agent: physical/ chemical	Physical: A nonreactive volatile material that causes the foam to rise Chemical: A compound such as water that chemically reacts with isocyanate to produce carbon dioxide as a byproduct	Increases the exotherm or heat generated by the reaction, and decreases the final density of the foam
Chain extender	Typically a small difunctional molecule that lengthens (extends) the chain of the hard segment	Gives foam enhanced mechanical properties and greater hardness, and may accelerate the foam curing process
Surfactant	An emulsifier that helps stabilize the mixture of two or more immiscible compounds (PU surfactants are generally silicones)	Used to manipulate the cell size of the foam, from very fine celled to very open celled

until fully cured. The ratio and amount of a catalyst can be used to tune how quickly foam cures and how dense it is.

The blowing agent, like the catalyst, is an ever present additive in foam formulations. Water is an example of a reactive blowing agent, as it reacts with isocyanate to form an amine and carbon dioxide. The carbon dioxide causes the expansion of PU foam as it is concurrently gelling and, in part, determines its final density. However, the reaction is not over: the produced amine reacts with another isocyanate group to form a urea bond, which becomes part of the final PU foam. These two reactions are highly exothermic, i.e., they release heat. Therefore the amount of water should not create excess heat; otherwise, runaway reactions can damage manufacturing equipment and begin to smoke or present a fire hazard. For this reason, physical blowing agents can be used to lower the density even further; these are typically nonreactive liquids that evaporate quickly and expand into a gas, such as fluorocarbons.

In some foams, a chain extender is added to increase the mechanical properties. Chain extenders are small molecules such as diols and diamines that react with the hard segment as polyols do. Additionally, they extend the length of the hard isocyanate segment, increasing the rigidity of the final foam. Increased rigidity is due to the increased isocyanate requirement relative to the total formulation.

To generate foam that does not collapse, a surfactant is required to stabilize the complex mixture that forms the B-side: polyol, water, chain extenders, blowing agents, catalysts, and pigments, as it reacts with the A-side or isocyanate. Many of these components are not miscible with each other, but through surfactants, the individual foam cells can expand and cure. An everyday example of surfactant stabilization is the phenomenon of blowing bubbles. Water itself will not form bubbles, but when combined with soap, which contains surfactants, a stable bubble can be blown and expanded.

With all this information in mind, we will revisit the decision to use polyether polyols to make most of the aforementioned foams. For each of those foams, water is likely the most commonly encountered liquid during everyday usage. Moreover, durability is a primary design criterion. Polyester-based foams tend to wear out faster, so polyether polyol seems the right choice. But there is one caveat: none of the products mentioned thus far *need* to last forever. The United States Department of the Interior estimates that an average of 300 million pairs of shoes are thrown away each year.[9] The average lifespan of a car is about 12 years.[10] Furniture, mattresses, and insulation are continually replaced.

Suppose these products are all made from polyether-based PU. In that case, once they are thrown away, they will contribute to the growing amount of waste in landfills or make their way into oceans and, in certain conditions, produce microplastic particles.[11–13] Because polyether-based PU products resist breakdown, they are prone to becoming persistent pollutants. From this perspective, the hydrolytic susceptibility of polyesters may be not a detriment but a boon. Another consideration is the use of petroleum to make the starting materials for polyether polyols: the manufacturing steps of chemical and petroleum industries cause over a billion tons of carbon dioxide to be emitted annually.[14] While the raw materials for polyether and polyester polyols currently come from petrochemicals, polyester polyols can easily be made from biological sources, thus offsetting some of the carbon emissions involved in their production. As a bonus, photosynthetic biological organisms will uptake atmospheric CO_2 and negate carbon emissions entirely when they produce polyester precursors.

Since there are many different biological sources to choose from, it is vital to select starting components that are conducive to the needs of the environment and the needs of PUs. While it is natural to think that biobased feedstock is an improvement over petroleum-based feedstock, one must consider how the starting materials fit into the existing supply chain. For example, vegetable oils derived from palm or soybean can be used to produce polyols.[15–18] Both of these oils are widely available and low cost, and their production could satisfy the needs of the PU industry. However, palm oil production contributes to deforestation and wildlife displacement.[19] Using soy puts a strain on its production as food, and there are many drawbacks to using food as fuel.[20] Besides, soy is a commodity, which means it is subject to price volatility and international tariffs. Even nonfood commodities such as castor beans, a common source of natural oil polyols, are occasionally subject to such tariffs.

Algae, however, are a suitable biological source of polyols because they avoid many of the problems traditionally associated with biosourced production. They can be grown in ponds or bioreactors, so they need not use arable land or large quantities of fresh water. Many algal species can grow in brackish water. Also, algae are neither staple food crops nor commodities, so they are insulated from traditional supply and demand shocks that can plague business cycles. Algae can replicate quickly and produce high yields of PU precursors. By contrast, plants have to first germinate, establish roots, and grow leaves and flowers to produce seeds, and many are sensitive to growing conditions, making them sensitive to climate change. Thus future yields of many of these plants are not guaranteed, as the conditions in which they grow may not be as favorable as they are today.

Algae-sourced polyol has some drawbacks. A significant cost in producing the target PU precursor in an aqueous medium is that the precursor must then be separated from the medium. When comparing the refining process to that of petrochemical precursors, algae-based PU precursors have extra steps associated with their unique production. For instance, the target molecule may not be produced efficiently by the algae. In this case, genetic engineering may be required to make the algae secrete enough of the precursor to be cost effective.

In some cases, the target molecule needs to be separated from the media by precipitation or separated from water by reverse osmosis. In other cases, distillation might be necessary. Another complication arises when pigments and other algae byproducts persist in the isolated precursor. These are sometimes difficult to remove without additional processing. And algae polyols may have a higher viscosity than petroleum counterparts; this is true of polyester polyols in general when compared to polyether polyols. These drawbacks do not outweigh the benefits, but the proper infrastructure must exist to process algae at a global scale, which will enable mass adoption for use in PU.

Thermoplastics

The chemistry that governs foams is also found in thermoplastic polyurethanes (TPUs). TPU is a polymer that can be melt processed and formed into its final shape in an infinitely repeatable cycle. A TPU product can be first made from liquid polymer and cured, then melted down and cast into different shapes. The advantage of TPUs is that the reactive process is separate from the final part of production. The result is that large quantities of TPUs can be supplied to manufacturers prereacted instead of as reactive chemicals that can be moisture sensitive or have a shelf life.

TPUs are made by using many of the components in PU foams, notably polyether or polyester polyols, chain extenders, and isocyanates. These components are generally reacted in an extruder and cut into pellets to be supplied to manufacturers. To ensure that the resultant polymer is melt processable, all components must be difunctional, i.e., have a functionality of exactly 2. This ensures that individual strands of high molecular weight polymer are generated instead of an irreversibly crosslinked three-dimensional network. Fig. 7.5 visualizes this difference, which is crucial to understanding the essential structure-property relationship of TPUs. Because the interactions between individual strands are broken by high temperature, the solid polymer can be melted into a liquid.

In pellet form, TPUs are loaded into a hopper and melted into a viscous liquid. From here, injection molding is used to create the final product. The hot liquid polymer is injected into a cavity that has the shape of the final part machined into it. The cavity is typically made of a metal such as stainless steel or aluminum. As the liquid cools, it freezes into the shape of the cavity. The product is then removed (demolded) and is ready for use. Many intricate shapes and geometries can be easily fabricated by this method.

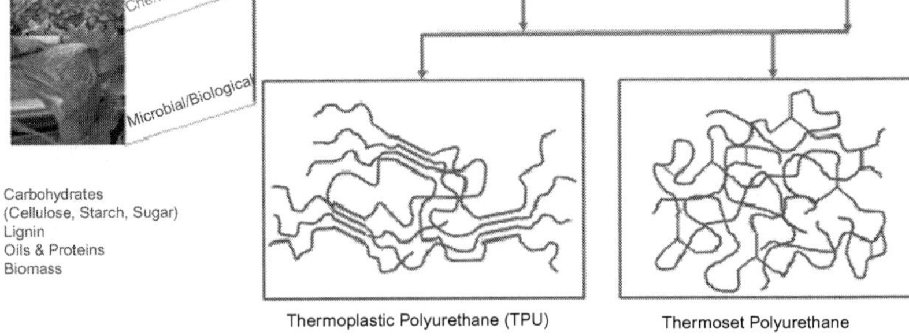

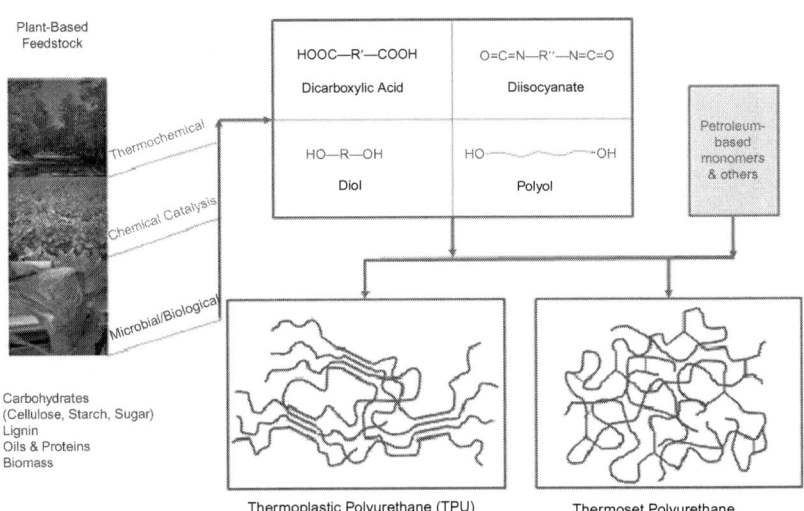

FIG. 7.5 Generic structure of a difunctional polymer (left) versus a multifunctional polymer (right). The polymer on the right cannot break any of its interactions. *Red groups* highlight the differences in crosslinking. *Reproduced with permission from Thien An PH, et al. Renewable polyurethanes from sustainable biological precursors.* Biomacromolecules *2021;22(5):1770–1794. Copyright 2021 American Chemical Society.*

The hardness of a part is tunable by precisely controlling the formulation, so TPUs can be made into a great variety of materials, from flexible rubber to hard resin. Applications range from abrasion-resistant shoe soles and wristwatch straps on the softer side to automotive body paneling and skateboard wheels at the rigid end of the spectrum. Fig. 7.6 shows examples.

Simply by changing the initial disposition of the TPU from pellets, even more applications can use this unique material. TPUs can also be extruded into thin films for use in breathable textiles and protective coatings. In the form of a filament or powder, they can be the starting material for 3D printing. A TPU commonly encountered in everyday life is spandex. It is uniquely produced by dissolving a polyol solution end-capped with unreacted isocyanate groups (a prepolymer) in a solvent. Then a separate solution containing a diamine is reacted with the first solution and spun into fibers as the solvent evaporates, creating the stretchy fabric.

At the cutting edge of urethane technology, hybrids of TPU and foams, called *E*-TPUs, have been made. They bring the low density of foams and combine them with the durable mechanical properties of TPU. The secret is the infusion of a blowing agent into the TPU pellets by a highly specialized machine. The blowing agent is typically carbon dioxide or nitrogen gas, but can also be a volatile hydrocarbon. When the pellets are melted and injection molded at high temperature, the gas expands and produces bead foam. The product is used in athletic running shoes to provide light weight, outstanding shock absorption, and energy return.

The typical "knobs" that are turned to produce a TPU for a specific application are similar to those for foams. The most impactful of these is the

FIG. 7.6 (A) Skateboard wheel, (B) foam case, (C) escalator handrail, (D) ski boot, (E) 3D printed flexible wrench, all made from thermoplastic polyurethane.

amount of hard segment in the given TPU. The hard segment consists of the chain extender and isocyanate, expressed as a weight percentage of the total composition. This percentage can be changed by increasing or decreasing the amount of chain extender, which respectively increases or decreases the amount of isocyanate needed to react the TPU fully. The type of chain extender can also influence the type of bonds in a TPU. For instance, amine chain extenders react with isocyanate to form urea bonds, whereas diol chain extenders simply form urethane bonds. Another effect of increasing the hard segment is that the temperature at which the TPU can be melt processed concurrently increases. It is essential to consider the molten state characteristics such as flow and viscosity, because this determines the ease with which manufacturers can make intricate molded parts.

At the cold end of temperature-based phase transitions is the glass transition temperature, defined as that at which a TPU changes from flexible to brittle. This is essential for cold-weather performance, such as the soles of ski boots. Some additives, known as plasticizers, can lower the glass transition temperature. They make TPUs softer and allow them to retain flexibility at colder temperatures than would be possible otherwise.

As with PU foams, formulators and manufacturers face the same choice of using water-resistant polyether polyols or oil and solvent-resistant polyester polyols. Either material gives good performance characteristics, but polyethers are preferred when hydrolysis resistance is paramount. Obtaining this performance characteristic with polyesters may seem an uphill battle. However, with increased consumer demand for biobased goods and consumer products that do not end up in landfills for hundreds of years, the argument can only grow in favor of polyesters. It may be tempting to continue using polyethers in TPU applications, since all TPUs, whether ether or ester based, can be infinitely melted, reprocessed, and reshaped into new articles. However, some estimates suggest that only 1% of PU products are ever recycled in this way.[21] This low rate reinforces the need for products on the market to clean up after *themselves* instead of expecting consumers and recyclers to do it for them.

Even if TPU recycling rates were drastically increased, the cost of sorting, collecting, and cleaning used articles, then remelting and reprocessing them into new materials, is not trivial, from either a monetary or a carbon emissions perspective. If the product itself can carry out the procedure of being decomposed, this could be perceived a benefit to the environment. Guaranteeing with certainty that a product will degrade instead of having only a small probability of being recycled is highly advantageous. It ensures that whether or not such a product is recycled, it will never contribute to the ever-rising amount of plastic pollution. As a result, the possibility of using algae as the starting block for the polyols for polyester-based TPUs must again be raised.

TPUs are an ideal use of algae-based monomers, because the building blocks are quite simple. All of them are difunctional polymers made of difunctional monomers. Algae are capable of generating these monomers and associated precursors.[22] TPU formulations can consist of over 60% polyol by weight. Using algae to make these polyols enables roughly the same percentage of the resultant TPUs to be biobased. While the products may not be as near impervious to water as their polyether counterparts, much like foams, none of the products made from TPUs need to last forever either.

Conclusions

If the evidence is so strongly in favor of using algae as a feedstock, then where are all the algae-based PUs? After all, there are myriad biobased polyols made from many different feedstocks,[23] yet there are almost no products on the market that currently use them. What will first determine the success or failure of algae-based PU is, first and foremost, its ability to match the performance characteristics of existing petroleum-based products. Public-private partnerships, such as collaborations between universities and companies, can bridge the gap between the expensive R&D needed to get prototype parts that meet product specifications and the large-scale manufacturers to mass-produce those parts.

The second challenge is that of mass production itself. It is a small matter to make a kilogram of algae-based PU, and another matter entirely to make 1 million parts with narrow tolerance and bring them to market. The infrastructure for algal monomer production requires a demand at the mass scale, but the mass production itself requires the infrastructure, causing a chicken-and-egg situation. To resolve this, multiple levels of pressure, from consumers to governments, must demand that companies rein in pollution and carbon-unfriendly products.

The third challenge is encountered when the first two are successfully navigated: algae PU must be integrated into existing factories with currently available machinery. Algae PU cannot require expensive and specialized equipment; it must behave as a drop-in solution for existing PU. Even if the output—the product—is identical to a petroleum product, the input, algae PU reactants, must behave as closely to petroleum PU reactants as possible. Imagine that the factory switches from a polyether polyol liquid to an algae polyester polyol solid at the same temperature. If the machinery used by the factory has no large-scale heating equipment to melt the polyester, then there is no pathway to production. Only when all three of these challenges are addressed can algae PU become a reality.

What we must do is important, but equally, so is what we must not do. It is easy to call something a green product because it is made from natural sources, without considering its environmental implications. One notorious example is bio-PET (polyethylene terephthalate) plastic bottles. These are derived partially from plants yet are chemically identical to commercial plastic bottles, meaning that they will still be persistent pollutants that never undergo biodegradation. Ironically, their production has higher carbon emissions than their petroleum-based counterparts and results in higher ecotoxicity and ozone depletion.[24,25] In an urge to label products as ecofriendly, it may be tempting to take half measures such as these. In this case, the results are more detrimental than taking no measures at all. Compared to all of the other ecofriendly options, algae foams and TPUs will have the upper hand in the marketplace if they can successfully show (1) equal performance, (2) reduced carbon footprint, and (3) timely biodegradation.

In summary, the material demands of many high-performance products can be met using algal PU. All of the currently available PU products will eventually be disposed of or replaced, so an end-of-life cycle must be planned from their inception. Through a deliberate selection of starting materials and formulation, we can begin to realize products that have a lifetime proportional to their useful service life. Rethinking polyester PUs means rethinking the idea that built to last needs to last forever.

Close-Up: Raw materials for future polyurethane manufacture

Polyurethanes are ubiquitous in our lives. Innovation in this field has led to an explosion of applications. Using biobased materials as feedstock for polyurethanes seems a natural choice of raw material, but there are several hurdles to mass adoption.

Biobased polyurethanes must match the performance of those made from fossil fuels. Manufacturers are hesitant to adopt new materials, because most applications require very tight tolerance in the physical properties of the polyurethane foam, along with high durability. Future Foam, Inc. currently manufactures a few specialty viscoelastic products that meet or exceed the performance of traditional polyurethanes and are well received by the market. Significant expansion of the portfolio to address a mass market will require the use of other raw materials that must meet equally high performance criteria. Polyurethane manufacturers have highly developed and efficient process equipment, manufacturing processes, and distribution networks. Biobased raw materials must be able to use these existing systems without major modification to remove a significant barrier to adoption and increase speed to market. These materials would ideally be "drop-ins" for their fossil-based alternatives in the manufacturing process.

There is increased awareness of biobased polyurethanes at the consumer level. This is driving demand, but it is still early in the innovator stage of the adoption cycle. Price parity between fossil-based and biobased material will be needed to move into the mass adoption stage.

These hurdles are typical early stage phenomena. They will be overcome and adoption will follow. The future of biobased polyurethanes is bright.

Anshul Gupta

Chief Operating Officer, Future Foam, Inc., Council Bluffs, IA, United States

References

1. Bayer O. Das di-isocyanat-polyadditionsverfahren (polyurethane). *Angew Chem.* 1947;59 (9):257–272.
2. DE728981C. Process for the production of polyurethanes and polyureas; 1942.
3. Heinrich R., Heinz S., Werner S. U.S. Patent No. 2,511,544. Washington, DC: U.S. Patent and Trademark Office; 1950.
4. Szycher M, ed. *Szycher's Handbook of Polyurethanes.* CRC Press; 1999.
5. Petrović ZS, Javni I. The effect of soft-segment length and concentration on phase separation in segmented polyurethanes. *J Polym Sci B Polym Phys.* 1989;27(3):545–560.
6. Herrington R, Hock K, eds. *Flexible Polyurethane Foams.* Dow Chemical; 1997.
7. Schollenberger CS, Stewart FD. Thermoplastic polyurethane hydrolysis stability. *J Elastoplast.* 1971;3(1):28–56.
8. Sonnenschein MF. Polyurethanes: Science, Technology, Markets, and Trends. John Wiley & Sons; 2021.
9. http://www.recyclingadvocates.org/do-just-one-thing-give-old-shoes-new-life-aug-2018/.
10. https://www.aarp.org/auto/trends-lifestyle/info-2018/how-long-do-cars-last.html.
11. do Sul JAI, Costa MF. The present and future of microplastic pollution in the marine environment. *Environ Pollut.* 2014;185:352–364.
12. Cao Y, Wang Q, Ruan Y, et al. Intra-day microplastic variations in wastewater: A case study of a sewage treatment plant in Hong Kong. *Mar Pollut Bull.* 2020;160:111535.
13. Sharma S, Chatterjee S. Microplastic pollution, a threat to marine ecosystem and human health: a short review. *Environ Sci Pollut Res.* 2017;24(27):21530–21547.
14. Ritchie H, Roser M. CO_2 and Greenhouse Gas Emissions; 2020. Published online at OurWorldInData.org. Retrieved from: https://ourworldindata.org/co2-and-other-greenhouse-gas-emissions.
15. Tanaka R, Hirose S, Hatakeyama H. Preparation and characterization of polyurethane foams using a palm oil-based polyol. *Bioresour Technol.* 2008;99(9):3810–3816.
16. Li Y, Luo X, Hu S. Polyols and polyurethanes from vegetable oils and their derivatives. In: *Bio-Based Polyols and Polyurethanes.* Cham: Springer; 2015. SpringerBriefs in Molecular Science; https://doi.org/10.1007/978-3-319-21539-6_2.
17. Ionescu M, Petrović ZS, Wan X. Ethoxylated soybean polyols for polyurethanes. *J Polym Environ.* 2007;15(4):237–243.
18. Otieno NE, et al. Palm oil production in Malaysia: an analytical systems model for balancing economic prosperity, forest conservation and social welfare. *Agric Sci.* 2016;7 (2):55–69.
19. Gatti RC, et al. Sustainable palm oil may not be so sustainable. *Sci Total Environ.* 2019;652:48–51.
20. Herrmann R, et al. Competition between biofuel feedstock and food production: empirical evidence from sugarcane outgrower settings in Malawi. *Biomass Bioenergy.* 2018;114:100–111.

21. Magnin A, Pollet E, Phalip V, Avérous L. Evaluation of biological degradation of polyurethanes. *Biotechnol Adv.* 2020;39:107457.
22. Zia KM, Zuber M, Ali M, eds. *Algae Based Polymers, Blends, and Composites: Chemistry, Biotechnology and Materials Science.* Elsevier; 2017.
23. Ionescu M. Chemistry and Technology of Polyols for Polyurethanes. iSmithers Rapra Publishing; 2005.
24. Chen L, Pelton RE, Smith TM. Comparative life cycle assessment of fossil and bio-based polyethylene terephthalate (PET) bottles. *J Clean Prod.* 2016;137:667–676.
25. Gursel IV, Moretti C, Hamelin L, et al. Comparative cradle-to-grave life cycle assessment of bio-based and petrochemical PET bottles. *Sci Total Environ.* 2021;793:148642.

CHAPTER

8

Coatings, adhesives, and sealants from polyester polyurethanes

Naser Pourahmady

University of California San Diego, La Jolla, CA, United States

Introduction

History of coatings and adhesives based on renewable resources

The use of renewable materials for painting and gluing goes back to the Stone Age, where cave dwellers decorated their walls with plant-based pigments and joined things together with sticky material from tree sap. The Egyptians glued papyrus paper with starch as early as 3500 BCE.[1] Around 700 BCE, the Chinese learned to use rice starch for surface treatment of paper and as a binder in paper production.[2] Improvements to plant- and animal-based glues and coatings continued until the nineteenth century. In 1867 the first patent for paints and coatings was awarded to the Averill Chemical Paint Company.[3]

Bio-based materials predominated until World War II and the invention of synthetic polymers. Synthetic coatings and adhesives were among the earliest applications, and petroleum-based raw materials proliferated as more monomers of consistent quality became available at a lower cost than plant materials, facilitating the development of new products. With advances in polymer science, products emerged ranging from architectural paints to aerospace and marine coatings, from household paper glue to construction and automotive adhesives and sealants.

Many transformational developments have occurred and are shaping the future of the coatings industry: powder coatings technology, UV-cure coatings, electrocoating, waterborne technologies, and recently nano-composite and thin-film coatings. Likewise in the adhesives industry: cyanoacrylates (known as krazy glue), epoxy adhesives, moisture-cure silicon-based adhesives, urethane adhesives, and many new developments

Rethinking Polyester Polyurethanes
https://doi.org/10.1016/B978-0-323-99982-3.00008-0

in the area of engineering hot-melt adhesives. Polyurethanes have been and will continue to be a significant part of all these advances,[4] which have resulted in constant improvement of the cost/performance balance. Most improvements used to be based on innovations in petroleum chemistry. In recent decades, environmental impact and cost competitiveness of renewable resources have been attracting industry leaders and researchers.[5]

Environmental regulations and renewed interest in bio-based feedstock

The push by environmental activist groups and regulatory agencies has made renewable resources the raw material of choice for coatings and adhesives.[6] The drive is threefold: the green product (low hazard to environment), biodegradability (waste management), and raw material availability. Renewable raw material does not automatically make a product more environmentally friendly or biodegradable. This topic is covered in other chapters, as are raw material availability and biodegradability. Here we focus on algae-based chemicals to make polyester polyurethane coatings, adhesives, and sealants, and comparing the options available from algae feedstock to current petroleum-based processes.

Polyester polyurethane coatings and adhesives from algae-based raw materials

From renewable resources to polyurethane coatings and adhesives

Converting any renewable feedstock to a coating or adhesive requires four steps.

(1) Chemical refining of the feedstock to separate and clean up the valuable chemicals.
(2) Conversion of the refined chemicals to functional monomers.
(3) Polymerization of the monomers by many existing industrial processes.
(4) Formulation of the polymer resin product with additives for the final application.

All steps must be considered to enable a fair evaluation of the impact of any product on the environment. Note that only the refinery step is dependent on the nature and form of the renewable feedstock. Fig. 8.1 shows some chemical details of the process.

The downstream processes become less dependent on the source of feedstock and usually lead to the formation of similar or identical chemicals and monomers from the different feedstocks. Therefore, discussing

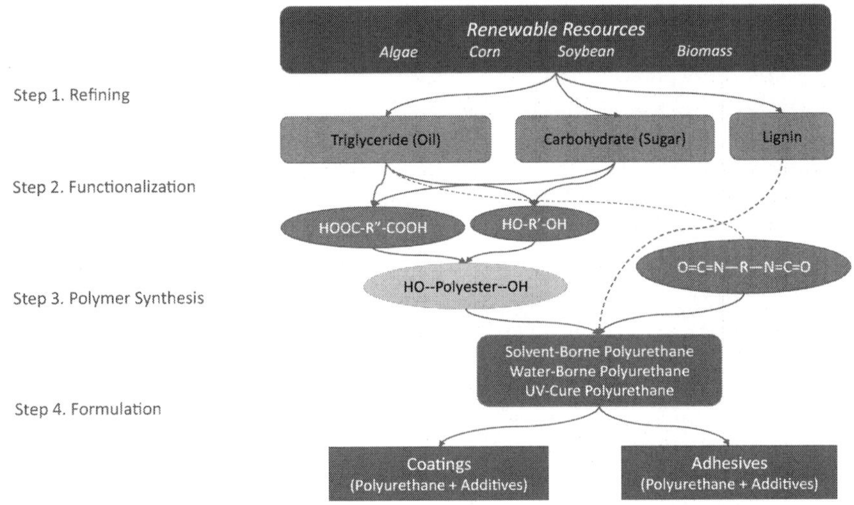

FIG. 8.1 Stepwise process from renewable resources to coatings and adhesives.

polyurethane coatings and adhesives based on renewable polyester polyols will have little or no relevance to specific biological sources.

The most common refinery products of bio-based feedstock useful for urethane manufacturing are oil (triglycerides) or sugar (carbohydrates).[7] The chemical composition of these products is very similar regardless of the precursor. Therefore, the polyester polyurethane chemistry discussion from algae feedstock is no different from polyurethanes made of raw materials sourced from soybean, corn, biomass, or any similar renewable resource.

The reason behind selecting the triglyceride or carbohydrate route in the refinery stage is the need for some reactive nucleophile functionality in the monomers, mainly hydroxyl groups, to react with acid-functional monomers to form polyester and with isocyanate monomers to form polyurethane. The triglyceride structure in vegetable and algae oils is suitable for converting diacids, diols, and polyols via hydrolysis, oxidation, or some other route.[8] Similarly, carbohydrates or alginate structures either contain hydroxyl functionality or can be easily converted to polyols. One exception is lignin extracted from wood. It has a complex chemical structure and usually contains functional groups that can be incorporated into urethane products.[9] Some coatings and adhesives are developed with the incorporation of lignin chemicals and are discussed in detail in the literature.[10] We are focusing on algae-based chemicals and so will cover the algae oil and alginate carbohydrate routes.

The refining, functionalization, and polymerization steps of Fig. 8.1 are discussed elsewhere in this book. Here we cover the work done on polymers specific to coatings and adhesives and formulations targeting these applications.

The algae oil route to polyurethane coatings and adhesives

The chemical process is essentially the same as that for vegetable oil or any other triglyceride feedstock. The minor differences in the composition of algae oil and other renewable resources are discussed by Petrovic[11] and summarized in Table 8.1. The most common triglycerides from microalgae contain fatty acid chains with 16 to 20 carbon atoms and up to three unsaturated bonds per chain. The unsaturation in the fatty acid chains starts from C-9, as shown in Fig. 8.2. Triglycerides provide a facile route for the production of azelaic acid from algae oil by hydrolysis of fatty acid chains followed by ozonolysis of the double bonds. Pure azelaic acid is then separated from the resulting mixture.[12]

The most common way of making polyester polyurethane is from the reaction of polyols and diols with diisocyanates. The algae-based diacid has been used to make polyester polyols by reacting with diols from other renewable resources.[13] The diacid product can also be reduced to 1,9-nonanediol or converted to diisocyanates to make polyurethanes entirely from algae-based feedstock.[14] The polyols and diol products from algae feedstock can also be used in combination with petroleum-based monomers in the production of polyurethane for coatings and adhesives.

A variety of polyurethanes based on azelaic acid polyesters are developed and used to formulate coatings and adhesive.[15,16] Awasthi and coworkers compared the performance of polyols from azelaic to those made of other difunctional acids in urethane coatings.[17] They found polyurethanes made of azelaic acid polyesters comparable to other linear aliphatic polyester polyurethanes in thermal and chemical properties, but somewhat inferior to polymers made of cycloaliphatic diacids. Another study found polyurethane coatings made of azelaic acid polyesters had good UV resistance and hydrolytic stability and were suitable for outdoor applications.[18] Polyester polyurethanes are commonly not recommended for outdoor applications due to their poor UV and hydrolytic stability.

An amidation method developed by Patil and coworkers[19] converts algae oil to fatty amid diols, as shown in Fig. 8.3. They used these diols combined with others such as butanediol, isosorbide, and bisphenol A to make polyether polyols and polyurethane coatings that have unique antimicrobial properties. Because of the ether backbone and the hydrophobicity of the fatty acid amide moieties, these coatings also show good outdoor stability, chemical resistance, and anticorrosion effects[20] when tested on steel panels. The authors also reported converting fatty amide diols to alkyds and polyester polyols and preparing polyurethane coatings from fatty amide polyols.[21] The antimicrobial effect of these coatings was also investigated. This approach is unique in introducing amide linkage in the backbone of coating and adhesive polyurethanes and the presence of fatty acid pendant groups, which enhances hydrophobicity and hydrolytic stability as expected. However, if biodegradability is of primary

TABLE 8.1 Fatty acid content of triglycerides from renewable sources.

	Composition (mole %) fatty acid chain length/unsaturation										
Bio-based oil	**14/0**	**16/0**	**16/1**	**18/0**	**18/1**	**18/2**	**18/3**	**20/0**	**20/4**	**22/0**	**22/1**
Microalgae		13.9	4.7	1.7	11.8	13.2	11.8		34		
Palm	1	44.4	0.2	4.1	39.3	10	0.4	0.3		0.1	
Soybean	0.1	10.6	0.1	4	23.3	53.7	7.6	0.3		0.3	
Linseed		6		4	22	16	53	0.5			
Rapeseed	0.1	3.8	0.3	1.2	18.5	14.5	11	0.7		0.5	41.1

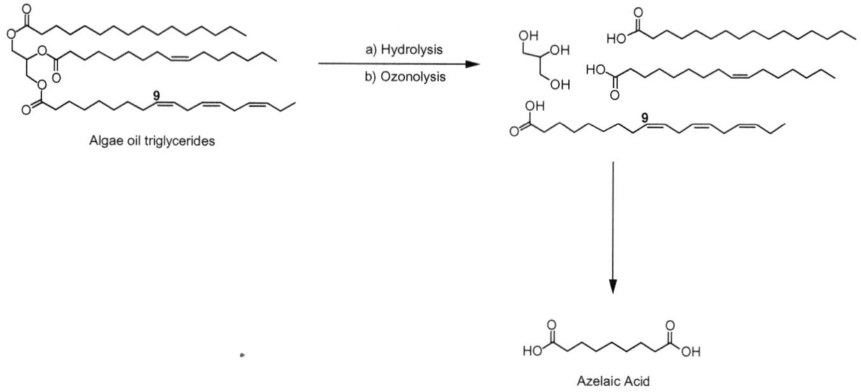

FIG. 8.2 Difunctional acid from triglycerides sourced from plants and algae.

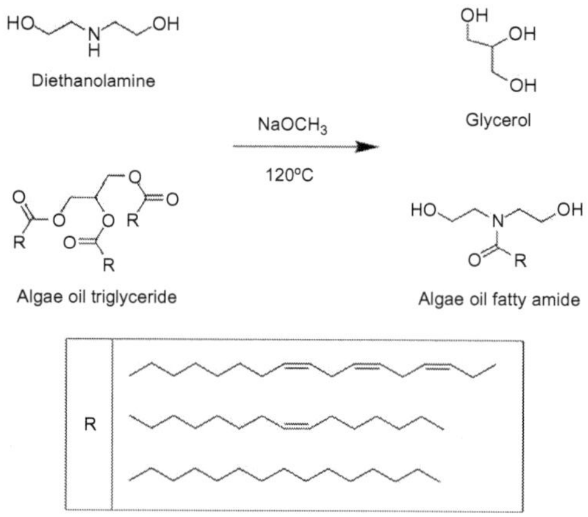

FIG. 8.3 Conversion of algae oil to fatty amide diols for polyurethane.

interest, this may not be a favorable route as compared to other polyester polyurethanes.

Kadam and coworkers reported polyurethane coating films from algae-based fatty amid polyester polyols and their activity against mosquito larvae and ants upon 2 to 12h of exposure.[22] They also reported good results from cytocompatibility studies.

Another common technology of converting vegetable and algae oils to polyols and polyurethanes is epoxidation followed by hydroformylation or alcoholysis to produce a bio-based polyol, as shown in Fig. 8.4. This route is discussed in detail by Petrovic,[11] and some modifications have also been reported.[23] Although more economical, there are several disadvantages to this route.

- The product is usually a poorly defined mixture of multifunctional polyols, therefore not suitable for preparing thermoplastic

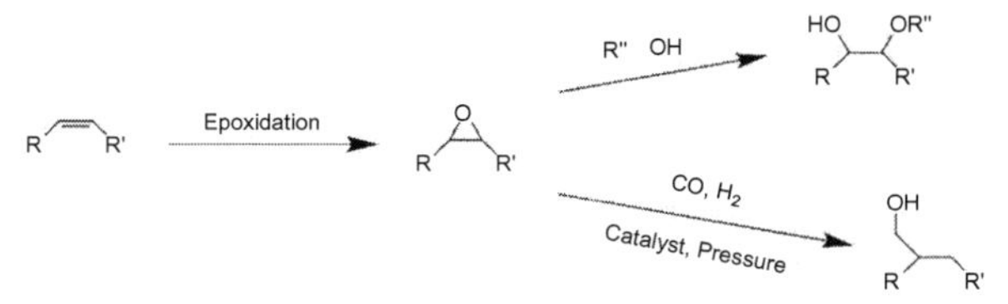

FIG. 8.4 Epoxidation route for conversion of algae oil to polyol.

polyurethane products that require difunctional monomers with high purity.
- The hydroxyl functionality produced by this route is usually a secondary hydroxyl and therefore lower in reactivity toward isocyanates. The hydroformylation route overcomes this problem; however, it requires a high-pressure carbonylation step, which is costly and diminishes the economic advantage over the azelaic acid route.
- Polyols with consistent quality are difficult to produce by this route because of the variation in the oil composition by region, seasonality, and the refining process.

Incorporation of natural oils into waterborne urethane coatings by epoxidation route was reported by Pajerski[24] using epoxidized vegetable oil, and Nudin and his group[25] using Jatropha oil. Because of the problems outlined before, it is impossible to make polyurethane coatings purely based on this route. Pajerski and his team could incorporate as high as 50% oil-based polyol in waterborne polyurethane dispersion coatings, the highest reported for any commercial product so far.[26] Salah and group studied the colloidal stability and mechanical properties of waterborne polyurethane dispersion (PUD) coatings made by this method and the change in the level of oil and acid functional stabilizing group.[27] They reported improved hydrophobicity and higher crosslink density and suggested potential application in wood and decorative coatings.

There is another route reported in the patent literature that provides a more environmental-friendly, nonisocyanate process to convert triglycerides to polyurethanes.[28] Fig. 8.5 shows the reaction scheme. Others previously reported this kind of nonisocyanate route to polyurethanes from cyclic carbonate and diamines. However, the patent by Wilkes and his team was the first to utilize the chemistry for polymerization of epoxidized vegetable oil. Interestingly, the group did not mention algae oil in their work. The specific examples started with epoxidized soybean oil to make cyclic carbonate, followed by an ethylene diamine reaction to make polyurethane. Later on, Mahenderan reported extending this work to other vegetable oils and polymerization with aromatic amines and polyamines.[29] Because of the presence of multiple epoxy groups in algae and vegetable oils, this route can only provide thermoset polyester polyurethanes with low biodegradability and no recyclability, which makes

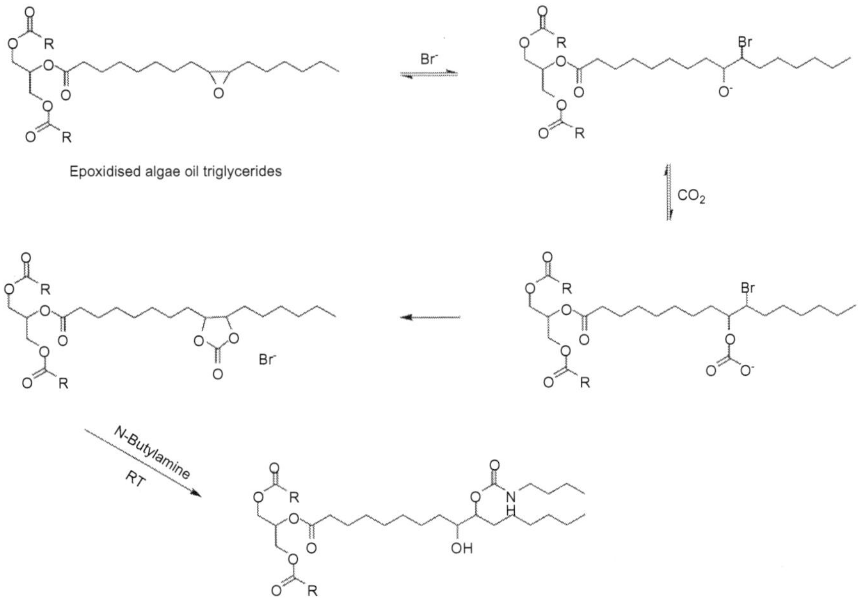

FIG. 8.5 Nonisocyanate route from triglycerides to polyurethane.

them less desirable for most common applications. Another drawback is the limitation in fine-tuning the polymer architecture for target performance until several short-chain carbonate monomers become readily available to copolymerize with oil-based carbonates. Therefore, despite this technology's simplicity and environmental friendliness, it has not found any commercial application.

Other chemistries are applied in the conversion of algae oils and vegetable oils into polyurethane and formulation to coatings and adhesives. Simmons and his group developed a two-component polyurethane adhesive using the product of transesterification of triglycerides with glycerol or trimethylolpropane as one component and polyisocyanates as the second component.[30] This type of chemistry is generally applicable to the triglycerides from any renewable feedstock, and it is a more economical way to a partially renewable adhesive.

Alginate route to polyurethanes for coatings and adhesives

Alginic acid, also called algin, is the polysaccharide form of bio-based materials found in the cell walls of brown algae. Due to its hydrophilic nature, it is usually found in a hydrated form as a viscous gum or in the form of neutralized sodium or calcium salt, which is called alginate. These polysaccharides have been extracted from algae and widely studied and converted to various functional monomers to make polymers for

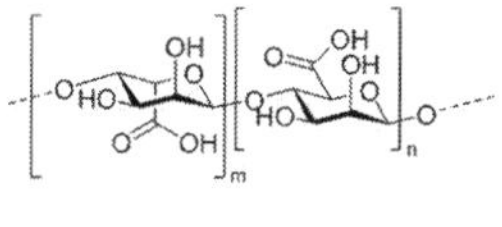

FIG. 8.6 Carbohydrate chain structure of alginate.

various applications. A typical structure reported for alginate consists of a linear chain of mannuronic acid and l-guluronic acid linked together at the 1,4 positions (Fig. 8.6). This natural polymer has found industrial application as it is processable and valuable without any significant structural modification. The hydrophilic polymer is a suitable film former by itself, and it has been used as edible coatings for food packaging.[31] Some studies have found antioxidant activity[32] in alginates. Others have shown enhancement of physicochemical and biological properties of fruit products by alginate-based coatings.[33] Improved gel strength was reported when alginate was used in combination with polyurethane coatings.[34]

Martin and Grossmann[35] have discussed the conversion of these sugar structures to furfural and its derivatives. One of the useful monomers in this family, furan-dicarboxylic acid, has been used to make polyols and polyurethanes. Several spirocyclic polyols were made by incorporation of furfural derivatives and converted to polyesters and polyurethanes.[36] Incorporation of these monomers in the backbone of polymers significantly increased the glass transition temperature. This use can be important for the manufacture of hard coatings and composite materials.

A nonisocyanate route using mono- and disaccharides has also been reported in making polyurethanes from renewable resources.[37] Although applicable to alginate carbohydrates, the authors did not explicitly apply this to algae polysaccharides.

Formulation of coatings and adhesives from renewable polyester polyurethane

From polymerization tank to paint can

We have discussed major advances in chemistries that can lead to the development of new environmentally friendly adhesives and coatings. Now we turn to the last step in Fig. 8.1 and ask how many of the chemistries are currently being used to make products. The answer is very few. Why?

It is very rare that a polyurethane manufactured by a polymer supplier ends up on a store shelf as paint or an adhesive without any formulation. Regardless of the shape and form of the polyurethane resin and the manufacturing process, it will need formulation additives to make it suitable for specific applications. Although polyurethane resin is the main ingredient in some of the coatings and adhesives, the formulation additives can make up as much as 50% of most pigmented applications. Adhesives typically require fewer additives than coatings and sealants. Some of the additives in coatings and adhesives, such as plasticizers, dispersants, rheology modifiers, compatibilizers, tackifiers, adhesion promoters, defoamers, and crosslinkers are polymeric in nature. Therefore, even if the polyurethane resin in the formulation is bio-based and/or biodegradable, the product cannot be claimed as such.

Many additive manufacturers are trying to develop more environmental friendly alternatives. Even nonpolymeric additives such as fillers, pigments, coalescent agents, matting agents, dryers, biocides, and antioxidants could alter the ecological properties of the coating or adhesive. Most commercial industrial coatings contain 8 to 10 ingredients in addition to the base polymer. Therefore, although the substitution of a bio-based or biodegradable resin may considerably improve the product's environmental profile, it does not entirely solve the problem.

How much of the problem is caused by formulation additives and how to overcome these issues depend on the manufacturing process and the intended application. Standard processes for polyurethane coatings and adhesives are discussed later, and the ecological issues associated with each and potential solution with algae or other bio-based chemicals are considered. Processes for coatings and adhesives share the most common monomers and polymers.[38] The main difference lies in the choice of monomer composition and formulation additives. In general, adhesives are made of monomer composition to yield polymers with a lower glass transition temperature (softer) than those used for coatings.

Solvent-borne coatings and adhesives

This is the oldest and most common technology for the production of coatings and adhesives. The polyurethane resin is made in a solvent and formulated with solvent-borne or solvent-compatible additives. Most makers of high-performance coatings such as automotive topcoat and specialty industrial coatings still use this process, but they are under constant regulatory pressure to change. A large portion of the adhesive industry is also solvent borne but is gradually switching as new methods become available. The solvent-borne process is the least environmentally friendly because of the volatile organic compounds released into the environment during application. Despite the effort by solvent manufacturers to develop less hazardous alternatives,[39] solvent-borne coatings and adhesives remain the least favored, and the market is shrinking. Existing bio-based resins and additives will not change the picture. A new nontoxic solvent is needed. Years of research have shown that finding a more environmentally friendly solvent than water is unlikely. Therefore, the waterborne route has become the most obvious alternative.[38]

Before we discuss the alternate techniques, it is crucial to understand the role of polymer crosslinking in these products. Unlike many other applications such as foams, extrusion, and molded plastic parts, the base polymer system used for coatings and adhesives must be thermoplastic before application and thermoset after application to the substrate. Before the application, the need for a thermoplastic state is that the coating or adhesive needs to flow into different shapes, fill the gaps, cover the substrate, then fuse into a continuous film or filling to adhere and protect. A thermoset polymer will be neither melt flowing nor soluble in any solvent to be formulated and used as a coating or adhesive. However, once it is applied to the substrate, the polymer needs to be crosslinked (thermoset) to gain strength, durability, and solvent resistance. An enormous amount of research has gone into the development of crosslink technologies. Some chemistries are built into the polymer backbone to trigger crosslinking after application via heat curing or other mechanisms. These are called self-crosslinking or internal crosslinking and enable the suppliers to provide ready-to-use coatings in a 1-pack (called 1K) system. The systems using external crosslinking come in two separate packages (2K) and need to be mixed in specified ratios right before application.

Waterborne coatings and adhesives

Although water seems an obvious choice for least hazardous solvent, switching polymerization and formulation from organic solvents is not straightforward.[40] Most starting materials for polymerization, whether from renewable resources or petroleum, are not soluble in water. The same is true for most formulation additives and polymers. Resistance to water and moisture is one of the desired features of all coating components for most applications. Therefore, new methods were needed for waterborne products. That is where heterogeneous polymerization methods started, with oil-in-water emulsions.[40] For many years, almost all waterborne polymers were made by emulsion polymerization of acrylic, styrene, and other vinyl monomers. However, this technology is only suitable for addition polymerization for monomers containing active double bonds. Waterborne polyurethane dispersion (PUD) started in the early 1970s when Dieterich introduced the concept of polyurethane ionomers,[41] which enabled stabilization of polymer particles in aqueous dispersion. Since then, many PUD products have been developed by polymer manufacturers and have found applications in various coatings, adhesives, sealants, and binders.

The technology of waterborne PUD products is versatile and constantly improving.[42] The stabilization of organic particles in aqueous media has been done by anionic, cationic, or nonionic routes, and commercial products in all three categories are on the market. Several new methods have also been developed to promote self-crosslinking in waterborne PUD polymers.[24,43] The elimination of organic solvents enables the coatings industry to formulate low-volatility organic compounds and environmentally friendly products for this market. Furthermore, since waterborne PUD is a heterogeneous system, both water-soluble and oil-soluble additives can be formulated into waterborne coatings and adhesives. This has facilitated the development of 1 K coatings (reactive components in one package) where the active reactants are placed in different phases, which keeps them from reacting until application and drying. This is particularly important in the self-crosslinking systems.

Because of the ecological advantage of PUD, the market is growing, and solvent-borne urethanes are being replaced in every application category as long as they can meet the target performance.[44] Therefore, the introduction of bio-based and/or biodegradable technologies in this market provides a higher chance of success.[45] Heinrich[46] has recently reviewed the opportunities for the incorporation of bio-based technologies in adhesives. Development of waterborne polyurethane adhesives from algae chemicals is currently underway in our labs at the University of California San Diego.

Solvent-free processes for coatings and adhesives

There are manufacturing processes that eliminate the need for a carrier solvent.[47] The powder coating and UV-cure methods are developed to apply coatings without the use of any solvent. There are also nonsolvent methods developed for 2 K adhesives where the reactants are mixed and applied at the application site. The potential application of polyurethanes from renewable resources and algae chemicals in all of these areas is also promising. However, the limitation in choosing materials suitable for these processes presents an added challenge in substituting bio-based feedstock for petroleum-based chemicals.

Cost/performance dilemma

The main criterion for the selection of raw materials and technology for a product in any new or existing application is cost vs. performance. All other factors such as safety, environmental profile, customer appeal, sustainability, and regulatory issues are always considered as part of the cost/performance balance or, in some cases, as tiebreakers. Keep in mind that renewable feedstock is competing with petroleum-sourced chemicals for both cost and performance. An evaluation of cost/performance may be helpful to divide bio-based raw materials into two categories: those that are exact replacements for petroleum-based chemicals, such as acrylic acid and butanediol; and those that are different from petroleum chemicals and only available from renewable resources, such as lactic acid and azelaic acid. With this in mind, we can also divide the value proposition for bio-based raw materials to cost advantage or performance advantage over their petroleum-based counterparts. Therefore, all bio-based monomers fall in one of the four quadrants of Fig. 8.7. An exception to this classification would be a method that offers both cost and performance advantages. Based on all R&D work to date it is unlikely, if possible at all, to obtain both cost and performance advantage at the same time by substitution of bio-based alternatives for petroleum products. The situation is unlikely to change until we really start running out of petroleum resources and experience a huge jump in the price of crude oil. Therefore, we have excluded such a category from our evaluation and only compare strategies targeting either cost or performance advantage by substitution to renewable feedstock.

Consider the renewable-based technologies within quadrants 1 and 4 of the figure. This group's value proposition is based on cost. How likely is it for a bio-based feedstock to offer a cost advantage over a petroleum-based feedstock? The chemical industry has been seeking an answer for 30 years. Because less than 10% of world crude oil production ends up in chemicals and polymers, the energy sector, which uses over 90%, determines the price. The increase in the price of crude from \$20 to \$120 a barrel in 1970s created a notion that the world was running out of fossil fuel and that biofuel might be the choice of sustainable energy. Therefore, during the 1980s and 1990s, many biofuel refineries were built in Europe and the United States with the presumption of a cost advantage in the future. With the biorefinery fever came the promise of many functional chemicals and monomers from renewable resources for the polymer industry at prices

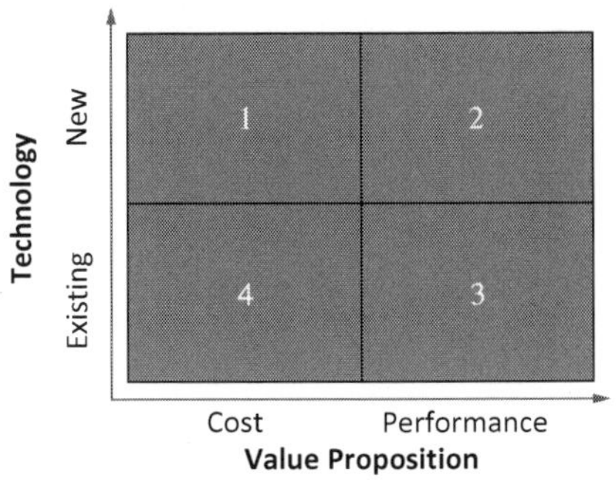

FIG. 8.7 Classification of value proposition for bio-based technologies.

comparable to petroleum-based chemicals. A long list of functional monomers were identified as target chemicals to be produced from bio-based feedstock by the biorefineries. At the top of the list was acrylic acid and acrylic monomers, adipic acid and its derivatives, propanediol, butanediol, and a few other difunctional monomers.

Two decades into the 21st century, more stability and the occasional drop in the price of crude on the one hand and the food vs. feed argument for bio-based feedstock on the other considerably dim the hope of bio-based achieving a cost advantage. The long-awaited governmental regulations to push the chemical industry in this direction did not materialize either. Therefore, biorefineries have been shutting down, and many start-up businesses that produced monomers from renewable resources have closed or declared bankruptcy. A few companies with deeper pockets continue within quadrant 4 with the hope of winning the cost proposition in the future with volume growth and an increase in the price of crude. The situation is not much better for quadrant 3. Finding a performance advantage with a drop-in substitution in the feedstock is not easy. Production of higher purity grade products with consistent quality appears to be more complex with bio-based raw materials. Therefore, the most promising area for polymers from bio-based feedstock seems to be quadrant 2, higher performance polymers from new renewables. This is true for polyurethane chemicals and particularly important for bio-based polyester polyurethanes. Therefore, in addition to some common monomers such as azelaic acid, much research is now devoted to the development of commercially viable routes for new monomers such as furan dicarboxylic acid and derivatives from algae and other bio-based feedstock.

One should also consider the concept of biodegradation in the cost/performance balance. However, biodegradability is desired only for specific applications that are addressed in the application outlook section later. In most applications, the consumers still expect the product to retain all desired properties associated with nonbiodegradable counterparts. Whether they are willing to pay a premium for biodegradability is still questionable.

Status and future

The state of technology for polyurethanes from algae feedstock

Among the several routes discussed for making polyurethanes from algae-based chemicals, the one with the highest commercial merit seems to be azelaic acid. Although it has been commercially available for many years, it competes with other similar monomers such as sebacic acid, adipic acid, and succinic acid, which are also available commercially at lower prices. The value proposition for all of these diacid monomers so far is their ability to provide polyester polyols leading to polyurethane products with good physical and mechanical properties, with the added advantage of biodegradability, which is desired in some applications. So far, no distinct performance advantage is reported for polyester polyols made of azelaic acid over their counterparts to justify the premium price. Therefore the use of this algae-based chemical will remain limited until the production volume of algae feedstock grows to the point of making it more cost-competitive.

A similar scenario is true for the alginate route and sugar-based products. That is why the focus on new monomers and polymers, which may not be readily available from other renewable resources and petroleum-based chemicals, becomes very important. Surprisingly, the fact that algae oil offers the highest degree of unsaturation in the backbone of triglycerides compared to other plant-based oils (see Table 8.1) has not been exploited commercially. This makes the epoxidation and functionalization route more suitable for incorporation of algae oil in polyurethane products. A similar product, epoxidized soybean oil, has been available in large volume at low price for many years because of its primary use as a plasticizer by the PVC industry. Soy-based polyols from this route are now commercial and have found applications in urethane foams and a few other products.

Application outlook

Polyurethane coatings[48] and adhesives[49] have many applications, from construction to automotive industries to medical and electronics, furniture, appliances, and industrial equipment, both indoor and outdoor. For coatings, automotive topcoat, industrial machinery coating, wood floor finish, furniture and cabinetry, and synthetic leather are among the major users. For urethane adhesives, textile lamination and carpet underlay are traditional applications. The packaging industry uses polyurethane adhesives in all shapes and forms for laminating film to film, film to foil, or film to paper. Most polyurethane adhesives for packaging meet US Food and Drug Administration approval for food packaging. Urethane adhesives are used in composite sheet molding compounds and bulletproof glass in the automotive and transportation market. In furniture and shoes, polyurethane adhesives hold a significant share of the market.

The interest in using renewable resources or biodegradable/compostable polymers in coatings and adhesive is not the same for all market segments. Every market and every specific application has its list of must-haves and like-to-haves. Several big industrial consumers such as Walmart and Procter & Gamble have included plastics from renewable resources in their sustainability initiatives. However, customers are often unwilling to sacrifice any performance in the products they buy or pay any premium for switching from petroleum-based raw materials to more environment-friendly alternatives. Therefore, to win the market, the renewable feedstock will have to offer better performance or equal performance at the same price as their petroleum-based counterparts. Among renewable sourced feedstock, algae-based chemicals present a great potential to meet this requirement.

However, biodegradability is very much application driven and does not follow the same rules as renewable resources. The ever-increasing waste management problems have created a strong push for several packaging industry sectors to demand biodegradable materials and willingness to pay a premium for them. This is true for coatings such as those in disposable coffee cups and film and sheet products, in adhesives used in food packaging, and in many other disposable products. Polyurethane manufacturers are active in developing polymers from renewable raw materials and addressing biodegradability issues for selected applications. Polyester polyurethanes are a class of polymers that offer an excellent route to address both ecological concerns of consumers and waste management problems. Therefore it is not unreasonable to expect a higher growth rate for coatings and adhesive products based on polyester polyurethanes for many years to come.

Close-Up: Biodegradable minor components in packaging

Packaging plays an essential role in safely delivering food and beverage products to customers. Packaging materials are designed to balance several critical criteria, including compliance with food safety regulations, maintenance of freshness and quality, environmental sustainability, affordability, and consumer preferences, including convenience.

From a sustainability point of view, many countries and companies have lofty goals to reduce their carbon footprint as well as design packaging to be recyclable, compostable, or reusable by 2030. Packaging contributes significantly to greenhouse gas emissions, and finding renewable alternatives would lower the carbon footprint significantly.

Flexible plastic packaging is probably one of the most sustainable formats of single-use packaging in terms of low material use. Flexible bags are multilayer and multimaterial to get a balance of properties such as barrier provision and puncture resistance, which makes recycling a challenge. The current recycling infrastructure is not equipped to handle these packages. Development of bio-based and biodegradable materials for alternative packaging will ensure managed disposal along with organic waste to composters and will reduce greenhouse gas emissions considerably.

Thin films are made from bio-based materials that are also biodegradable, but a few minor nondegradable components such as coatings, inks, and adhesives are needed to provide functionalities such as structural integrity, sealing, and graphics. Today, we are allowed to use small quantities of these components (<1% of the total packaging weight) in certified compostable packaging, but these limits will become more stringent by 2025, so we must look for biodegradable alternatives. Polyurethane adhesives and coatings based on algae polyols would satisfy both the bio-based and biodegradability conditions. Although the inks used in compostable packages are tested for toxicity and heavy metals, it would be great to get them formulated in compostable resin carriers.

Sridevi Narayan-Sarathy, Ph.D.
PepsiCo, Plano Texas, United States

References

1. Pliny E. The Natural History. Vol Book XIII; 2022 [Chapter 17].
2. Hunter D. Papermaking. Dover Publications; 1947.
3. Averill Chemical Paint Company. Seeley & Stevens; 1890.
4. Szycher M. In: Szycher M, ed. *Szycher's Handbook of Polyurethanes*. 2nd ed. Taylor & Francis Group; 2012.
5. Gandini A, Lacerda TM. From monomers to polymers from renewable resources. *Prog Polym Sci*. 2015;48.
6. Mucci VL, Hormaiztegui MEV, Aranguren MI. Plant oil-based waterborne polyurethanes: a brief review. *J Renew Mater*. 2020;8(6). https://doi.org/10.32604/jrm.2020.09455.
7. Vijayendran B. Bioproducts from biorefineries—trends, challenges and opportunities. *J Bus Chem*. 2010;7(3).
8. Çelebi M, Yazici T. Synthesis and characterization of bio-based polyester polyol. *J Turkish Chem Soc Sect A Chem*. 2016;3(3). https://doi.org/10.18596/jotcsa.287306.
9. Alinejad M, Henry C, Nikafshar S, et al. Lignin-based polyurethanes: opportunities for bio-based foams, elastomers, coatings and adhesives. *Polymers*. 2019;11(7). https://doi.org/10.3390/polym11071202.
10. Klein SE, Alzagameem A, Rumpf J, Korte I, Kreyenschmidt J, Schulze M. Antimicrobial activity of lignin-derived polyurethane coatings prepared from unmodified and demethylated lignins. *Coatings*. 2019;9(8). https://doi.org/10.3390/coatings9080494.
11. Petrović ZS. Polymers from biological oils. *Contemp Mater*. 2010;1:1. https://doi.org/10.5767/anurs.cmat.100101.en.039p.
12. Phung Hai TA, Neelakantan N, Tessman M, et al. Flexible polyurethanes, renewable fuels, and flavorings from a microalgae oil waste stream. *Green Chem*. 2020;22(10). https://doi.org/10.1039/d0gc00852d.
13. Mohd Noor N, Sendijarevic A, Sendijarevic V, et al. Comparison of adipic versus renewable azelaic acid polyester polyols as building blocks in soft thermoplastic polyurethanes. *JAOCS J Am Oil Chem Soc*. 2016;93(11). https://doi.org/10.1007/s11746-016-2903-9.
14. Hojabri L, Jose J, Leao AL, Bouzidi L, Narine SS. Synthesis and physical properties of lipid-based poly(ester-urethane)s, I: effect of varying polyester segment length. *Polymer*. 2012;53(17). https://doi.org/10.1016/j.polymer.2012.06.011.

15. Tuan Ismail TNM, Ibrahim NA, Sendijarevic V, et al. Thermal and mechanical properties of thermoplastic urethanes made from crystalline and amorphous azelate polyols. *J Appl Polym Sci.* 2019;136(34). https://doi.org/10.1002/app.47890.
16. Tuan Ismail TNM, Ibrahim NA, Poo Palam KD, et al. Improved dynamic properties of thermoplastic polyurethanes made from co-monomeric polyester polyol soft segments based on azelaic acid. *J Appl Polym Sci.* 2021. https://doi.org/10.1002/app.50815. Published online.
17. Awasthi S, Agarwal D. The effect of difunctional acids on the performance properties of polyurethane coatings. *J Coatings Technol Res.* 2009;6(3). https://doi.org/10.1007/s11998-008-9121-9.
18. Awasthi S, Agarwal D. Preparation and characterisation of polyurethane coatings based on polyester polyol. *Pigm Resin Technol.* 2010;39(4). https://doi.org/10.1108/03699421011055518.
19. Patil CK, Jirimali HD, Paradeshi JS, et al. Chemical transformation of renewable algae oil to polyetherimide polyols for polyurethane coatings. *Prog Org Coat.* 2021;151. https://doi.org/10.1016/j.porgcoat.2020.106084.
20. Patil CK, Jirimali HD, Paradeshi JS, Chaudhari BL, Gite VV. Functional antimicrobial and anticorrosive polyurethane composite coatings from algae oil and silver doped eggshell hydroxyapatite for sustainable development. *Prog Org Coat.* 2019;128. https://doi.org/10.1016/j.porgcoat.2018.11.002.
21. Patil CK, Jirimali HD, Paradeshi JS, et al. Synthesis of biobased polyols using algae oil for multifunctional polyurethane coatings. *Green Mater.* 2018;6(4). https://doi.org/10.1680/jgrma.18.00046.
22. Kadam A, Pawar M, Thamke V, Yemul O. Polyester amide based polyurethane coatings from algae oil and their larvicidal, anti-ant properties. *Prog Org Coat.* 2017;107. https://doi.org/10.1016/j.porgcoat.2017.03.013.
23. Zhao H, Herrington R, Rodriguez F. US20110118432A1. Published online; 2011.
24. Pajerski A, Lerner S. Aqueous Polymer Compositions from Epoxidized Natural Oils. US Pat 8,658,889. Published online; 2013.
25. Nurdin NS, Salah S, Lim AT, Francis AY, Abdullah L, Saifulazry S. Effect of DMPA content on colloidal stability of jatropha oil-based waterborne polyurethane dispersion. *IOP Conf Ser Mater Sci Eng.* 2020;778. https://doi.org/10.1088/1757-899X/778/1/012107.
26. Pourahmady N. Growing performance. *Eur Coat J.* 2016;4.
27. Salah S, Abdullah LC, Aung MM, et al. Chemical and thermo-mechanical properties of waterborne polyurethane dispersion derived from jatropha oil. *Polymers.* 2021;13(5). https://doi.org/10.3390/polym13050795.
28. Wilkes G, Sohn S, Tamami B. US Patent 7,045,577. Published online; 2006.
29. Mahendran AR, Aust N, Wuzella G, Müller U, Kandelbauer A. Bio-based non-isocyanate urethane derived from plant oil. *J Polym Environ.* 2012;20(4). https://doi.org/10.1007/s10924-012-0491-9.
30. Simons JB. US20060182957A1. Published online; 2006.
31. Parreidt TS, Müller K, Schmid M. Alginate-based edible films and coatings for food packaging applications. *Foods.* 2018;7(10). https://doi.org/10.3390/foods7100170.
32. Rastegar S, Hassanzadeh Khankahdani H, Rahimzadeh M. Effectiveness of alginate coating on antioxidant enzymes and biochemical changes during storage of mango fruit. *J Food Biochem.* 2019;43(11). https://doi.org/10.1111/jfbc.12990.
33. Medina-Jaramillo C, Quintero-Pimiento C, Díaz-Díaz D, Goyanes S, López-Córdoba A. Improvement of andean blueberries postharvest preservation using carvacrol/alginate-edible coatings. *Polymers.* 2020;12(10). https://doi.org/10.3390/polym12102352.
34. Iijima S, Mano T, Taniguchi M, Kobayashi T. Immobilization of hybridoma cells with alginate and urethane polymer and improved monoclonal antibody production. *Appl Microbiol Biotechnol.* 1988;28(6). https://doi.org/10.1007/BF00250414.

35. Martín M, Grossmann IE. Optimal production of furfural and DMF from algae and switchgrass. *Ind Eng Chem Res.* 2016;55(12). https://doi.org/10.1021/acs.iecr.5b03038.
36. Warlin N, Garcia Gonzalez MN, Mankar S, et al. A rigid spirocyclic diol from fructose-based 5-hydroxymethylfurfural: synthesis, life-cycle assessment, and polymerization for renewable polyesters and poly(urethane-urea)s. *Green Chem.* 2019;21(24). https://doi.org/10.1039/c9gc03055g.
37. Xi X, Pizzi A, Delmotte L. Isocyanate-free polyurethane coatings and adhesives from mono- and di-saccharides. *Polymers.* 2018;10(4). https://doi.org/10.3390/polym10040402.
38. Golling FE, Pires R, Hecking A, et al. Polyurethanes for coatings and adhesives—chemistry and applications. *Polym Int.* 2019;68(5). https://doi.org/10.1002/pi.5665.
39. Zhenova A. Challenges in the development of new green solvents for polymer dissolution. *Polym Int.* 2020;69(10). https://doi.org/10.1002/pi.6072.
40. Aguirre M, Hamzehlou S, González E, Leiza JR. Renewable feedstocks in emulsion polymerization: coating and adhesive applications. *Adv Chem Eng.* 2020;56. https://doi.org/10.1016/bs.ache.2020.07.004.
41. Dieterich D, Keberle W, Witt H. Waterborne polyurethanes. *Angew Chem Internat Ed.* 1970;9:40–50.
42. Honarkar H. Waterborne polyurethanes: a review. *J Dispers Sci Technol.* 2018;39(4). https://doi.org/10.1080/01932691.2017.1327818.
43. Tillet G, Boutevin B, Ameduri B. Chemical reactions of polymer crosslinking and post-crosslinking at room and medium temperature. *Prog Polym Sci.* 2011;36(2). https://doi.org/10.1016/j.progpolymsci.2010.08.003.
44. Weiss KD. Paint and coatings: a mature industry in transition. *Prog Polym Sci.* 1997;22(2). https://doi.org/10.1016/S0079-6700(96)00019-6.
45. Tenorio-Alfonso A, Sanchez MC, Franco JM. A review of the sustainable approaches in the production of bio-based polyurethanes. *J Polym Environ.* 2020;28:749–774.
46. Heinrich LA. Future opportunities for biobased adhesives. *Green Chem.* 2019;21:1866–1888.
47. Bellido-Aguilar DA, Zheng S, Huang Y, Zeng X, Zhang Q, Chen Z. Solvent-free synthesis and hydrophobization of biobased epoxy coatings for anti-icing and anticorrosion applications. *ACS Sustain Chem Eng.* 2019;7(23). https://doi.org/10.1021/acssuschemeng.9b05091.
48. Avar G, Meier-Westhues U, Casselmann H, Achten D. Polyurethanes. In: *Polymer Science: A Comprehensive Reference, 10 Volume Set.* vol 10; 2012. https://doi.org/10.1016/B978-0-444-53349-4.00275-2.
49. Lay DG, Cranley P, Pizzi A. Handbook of Adhesive Technology. 3rd ed. CRC Press; 2017. https://doi.org/10.1201/9781315120942.

CHAPTER

9

Biodegradable biocomposites

Robert S Pomeroy

University of California San Diego, La Jolla, CA, United States

Introduction

There are entire textbooks[1–3] and excellent reviews[4–6] on composites and biocomposites.[7–12] The goal of this chapter is to address these materials within the context of an algae-based renewable, biodegradable material. While glass and carbon fibers are chemically inert with a substantial formulation history,[10,11] they have environmental drawbacks with respect to the energy needed in their production[12] and the emission of volatile organic compounds (VOCs).[13] Here we provide a brief introduction to composites and biocomposites and the components of which they are made (Fig. 9.1). Then we discuss the advantages and concerns associated with biocomposites.

Composites consist of two components, which can be natural or artificial. By combining them, each with different physical or chemical properties, a synergistic interaction occurs, resulting in materials that possess superior characteristics neither achieves on its own. The component materials do not dissolve in one another. They blend and do not lose their identities. Typically, the composite has added strength, efficiency, or durability. The lightweight and insulation capabilities for transportation can yield significant savings of fossil fuels that partly offset their environmental impact. Biocomposite research needs to move toward new polymers derived from renewable sources and environmentally friendly fillers that permit safer and more sustainable production of plastics.

Manufactured composites date back over 5000 years to the Sumerians of Mesopotamia.[14,15] About 3400 BCE, they made a version of plywood. The Egyptians created an analog to papier-mâché as a composite of linen soaked in plaster. Sumerians and Egyptians are credited with adding straw to strengthen mud bricks and pottery.

Rethinking Polyester Polyurethanes
https://doi.org/10.1016/B978-0-323-99982-3.00006-7

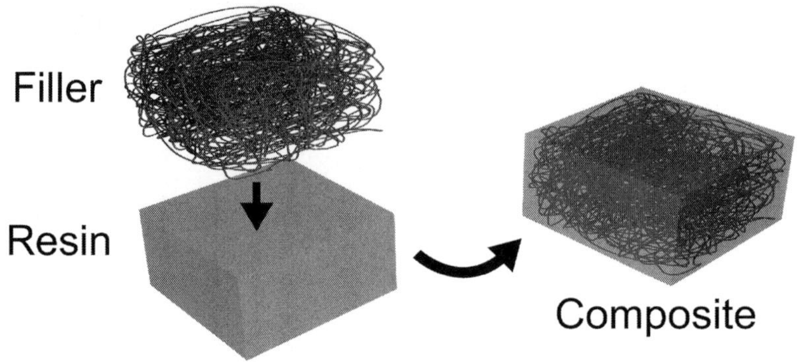

FIG. 9.1 Composite materials: the matrix+the fiber or filler = the composite.

Compositional advances have played a significant role in shaping historical events. From the start of the chemical revolution in the late 19th century, composites of earth materials were replaced by the creation of synthetic resins. Early chemists made solid materials by crosslinking lower molecular weight liquid materials through polymerization. In the early 20th century, they synthesized new polymers such as vinyl, polystyrene, phenolic, and polyester. Many of today's composites have their roots in fiber-reinforced polymers (FRPs) developed initially by Corning in 1935, who combined glass fibers with synthetic resins to make materials with superior strength and rigidity. From this point on, there were many advances on the chemical side, developing new unsaturated polyesters and epoxies, and composites with superior strength-to-weight ratios. Given that these materials were also nonconductive of heat and electricity and transparent to radio frequencies, many found their way into applications after WWII.

The automotive and aerospace industries are drivers of composite innovation. The 1953 Corvette used preformed fiberglass combined with resin and molded in matched metal dies to create the car's body. In addition to the materials, the manufacturing method plays a role in new developments. The most common is compression molding, which can be divided into sheet molding and bulk molding. Other methods are pultrusion, vacuum bag molding, and large-scale filament winding.

The introduction of carbon fiber was a significant change, greatly enhancing strength-to-weight ratios and creating superior materials for the aerospace, marine, automotive, and high-performance sporting goods industries. In the 1970s, ultra-high molecular weight polyethylene pushed the technology to new heights with products that became part of everyday life. By the mid-2000s, the development cycle of fibers, resins, and methodology resulted in high-strength and rigid composites, expanding the market to appliances and other consumer goods.

Composites research attracts grants from governments, manufacturers, and universities. These investments accelerate innovation. Specialized companies can find a place in the industry. Two applications that continue to experience innovative growth are airplane materials and composite sheets for marine use. Other materials, such as environmentally friendly resins incorporating recycled plastics and bio-based polymers, meet the demand for more robust, lighter, environmentally friendly products. In the future, still-to-be-developed fibers and resins will create even more applications for everyday and specialized use.

Biocomposites: Bio-based biodegradable fiber-reinforced polymers

The resin matrix

Biocomposites[16–18] are made from at least one bio-sourced component—matrix, filler, or both. FRPs are fibers embedded in a resin matrix. Bio-based means the resin matrix or the fiber is from biologically sourced material. For our purposes, bio-based also needs to be biodegradable. Resins, even if bio-based, that are polyethylene, polypropylene, polyvinyl chloride, polystyrene, ethylene vinyl acetate, and epoxy are excluded from this discussion since they do not wholly or readily biodegrade. Polyester-based polyols, created with bio-based diols and diacids, omega hydroxy acids, or lactic acid, will biodegrade. Resins used in FRP composites are either thermoset or thermoplastic.

Most composites are made with thermoset resins. They are converted from liquid to solid through polymerization or crosslinking. When finished goods are produced, thermosetting resins are "cured" using a catalyst, heat, or a combination of the two. Once cured, solid thermoset resins cannot be converted back to their original liquid form. Polyesters and polyurethanes (PUs) are common thermoset plastics. The fiber material can be either bio-based like cellulose or chemically inert like silica. Carbon fiber is excluded from this discussion because of its cost and the difficulty of recycling, often resulting in landfill disposal.

Thermoplastic resins are not crosslinked and can be melted, formed, remelted, and re-formed. This desirable property comes with the price of using monomers such as acrylonitrile butadiene styrene, polyethylene, polystyrene, and polycarbonate, which are not biodegradable.

Material scientists can tailor the properties by choosing the matrix resin, the fiber, its orientation, and the manufacturing method to meet the desired specifications. The choice of resin selects for properties such as resistance to heat, chemicals, and weathering by choosing an appropriate matrix material. By aligning fibers in a particular orientation, one can create a composite that is stronger in one dimension than in another. This combination of plastic and reinforcement fiber produces some of the most robust and versatile materials (for their weight) ever developed. Due to its strong yet flexible properties, FRP can replace materials such as wood, aluminum, granite, and even steel.

Nonisocyanate polyurethane (NIPU)

Nonisocyanate PU applications have won the Presidential Green Chemistry challenge twice, in 2015[19] and 2021.[20] NIPU is a safer plant-based solution to PU resin by creating the urethane bond without the use of diisocyanates. The plant-based source removes petroleum, and the elimination of the diisocyante chemistry removes concern about the toxicity of isocyanates. Diisocyantes are designated carcinogenic, mutagenic, and harmful to reproduction. The main threat to factory workers from isocyanates is respiratory or allergic reactions. As potent sensitizers, they are the leading cause of occupational asthma.

NIPU was initially described in 1957 by Dyer and Scott.[21] They were exploring a means of making PUs that eliminated the use of the moisture-sensitive isocyanates. They reacted amines with cyclic carbonates to form the urethane group. Due to its 100% atom economy, this process remains the preferred route to making NIPUs. Until recently, the research was pretty limited. Advances in synthesizing cyclic carbonates using the chemical insertion of CO_2 into epoxy resins have spurred increased interest because of the growing market demand for PUs with outstanding performance and sustainability.[22–26]

NIPUs are formed through the polyaddition of cyclic carbonates and polyfunctional amines (Fig. 9.2). NIPUs eliminate diisocyanates and lead to materials with unique and tunable properties. They exhibit better chemical, mechanical, and thermal resistance, and the process is invulnerable to moisture, an essential technological feature.

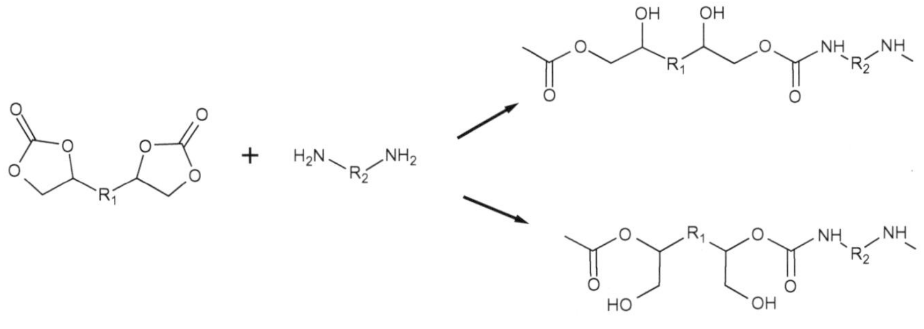

FIG. 9.2 NIPU chemistry: the reaction of bis(cyclic carbonate)s and diamines.

Based on the similarities in resulting urethane bond formation, NIPUs might appear to be direct drop-ins for conventional PUs. However, this is not the case, especially in manufacturing practice. As a result, the market size of NIPUs remains negligible. Conventional PUs can be formulated to create an array of commercial products, from thermoplastics to elastomeric materials (phase-separated PUs) to thermoset plastics. Moreover, a significant portion of PU production is in manufacturing foams. Conventional PUs using isocyanate-based chemistry are self-blowing, which is not the case for NIPUs. Perhaps this is a factor in limiting NIPU's industrial acceptance.

Their translation to commercialization has been slowed by the inability to synthesize activated carbonates with the desired molecular weights at industrial scale. Recent developments have achieved high molar masses of NIPU polymers when large cyclic carbonates are employed. Formulating these carbonates with suitable chain extenders creates phase-separated materials and allows the preparation of self-blowing foams. Improvements in the commercial viability of a more sustainable and greener PU while reducing the inherent risk of the isocyanate-based chemistry create new opportunities for these materials.

A solvent-free and catalyst-free method for preparing PUs without harmful isocyanate has been reported.[27] Sebacic bis(cyclocarbonate) and bio-based dimer diamines were used to make the NIPU. The NIPUs are amorphous and have a glass transition temperature suitable for several applications. Thermal and rheological analyses of a series of NIPUs using various amine functionalities of bio-based dimer diamines created tunable structure-property relationships of the NIPU. Adjusting the degree of crosslinking of the chains changes the elasticity. The NIPU formulations with the highest crosslinking displayed superior elasticity.

NIPU as a polymer matrix for nanocomposites

NIPUs show promising possibilities through the simple manipulation of the side chain chemistry. These modifications will affect mechanical and thermal properties, expanding applications in biocomposite formulation. Fillers may be introduced as blends so that only physical interactions occur between them and the matrix. Although this method ensures substantial improvements in mechanical and thermal properties of final products, the possibility of chemically binding the fillers used with NIPU matrices plays a significant role in nonisocyanate PU composites. Furthermore, organic and inorganic fillers modified prior to their introduction into the NIPU side chain modifications, such as adding carbonate, amine, or hydroxyl groups, will open new opportunities in combination with chemically bound fillers or nanofillers. This chemistry will expand the achievable properties of NIPU composites and open a wide range of applications. NIPU nanocomposite materials are a significant area of research activity. They show improved properties when adding the nanofiller at low levels, amounts of 1 to 5%, compared to traditional microstructured composites that require higher loading levels.

Fillers and fibers

Fig. 9.3 illustrates the construction of a composite.

Inert fillers

A filler is a substance added to fill space. They are typically particulates, where dimensions of three or two of the geometric axes are of a similar order of magnitude, approximately spherical. The addition of the filler often affects the bulk properties and increases the resulting material's density, viscosity, or strength. Fillers can account for 40 to 65% by weight in the finished composite. The use of fillers reduces the cost of composites and imparts performance improvements that might not otherwise be achieved by the resin or reinforcement (fiber) alone. Compared to the resins and reinforcements, fillers are the least expensive of the major components. As such, they are often referred to as extenders. Fillers improve mechanical properties and reduce flammability by simply reducing the amount of resin present. Adding fillers results in less consistent dimensional control of molded parts. Other properties are water resistance, weathering, surface smoothness, stiffness, dimensional stability, and temperature resistance. The addition of fillers has a significant impact on the price. Typical inorganic fillers are chalk (calcium carbonate), clay (kaolin), gypsum (calcium sulfate), alumina trihydrate (gibbsite), and silica. The chemical compatibility of the filler and the resin and how well they mix and compact affect the permeability. If the filler is compatible, it will adhere well to the resin. The resulting composite's polymer matrix will be impermeable. If less compatible, voids will appear, allowing other substances to permeate the polymer. The formation of voids is an essential concern if one is designing polymers to be barriers or to allow specific molecules to pass through. The formation of voids in the fiber-reinforced

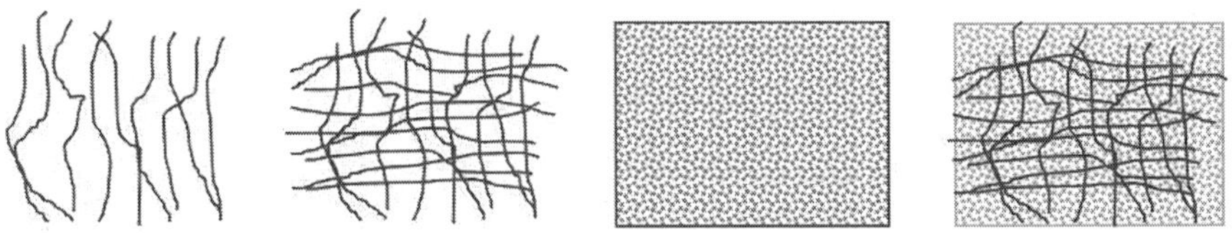

FIG. 9.3 Steps in construction of a composite, from left to right: fibers aligned, fibers matted, filler, fibers and filler.

polymer composite is the main factor that affects its performance. For structural strength, the percentage of voids should be kept to a minimum.

Fibers

Fibers have one dimension that is several orders of magnitude larger than the other two dimensions. This difference affects properties like strength, stiffness, and impact (energy required to initiate the onset of damage to material failure). The addition of fiber is not so much about acting as a filler and lowering the price; instead, its presence increases strength and durability. The strength of a fiber-reinforced plastic is much greater than that of the base resin. The reinforcement comes from fibers, fabrics, or mats made of fibers. Plastic laminate is the most common and most robust type of reinforced plastic.

Natural fibers

Recently, plant-derived sustainable materials such as cellulose nanocrystals, natural fibers, lignin, biochar, and polysaccharides have been extensively investigated as substitutes for glass fibers and fillers such as carbon black. Natural fibers have a wide array of physical and chemical properties and can also be modified. Using natural fibers allows the polymer chemist to tune the composite's properties to meet the application's required specifications. Standard natural fibers are cellulose, hemicelluloses, pectin, and lignin. Natural fibers have differing levels of these components, and the ratios depend on the types of fibers and variability caused by growing and harvesting conditions. Cellulose and hemicellulose are similar. Cellulose is a semicrystalline polysaccharide and imparts the hydrophilic nature of natural fibers. Hemicellulose, by comparison, has shorter chain lengths and lower molecular weight and is an amorphous polysaccharide. The amorphous nature of hemicellulose results in partial solubility in water and alkaline solutions. Pectin is also a polysaccharide and holds fibers together. Lignin is an amorphous polymer with a higher content of aromatic functionalities. As a result, it has less hydrophilicity and provides structural integrity.

Natural fibers are lighter, cheaper, and environmentally smarter substitutes for glass. The choice to use them for composites starts with the production of natural fibers, reducing environmental impact. Many are lighter in weight, leading to materials with lower density. In the transportation sector, reducing weight increases gas mileage. And natural fibers can be broken down in the environment aerobically and anaerobically.

Natural fiber composites are realistic alternatives to glass-reinforced composites in many applications. In some cases, glass is superior; however, with proper formulation and manufacturing, many believe that equivalent materials with lower cost and reduced environmental impact are achievable. For example, hemp fiber-epoxy, flax fiber, and China reed fiber have been used and compared to glass in automotive applications.

The most significant advantages of natural fibers are cost of materials, sustainability, and density. They are renewable on an 8-month timescale, are sustainable, and cost approximately \$0.25 a pound. Glass fibers cost \$0.55 to \$0.90 a pound. The density of natural fibers ranges between 1.15 and 1.50 g/cm^3 compared to 2.4 g/cm^3 for glass, making natural fibers on average 55% lighter. They are used primarily to reinforce thermoset resins; however, they are moving to use in thermoplastics, as it makes them potentially recyclable. Altogether it would appear that natural fiber FRP results in less pollution, including greenhouse gases, and biodegradability, combined with a reduction in overall fossil fuel energy consumption. Synthetic fibers' strength is much higher than that of natural fibers. The loss of strength limits the replacement of synthetic fibers with natural fibers for environmental protection. A complete life cycle assessment has not been published.

Natural fiber feedstocks for fiber-reinforced composites can often be sourced from wood or agricultural materials. This makes them advantageous, as they are essentially waste streams from other activities. The utilization of these waste streams, wood flour from sawdust or fibers from plant stalks such as jute and hemp, helps to valorize the activity and reduces waste. The fibers need to be dried and sized prior to use in composites. As a repurposed waste stream, these materials are low in cost and provide additional economic activity.

Biocomposites

Biocomposite sales are expected to reach \$41 billion globally by 2025.[28] Automotive, 3D printing, packaging, and biomedical top the list of emerging applications. Nanoscience and nanotechnology introduce exciting research areas by exploiting renewable nanomaterials as reinforcements to advanced materials with innovative properties that conventional materials cannot meet. Nanoscience is material science at the submicron scale. For biocomposites, this means using nano-sized filler materials such as nanoclays and nanocellulose. Research exploring the physical, mechanical, flame, and gas barrier properties of polymers made with nanomaterials is underway. The potential of natural fiber nanocomposites is to overcome the reduction in mechanical properties typically experienced with natural fibers and to eliminate the use of the traditional fibers such as glass, carbon nanotubes, carbon black, silica, and metal oxides. This is expected to result in high-strength, nontoxic, biodegradable, low-density, high-specific-strength composites from abundant low-cost materials.

Incorporating renewable nanomaterials into biocomposites enhances bio-based resins such as polylactic acid's physical, mechanical, chemical, and barrier properties. Nanomaterial-reinforced polymer nanocomposites have applications from packaging to medical and high-performance engineered materials, as the nanoparticle filler provides performance associated with micron-sized glass and carbon additives.

Biocomposite recycling

Composite materials will be subject to the four Rs, like any other resource on the planet. We can no longer dispose of objects and the energy invested in them to landfills; the investment is too great. They will need to be reused, repurposed, or recycled, or have the energy recovered. The challenge, then, is to design and create high-quality materials that can be recycled cost-effectively in an environmentally responsible way. Recycling requires clean, chemically uniform waste streams. Examples do exist. Fibrex is a wood polyvinyl chloride composite developed by Anderson Windows and used to construct new window frames. This example works because windows frames are not routinely replaced for long periods. The problem with most recycling methodologies is that the resulting material is inferior to the original. This loss of critical properties is termed downcycling.

Over time and recycling, the physical and chemical properties change, and not for the better. After recycling multiple times, the tensile strength and tensile modulus of natural fiber biocomposites are degraded and significantly reduced. Exposure to ultraviolet light (UV) can change the chemical composition, so even with the most efficient recycling, downcycling is likely to result. Lastly, recycling involves a heating step to soften and melt the polymer. Natural fibers will not remain unaltered through a temperature excursion. The fibers will become more brittle, affecting their flexibility and impact. The chemistry of the new material will result in changes to the glass transition temperature, hygroscopicity, viscosity, rheology, and dimensional stability of the resin matrix. Changes will also occur in the fibers.

Could these outcomes be mitigated? Could additives slow the impact of the recycling process? An active area of investigation is functional additives, compounds that are added in the manufacturing process to enhance product characteristics or suppress undesirable ones. These include coupling reagents, plasticizers, chain extenders, and molecules to provide UV protection.[29–31]

Coupling agents increase the miscibility of the typically hydrophilic fibers to mix more uniformly with the hydrophobic matrix resin. The surface chemistry of the fibers can be changed by washing them in NaOH, alcohol, or benzene. Another approach is to modify the surface via reactions with free hydroxyl groups to esterify, silanize, or form amide and urethane (carbamate) linkages. Instead of modifying the fibers, a plasticizer can be added to lower the melting temperature of the resin, reducing changes to the fibers. Chain extenders can be added to increase the matrix resin's molecular weight, usually reduced by UV exposure, weathering, or heating. Additives that act like sunscreen sacrificially protect the resin's molecular bonds from UV damage and slow the rate of downcycling.

The properties that make composites and biocomposites valuable—their strength-to-weight ratio and durability—are the features that make downcycling unavoidable. The future challenge is to determine the best end-of-life outcome for these unique materials.

Biocomposite biodegradation

Bio-based polymers must be distinguished from biodegradable polymers. Biodegradables can be made from fossil resources, while some plastics made from biomass are nonbiodegradable. It is important to delineate the specific conditions and the time frame under which a "biodegradable" polymer degrades. Polylactic acid is a plastic often advertised as biodegradable, but this is true only if composted in industrial units, and they will most probably have limited biodegradation in a landfill. Biodegradation was covered in Chapter 4, and the instrumentation and methodology were covered in Chapter 5.

Let us define green composites as FRPs based on a biodegradable biopolymer matrix and natural fillers. Natural fillers lower the thermal stability of the resin. Therefore their use is limited to plastic materials with low melting temperatures. In fiber-reinforced composites, the matrix protects the fibers from external environmental damaging factors and transmits the externally applied loads to the reinforcement fibers, which can absorb the energy stress without deterioration of the material.

Two approaches are usually considered in designing environmentally degradable polymers to reduce the global problem of inert polymer waste. One is to design with inherent biodegradability, and another is to enhance the biodegradation of recalcitrant petroleum-based polymers by modifying them (e.g., in blends or composites) with degradable components, usually bio-based ones. Examples of biodegradable polymeric materials are starch, chitosan, chitin, cellulose, lignin, polylactic acid, poly(3-hydroxybutyrate), poly(3-hydroxybutyrate-3-hydroxyvalerate), poly (butyrate adipate-*co*-terephthalate), and poly-t-caprolactone.[32–38]

Biodegradability of polymer composites with natural fibers and inert fillers

The biodegradation rate of composite materials depends on several factors associated with the material: the presence of functional group microbes that can attack; surface area; permeability; and environmental conditions such as temperature, humidity, nutrients, pH, and microbial consortium. A rough surface with many hydrophilic functional groups biodegrades more readily than a smooth, hydrophobic surface. In principle, the choice of natural fibers and resin allows the chemist to tune the physical and biodegradation properties. Since the interface acts as an access point, controlling it dictates the physical, chemical, and biodegradation characteristics of the resulting material. Leveraging the interfacial interactions opens the door to creating products where the lifetime of the product is more closely matched to the lifetime of the material. The degradation of plastic composites under the action of aggressive environmental factors combined with cataloging the microbial diversity is the future trend of materials research.

Composite materials are energy intensive and difficult to recycle. The lack of clean, uniform waste streams and an efficient process for separating and recovering the resin from the fiber or filler limits economical recycling, given the cost of virgin materials. Biodegradation offers an environmental advantage, because a biocomposite will not last forever; microbes mineralize the components, which makes them available to be reabsorbed. This is what happens in aerobic degradation. Incorporation of an anaerobic digestion step can help recover some of the captured energy (Fig. 9.4). Anaerobic digestion takes place in the absence of oxygen. The result is that anaerobic microbes convert the carbon into methane and carbon dioxide, not solely carbon dioxide as in aerobic digestion. This methane, also called biogas, can be extracted and its energy content can be recovered through combustion.[39,40]

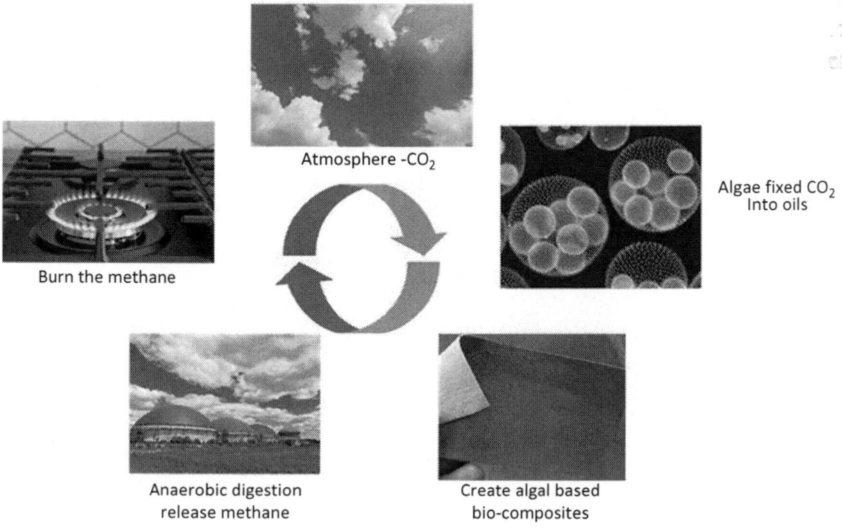

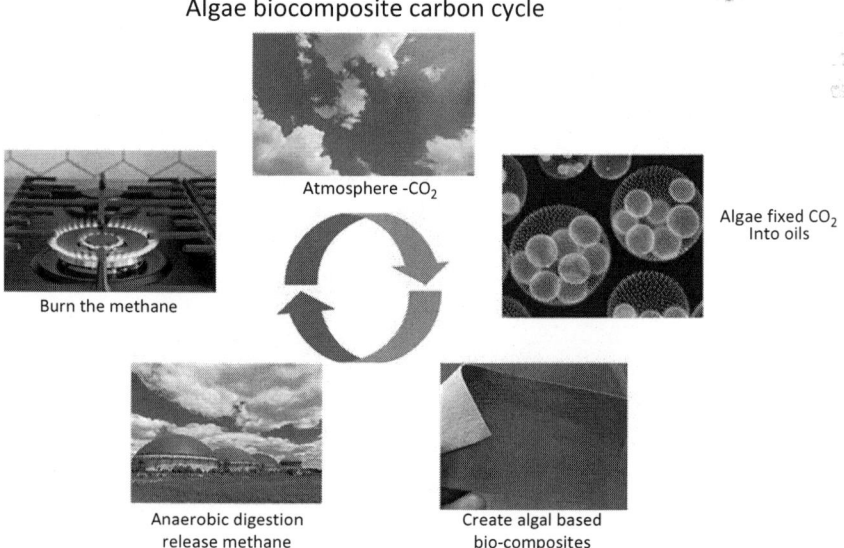

FIG. 9.4 The algae-based carbon cycle of an anaerobically biodegraded biocomposite.

Challenges to biocomposites

There are two factors limiting the large-scale adoption of natural fiber composites.[41–45] One is their lower strength compared to glass. The loss is often a result of the incompatibility between the fiber and the resin matrix. The other is their water absorption. It increases the weight, alters the surface, and creates voids. Mitigation of these two issues is receiving significant research attention. Water absorption is associated with the hydroxyl groups on the cellulose chains, and chemically modifying these groups can reduce hydrophilicity. Several promising techniques have been investigated, including washing with NaOH, silanizing, and esterification.

Fundamentally, more research to elucidate the molecular structure and interfacial chemistry between the matrix and the fibers is needed. Understanding the relationship between the chemistry, structure, and properties in greater detail is necessary.

There will be more discoveries associated with green polymers. Biotechnology can create an array of chemicals that exceed the chemical diversity of current petroleum-based monomers. Functional additives are another avenue to further enhance the viability of biocomposites. What is certain is that sustainable, biodegradable biocomposites already exist to fill the needs of many applications. Future research will only expand them into materials with very stringent expectations.

References

1. Chawla KK. Composite Materials: Science and Engineering; 2019. https://doi.org/10.1007/978-3-030-28983-6.
2. Mallick PK. Fiber-Reinforced Composites: Materials, Manufacturing, and Design. 3rd ed; 2007.
3. Gay D, Hoa SV, Tsai SW. Composite Materials: Design and Applications; 2002.
4. Ferry JD. Viscoelastic Properties of Polymers; 1980. https://doi.org/10.1149/1.2428174.
5. Thostenson ET, Ren Z, Chou TW. Advances in the science and technology of carbon nanotubes and their composites: a review. *Compos Sci Technol.* 2001;61(13). https://doi.org/10.1016/S0266-3538(01)00094-X.
6. Coleman JN, Khan U, Blau WJ, Gun'ko YK. Small but strong: a review of the mechanical properties of carbon nanotube-polymer composites. *Carbon.* 2006;44(9). https://doi.org/10.1016/j.carbon.2006.02.038.
7. Faruk O, Bledzki AK, Fink HP, Sain M. Biocomposites reinforced with natural fibers: 2000-2010. *Prog Polym Sci.* 2012;37(11). https://doi.org/10.1016/j.progpolymsci.2012.04.003.
8. Mohanty AK, Misra M, Hinrichsen G. Biofibres, biodegradable polymers and biocomposites: an overview. *Macromol Mater Eng.* 2000;276–277. https://doi.org/10.1002/(SICI)1439-2054(20000301)276:1<1::AID-MAME1>3.0.CO;2-W.
9. John MJ, Thomas S. Biofibres and biocomposites. *Carbohydr Polym.* 2008;71:3. https://doi.org/10.1016/j.carbpol.2007.05.040.

10. Singh J, Kumar M, Kumar S, Mohapatra SK. Polymer-plastics technology and engineering properties of glass-fiber hybrid composites: a review properties of glass-fiber hybrid composites: a review. *Polym Plast Technol Eng.* 2017;56(5):455–469. https://doi.org/10.1080/03602559.2016.1233271.
11. Chand S. Review carbon fibers for composites. *J Mater Sci.* 2000;35:1303–1313.
12. Das S. Life cycle assessment of carbon fiber-reinforced polymer composites. In: *LCA For Energy Systems and Food Products*; 2011. https://doi.org/10.1007/s11367-011-0264-z.
13. Sands JM, Fink BK, Mcknight SH, et al. Environmental Issues for Polymer Matrix Composites and Structural Adhesives. vol. 2. Springer-Verlag; 2001.
14. Mar-Bal, Inc. History of Composite Materials. Mar-Bal, Inc; 2022. https://www.mar-bal.com/language/en/applications/history-of-composites/. Accessed 9 August 2022.
15. Caltech. History of Composites—Overview; 2022. https://authors.library.caltech.edu/5456/1/hrst.mit.edu/hrs/materials/public/composites/Composites_Overview.htm. Accessed 9 August 2022.
16. Mustapha R, Rahmat AR, Majid RA, Noor S, Mustapha H. Polymer-plastics technology and materials vegetable oil-based epoxy resins and their composites with bio-based hardener: a short review vegetable oil-based epoxy resins and their composites with bio-based hardener: a short review. *Polym Plast Technol Mater.* 2019. https://doi.org/10.1080/25740881.2018.1563119. Published online.
17. Zini E, Scandola M. Green composites: an overview. *Polym Compos.* 2011;32:12. https://doi.org/10.1002/pc.21224.
18. Mark JE. Bio-based polymers and composites. By Richard P. Wool and Xiuzhi Susan Sun. *Angew Chem Int Ed.* 2006;45(37). https://doi.org/10.1002/anie.200585443.
19. EPA. Presidential Green Chemistry Challenge: 2015 Designing Greener Chemicals Award. US EPA; 2022. https://www.epa.gov/greenchemistry/presidential-green-chemistry-challenge-2015-designing-greener-chemicals-award. Accessed 9 August 2022.
20. EPA. Green Chemistry Challenge: 2021 Academic Award. US EPA; 2022. https://www.epa.gov/greenchemistry/green-chemistry-challenge-2021-academic-award. Accessed 9 August 2022.
21. Dyer E, Scott H. The preparation of polymeric and cyclic Urethans and Ureas from ethylene carbonate and amines. *J Am Chem Soc.* 1957;79(3). https://doi.org/10.1021/ja01560a045.
22. Tamami B, Sohn S, Wilkes GL. Incorporation of carbon dioxide into soybean oil and subsequent preparation and studies of nonisocyanate polyurethane networks. *J Appl Polym Sci.* 2004;92(2). https://doi.org/10.1002/app.20049.
23. Javni I, Doo PH, Petrović ZS. Soy-based polyurethanes by nonisocyanate route. *J Appl Polym Sci.* 2008;108(6). https://doi.org/10.1002/app.27995.
24. Deepa P, Jayakannan M. Solvent-free and nonisocyanate melt transurethane reaction for aliphatic polyurethanes and mechanistic aspects. *J Polym Sci Part A Polym Chem.* 2008;46(7). https://doi.org/10.1002/pola.22578.
25. Carré C, Bonnet L, Avérous L. Original biobased nonisocyanate polyurethanes: solvent- and catalyst-free synthesis, thermal properties and rheological behaviour. *RSC Adv.* 2014;4(96). https://doi.org/10.1039/c4ra09794g.
26. Cornille A, Dworakowska S, Bogdal D, Boutevin B, Caillol S. A new way of creating cellular polyurethane materials: NIPU foams. *Eur Polym J.* 2015;66. https://doi.org/10.1016/j.eurpolymj.2015.01.034.
27. He X, Xu X, Wan Q, Bo G, Yan Y. Solvent- and catalyst-free synthesis, hybridization and characterization of biobased nonisocyanate polyurethane (NIPU). *Polymers.* 2019;11(6). https://doi.org/10.3390/polym11061026.
28. Zwawi M. A review on natural fiber bio-composites, surface modifications and applications. *Molecules.* 2021;26:404. https://doi.org/10.3390/molecules26020404.
29. Biocomposite recycling project. *Reinf Plast.* 2021;65:1. https://doi.org/10.1016/j.repl.2020.12.070.

30. Shanmugam V, Mensah RA, Försth M, et al. Circular economy in biocomposite development: state-of-the-art, challenges and emerging trends. *Compos Part C Open Access.* 2021;5. https://doi.org/10.1016/j.jcomc.2021.100138.
31. Martina RA, Oskam IF. Practical guidelines for designing recycling, collaborative, and scalable business models: a case study of reusing textile fibers into biocomposite products. *J Clean Prod.* 2021;318. https://doi.org/10.1016/j.jclepro.2021.128542.
32. Jandas PJ, Prabakaran K, Mohanty S, Nayak SK. Evaluation of biodegradability of disposable product prepared from poly (lactic acid) under accelerated conditions. *Polym Degrad Stab.* 2019;164. https://doi.org/10.1016/j.polymdegradstab. 2019.04.004.
33. Ali RR, Rahman WAWA, Kasmaini RM, et al. Pineapple peel fibre biocomposite: characterisation and biodegradation studies. *Chem Eng Trans.* 2017;56. https://doi.org/10.3303/CET1756223.
34. Jandas PJ, Mohanty S, Nayak SK. Renewable resource-based biocomposites of various surface treated banana fiber and poly lactic acid: characterization and biodegradability. *J Polym Environ.* 2012;20(2). https://doi.org/10.1007/s10924-012-0415-8.
35. Raju G, Mas Haris MRH, Azura AR, Eid AM, AM. Chitosan epoxidized natural rubber biocomposites for sorption and biodegradability studies. *ACS Omega.* 2020;5(44). https://doi.org/10.1021/acsomega.0c04081.
36. Zhao L, Huang H, Han Q, et al. A novel approach to fabricate fully biodegradable poly(butylene succinate) biocomposites using a paper-manufacturing and compression molding method. *Compos Part A Appl Sci Manuf.* 2020;139. https://doi.org/10.1016/j.compositesa.2020.106117.
37. Chan MY, Koay SC, Husseinsyah S, Sam ST. Cross-linked chitosan/corn cob biocomposite films with salicylaldehyde on tensile, thermal, and biodegradable properties: a comparative study. *Adv Polym Technol.* 2018;37(4). https://doi.org/10.1002/adv.21784.
38. Kalita NK, Nagar MK, Mudenur C, Kalamdhad A, Katiyar V. Biodegradation of modified poly(lactic acid) based biocomposite films under thermophilic composting conditions. *Polym Test.* 2019;76. https://doi.org/10.1016/j.polymertesting.2019.02.021.
39. Dilawar H, Eskicioglu C. Laboratory and field scale biodegradability assessment of biocomposite cellphone cases for end-of-life management. *Waste Manag.* 2022;138. https://doi.org/10.1016/j.wasman.2021.11.033.
40. Andrew JJ, Dhakal HN. Sustainable biobased composites for advanced applications: recent trends and future opportunities—a critical review. *Compos Part C Open Access.* 2022;7. https://doi.org/10.1016/j.jcomc.2021.100220.
41. Muthuraj R, Misra M, Defersha F, Mohanty AK. Influence of processing parameters on the impact strength of biocomposites: a statistical approach. *Compos Part A Appl Sci Manuf.* 2016;83. https://doi.org/10.1016/j.compositesa.2015.09.003.
42. Nagalakshmaiah M, Afrin S, Malladi RP, et al. Biocomposites: present trends and challenges for the future. In: *Green Composites for Automotive Applications;* 2018. https://doi.org/10.1016/B978-0-08-102177-4.00009-4.
43. Fitzgerald A, Proud W, Kandemir A, et al. A life cycle engineering perspective on biocomposites as a solution for a sustainable recovery. *Sustainability.* 2021;13(3). https://doi.org/10.3390/su13031160.
44. Al-Oqla FM, Omari MA. Sustainable biocomposites: challenges, potential and barriers for development. *Green Energy Technol.* 2017. https://doi.org/10.1007/978-3-319-46610-1_2 [9783319466095].
45. Iqbal HMN, Keshavarz T. The challenge of biocompatibility evaluation of biocomposites. In: *Biomedical Composites;* 2017. https://doi.org/10.1016/b978-0-08-100752-5.00014-7.

PART V

Reimagining polyester polyurethanes

CHAPTER

10

Bioloop: The circular economy

Robert S Pomeroy

University of California San Diego, La Jolla, CA, United States

Introduction

In 2012 Tim De Chant suggested that if everyone on the planet consumed as much as the average US citizen, four Earths would be needed to sustain them.[1] This is provocative. Developing countries will be on the move to grow and gain access to the North American lifestyle. The next logical step is to conclude that either North Americans will have to change their consumption habits remarkably or people in developing countries can never realize their aspirations. While there were critics of this evaluation, it evokes a looming inequality and impending planetary doom. Is this our future, or a false either/or proposition? The premise assumes only one or the other is true and disregards another possibility. It accepts a business as usual model. In this view, things look grim. But here is another possibility: a disruptive technology.

Nicola Spaldin's argument on overcoming the energy needs of current silicon technology: "The true breakthroughs that will change the course of history will not come from initiatives to improve existing materials or devices, or to advance technologies that have already been identified .... Instead, they will come from off-beat individuals or small teams of fundamental researchers pushing the boundaries of knowledge in directions for which there is not yet an application."[2] Algae-based biofuels have been in the news for decades. Nevertheless, so long as petroleum is inexpensive and the companies that sell fossil fuels are so large and wealthy, algae biofuels cannot get a foothold. How can we get the potentially disruptive technology of algae-based materials into actual use? One solution is to develop a more profitable application that can perhaps solve different problems, specifically, polyester polyurethanes (PUs). Historically they were known to degrade.[3–5] They need to be reconsidered with a

Rethinking Polyester Polyurethanes
https://doi.org/10.1016/B978-0-323-99982-3.00011-0

recycle-by-design aspect and the potential of biotech to break down the barriers. Perhaps the best way forward is to understand the past.

Background

Planned obsolescence

We need to rethink material use and eliminate planned obsolescence, the practice of designing a product with an artificially predetermined limited useful lifetime. This approach generates the time between purchases and increases the long-term sales for the manufacturer. The result is that the consumer purchases replacements more often. It works most effectively when the consumer has a limited choice of producers. Before introducing planned obsolescence, the producer must know that the customer is at least somewhat likely to buy a replacement. Customer loyalty is why producers work so hard to build a brand. This strategy is effective in the marketplace because the consumer has little knowledge of realistic lifetimes. When more manufacturers are competing in a market, the competition results in products with increased durability. The current solid-state, flat-screen television market appears that way. TVs are no longer repaired, and they break down about the time the technology requires an upgrade, making repair unreasonable. An excellent analysis of this (CRT to LED transition) and the resulting impact on electronic waste has been published by Kalmykova et al.[6]

In essence, the waste itself is a flaw that may or may not be purposefully designed into the product. Metals recycling can be one of the most efficient ventures in material and energy expenditure. The main obstacle in recycling modern metals is the alloying elements and additives used to create materials with desired properties. These modifications make the recycling of mixed material streams more difficult. There is an ample number of facilities that segregate timber and steel from comingled waste. However, the presence of copper and tin in steel makes it difficult to recycle. Perhaps we should examine the metal's microstructure to understand better how its grain and texture are linked to its specific properties. We need to think of recycling material at its design stage. Making materials biodegradable is a start, but is it enough?

How does biodegradation fall short?

The United States Environmental Protection Agency (EPA) defines biodegradation as "a process by which microbial organisms transform or alter (through metabolic or enzymatic action) the structure of chemicals introduced into the environment."[7] The process depends on the environmental conditions, the composition of the material, the organism, and the reaction rate. Biodegradation is a biologically driven system.

Biodegradation environments

The sanitary landfill

There are four active stages to a sanitary landfill: aerobic, acidic, anaerobic, and methanogenic anaerobic, followed by a steady-state phase.[8] Each is characterized by the gases evolved. The consumption of oxygen represents the first stage as aerobic organisms break down the waste. As the oxygen is diminished, the landfill transitions to the second stage, where acid-forming anaerobic organisms initiate anaerobic decomposition. This stage is characterized by acidic conditions and a buildup of carbon dioxide (CO_2). In the third stage, the anaerobic digestion transitions to methane (CH_4) forming organisms. CH_4 steadily increases, as does the temperature, reaching 55°C. The fourth stage begins when the evolution of CO_2 and CH_4 comes to a steady state. Now the landfill and its microbial activity have stabilized. The CO_2 and CH_4 gases generated are roughly 50% each.[9]

Compost

The EPA defines compost as organic material that can be added to soil to help plants grow. Approximately 30% of household waste generated principally from food and yard waste can be composted. Composting keeps these materials from the landfill, and a compost pile decomposes aerobically. The oxygen level is maintained by regularly turning the pile over or using another method of oxygen introduction. Composting eliminates the production of CH_4, a greenhouse gas whose heat-trapping impact is 23 times greater than that of CO_2. Making compost creates great material for growing plants and at the same time prevents waste from being disposed of in landfills, thereby preserving scarce disposal space and reducing the amount of methane released.

Composting can be done on a backyard personal scale or an industrial scale. Commercial composting regulates the temperature, moisture level, and aeration of large quantities of ground or sized material. In principle, all materials are biodegradable, given enough time. This element of time creates an important distinction between the terms compostable and biodegradable. The former is more specific, linked to the rate of degradation. For a material to be considered compostable according to ASTM D4600,[10] 90% of it must degrade in 180 days under the controlled humidity, temperature, and aeration conditions of an industrial composting facility. In 2014 the EPA reported that by preventing 89 million tons of municipal solid waste from entering landfills, composting reduced CO_2 emissions in the atmosphere by 181 million metric tons.[11]

The open environment

Biodegradable in an unregulated environment means the material will degrade to base chemicals. This process is called mineralization, and the material is reduced to CO_2, water, mineral salts, and biomass through air oxidation in the presence of moisture and microbes. None of the intermediate molecules should be toxic to the environment. Toxicity is linked to a substance's inherent negative impact on an organ or organism. All things are toxic at some level or dose, so toxicity needs to be defined quantitatively. Acute toxicity is often measured as an LD_{50}, the concentration (typically mg of substance per kg of body weight) at which exposure to that substance causes death in 50% of the population. In relative terms, most substances have low toxicity if the LD_{50} is greater than 5000 mg/kg.

A plastic is biodegradable if it is produced from monomer units with linkages that will break, resulting in pieces digested by microorganisms. If this process is done aerobically, it reduces the production of CH_4 as a by-product. The plastic can be enzymatically degraded and digested or mineralized under the right temperature and humidity conditions with the proper microorganism consortia. Enzymatic digestion is initiated by enzyme-catalyzed hydrolysis breaking the bonds of the polymer, resulting in smaller pieces with lower molecular weights that can then be oxidized in the environment until all that remains are CO_2, water, and residual biomass.

Different microorganisms inhabit each ecosystem. The consortia are usually a mixture that may or may not break down the plastics enzymatically. Each ecosystem, soil, river, and lake represents a variety of conditions. The degradation time of plastic varies according to individual combinations of environmental conditions and the microbial ecosystem. Because all relevant factors cannot be easily known beforehand, the broad claim that material biodegrades in the environment is problematic.

Bioplastics

The term bioplastics can describe two separate things, depending on aspects of the material. *Bio-based* plastics are made at least in part from biological sources, not petroleum. *Biodegradable* plastics can be wholly mineralized by microbes in a reasonable time frame under the right set of conditions. Not all bio-based plastics are biodegradable, and not all biodegradable plastics are bio-based. Many petroleum-based polyesters are biodegradable. Furthermore, even plastic designated as biodegradable may not biodegrade in every environment. As scientists, we often start from a definitional foundation to create clarity for the analysis. The consumer, on the other hand, can be manipulated by the clever use of scientific terms to create confusion. This is the basis for greenwashing, which is addressed in the next chapter.

EVA and PET

For some plastics, monomers from bio-based sources are chemically identical to and indistinguishable from their petroleum counterparts. An example of a nonbiodegradable, bio-based plastic is ethylene vinyl acetate (EVA). While the source of the base chemicals is biological (sugar cane),[12] the monomers and resulting polymers, while carbon negative (a carbon sink), do not biodegrade and will persist in the environment like petroleum-based EVA. A polyester PU, polyethylene terephthalate (PET), which is what most drinking water bottles are made of, can be made from plants.[13] Again, the resulting material is the same as the petroleum-based material. There are reports of some organisms capable of biodegrading PET under specific conditions. PET for water bottles has a crystalline structure that remarkably inhibits microbial action. So, while it can be bio-based and potentially biodegradable, PET is not a solution to the problem of improperly disposed of drinking water bottles. PET does not entirely break down quickly in the environment and still threatens wildlife and ecosystems.

PLA

Another bio-based plastic is polylactic acid (PLA). It is used in shopping bags, drinking straws, transparent cups, 3-D printing filament, and many other products.[14] Because it can be created from plant-based sources such as sugar cane, corn, or potatoes, it is renewable. Obtaining the base chemicals from plants means that CO_2 is sequestered from the atmosphere and the use of fossil fuels is reduced. PLA is listed as biodegradable. However, when plastics made solely from PLA end up in seawater, they do not biodegrade. Worse, mixed PLA and polyester plastics give the appearance of biodegradation. PLA segments do not biodegrade sufficiently in the open environment or in a backyard composter, and the result is microplastics. The proper description for PLA is that it is *conditionally compostable*. It requires *industrial composting*. The temperature of the composter needs to be at or above 58°C (136°F). The good news for PLA, if properly collected and processed under the right conditions, is that it can be mineralized to CO_2 and water within a couple of weeks.

In 2018 California introduced a limited ban on plastic drinking straws. The intent was admirable. Many businesses switched to PLA straws. They paid more, and many people probably felt PLA was better environmentally. However, biodegradation takes place only if these straws are industrially composted or recycled. If packaging or straws made from PLA wind up on a beach or in a waterway, there is no advantage; they do not readily degrade under those conditions.

PHA

Polyhydroxyalkanoates (PHAs) are monomers with two functional groups, a carboxylic acid that must be terminal and a hydroxyl group that is ideally terminal. A very active area of research exists around the microbial production of PHAs. While this monomer is currently a small part of the polyester PU market, demand is expected to grow based on its inherent biodegradability. To compare biodegradability, a thin film of PHA placed on the seafloor in the tropics will degrade in 1–2 months.[15] In the Mediterranean, the degradation can take 10 times as long. What might result in an environment where the temperature is near 0°C and the nutrients are in low concentrations? Such conditions occur at the poles and the bottom of deep oceans. Are the bacteria that live in these environments even equipped to degrade these materials? PHAs do degrade in some environments, but clearly not in all. Given enough time, they slowly accumulate. Consider the *Titanic*. It rests at a depth where the oxygen concentration is so low that no rusting occurs despite the iron and the water. Rusting requires iron, water, and oxygen simultaneously. Similarly, biodegradation requires water, nutrients, and suitable microorganisms. Otherwise, like the *Titanic*, the plastic will rest at the bottom of these reservoirs.

New solutions, new problems: Mixed plastics and microplastics

Those examples point to an inevitability: even with the incorporation of PLA and PHAs, not all plastics will degrade adequately in all environments within an acceptable time frame. Places where degradation is slow will become depositories for littered plastic. The resultant pollution is not limited to the polymer itself but includes the chemicals used to give it the desired properties needed for manufacture or performance. A single plastic product may contain several additional chemicals that modify the physical characteristics and give rise to two problems: they may harm the ecosystem, and they complicate recycling. In addition, the plastics of the mixture may have differential degradation characteristics that can affect an ecosystem. Even if "biodegradable," these longer-lived materials will be a source of microplastics, an environmental problem all its own.

Microplastics are defined by the US National Oceanic and Atmospheric Administration and the European Chemicals Agency as fragments less than 5 mm (0.20 in.) in length. They cause pollution by entering natural ecosystems from various sources, including cosmetics, clothing, and industrial processes. Two classifications are currently recognized. Primary microplastics are fragments or particles already that size before entering the environment, e.g., microfibers from clothing, microbeads, and plastic pellets (also known as nurdles).[16] Secondary microplastics arise from the degradation of larger plastic products through natural weathering processes after entering the environment, e.g., water and soda bottles, fishing nets, plastic bags, microwave containers, tea bags, and tire wear.[17]

Microplastic particles are becoming ubiquitous environmental pollutants, and there is growing concern over their presence in the food supply. Many studies have demonstrated their presence and toxicity in plants and animals.[18] They are very difficult to remove from the environment and to prevent from entering the food web. Ingested by one organism, they travel up the food chain, and there is no natural mechanism for elimination. Birds and fish ingest and accumulate microplastics in their digestive systems over time. Many microplastics enter the environment through wastewater treatment facilities. The use of plastic-degrading organisms and enzymes is a promising method to address the problem.

Products billed as enzyme-mediated degradable plastics are not entirely biodegradable. Other products such as mixed PLA materials and polyethylene plastics have metal compounds incorporated in them to speed degradation. Some of these are referred to as oxo-biodegradable.[19] European Union and United Nations reports have not concluded that the addition of enzymes or metals results in complete mineralization. The problem is that these additives simply give the appearance of biodegradation and actually result in increased microplastic production.

Biodegradation alone as a solution to plastic pollution is fraught with problems, staring with the varying definitions of biodegradation and how it is tracked. The terminology of bio-based bioplastics and compostables can conjure up an illusion that is more of a problem than a solution. The generation of microplastics and associated pollutants is alarming. So, is biodegradation a materially, environmentally, and energetically efficient solution?

The circular economy

A circular economy engages manufacturing, technology, and economics to create a system with a regenerative approach[20–24] to the "four Earths" problem described by De Chant. The current linear economy extracts a natural resource and manufactures it into a product that is disposed of when its practical use is exhausted. The "take, make, waste" approach has directly contributed to global challenges such as climate change, the loss of biodiversity, and improper waste disposal resulting in pollution. Many products are intentionally designed according to the linear model. The manufacturer assumes that when ownership is transferred to the consumer, proper disposal becomes the consumer's responsibility. The petroleum industry accepts no liability for the CO_2 released by consumption.

The linear economy has thrived because it provides high profits and low prices for consumers in developed nations as part of a global economy. Environmentally, the linear system's profitability arises partly from manufacturers not having to account for the cost of waste disposal and pollution to air and water that accompanies material production. It incentivizes waste and pollution to provide jobs, drive the economy, and enhance tax revenue.

The circular economy offers a wide range of benefits. Product design factors in the social, economic, and environmental impacts of production, distribution, and end of use recovery or disposal. It reduces the need for virgin raw material, thereby reducing the pollution of sourcing, and it increases the lifetime of the product, reducing waste accumulation. Increasing the use of recycled or reclaimed materials reduces energy use and the resultant pollution. The drawback is that the recycled material may not be available or at an acceptable price. The issue that needs to be made clear to consumers is that the linear system does not price in the environmental and social costs of virgin material. A properly designed circular economy reduces carbon emissions, air pollution, and toxicity exposure, and at the same time embraces actions such as renewable energy and the protection of ecosystems. The total impact is assessed. A key reason why developing and developed countries should adopt this economic structure is that it provides employment, particularly for low-wage unskilled workers in waste management and recycling.

Probably the most significant downside in moving to a circular economy is the lost profitability of planned obsolescence under the linear model. The switch would have to be planned, supported, and given time to be successful. Jobs and goods cannot be erased overnight; it is too disruptive. While technological innovation will enhance the change, technology alone will be unsuccessful. The habits and lifestyles of the people involved require changing. In developed nations, ideas such as fast fashion must be reevaluated, as it will make no sense to view products as disposable if they can still serve their intended purpose. The question "Do you need a new one?" must become an integral part of societal thought process. People will also have to alter how they currently recycle. Cleaning and sorting materials at the consumer level has a remarkable impact on recycled materials' waste stream value. Repair, reuse, sharing, and mass transit must be accepted on a larger scale in developed countries. The attitude needs to expand to include cars, household appliances, and consumer electronics. The waste from these industries is staggering. Many items are disposed of not because their lifetime is exhausted. Lots of materials disposed of in the United States are sent overseas for a new life. This is not to suggest stoppage of the production of cars, for example, but clearly they can last a long time. Look at Cuba.[25]

The idea of a circular economy is not unrealistic. We have examples of success, such as metals recycling. If the four Earths dilemma is to be avoided, these circular practices will have to be implemented at a large scale. Should our path be business as usual, it is difficult to imagine all people sharing in the lifestyle of the developed world. The linear economy assumes a planet of unlimited abundance and unchecked growth. Awareness of this falsehood is the basis for De Chant's analysis. A sustainable world where material, energy, and environmental demands do not exceed the planet's natural restorative capacity will have to be created. In the absence of sustainability, extinctions, famine, and conflict from climate-induced migration are likely to become the norm.

A circular economy starts by envisioning the most environmentally efficient acquisition of raw materials combined with manufacturing practices focused on reducing energy use, resource use (e.g., water), and waste production. The resulting product is designed to be repairable and its materials recycled to create an energy and resource-efficient, closed-loop system. The product is designed and manufactured for long-term use to reduce the material inputs and minimize changes to the production equipment and infrastructure: less waste in making the product and less waste in retooling the factory. Optimally, waste should retain value through recycling, recovery of the material resource, or reuse for another application. No more take, make, waste.

Designing a circular economy that is restorative, innovative, and effective is a challenge, especially in light of inexpensive and readily available oil. The problem is clearly that the oil will run out. Developing countries are more in tune with a circular economy, but the challenge exists in those with heavy emphasis on consumer spending and disposable materials.

The scientific literature on circular economic models is small but growing. The future focus is on developing practical solutions to create sustainability. Circularity might have its beginnings with Kenneth Boulding's *The Economics of the Coming Spaceship Earth* (1966). The idea of a steady-state economy and "cradle-to-cradle" were described by Daly and Peterson.[26,27] The *Journal of Cleaner Production* is a current platform for exploring the conceptual foundation and principles of a circular economy. Suarez-Eiroa et al.[28] propose seven operational principles:

- adjusting inputs to the system to regeneration rates
- adjusting outputs from the system to absorption rates
- closing the system
- maintaining the value of resources within the system
- reducing the system's size
- designing for a circular economy
- educating for a circular economy

A model circular economy: Aluminum

Aluminum is the third most abundant element in the Earth's crust, after oxygen and silicon. The world production exceeds 57 million tons a year.[29] Given its prevalence, why is it so widely recycled worldwide? Two reasons: it is one of the most accessible materials to recycle; and while the ore is readily available, turning it into metal is very energy intensive, which means expensive. Fourteen thousand kilowatt hour of electricity is required to produce a metric ton. The Aluminum Association reports that nearly 75% of all aluminum ever produced is still in use.[30] Most of the aluminum encountered has already been recycled, in some cases several times! Aluminum recycling is perhaps the best existing example of a circular economy, an almost perfect closed-loop solution.

The energy expense justifies recycling, but what makes aluminum so efficient to recycle? The first step is to shred it. The chips can be separated efficiently from other materials such as plastic and glass by using an infrared sorter followed by a magnet to separate ferrous metals. The resulting chips are melted down at over 1200°F in air. This high-temperature step makes aluminum, iron, and glass recyclable but is an issue for plastics, as will be discussed later. The high temperature required to melt aluminum vaporizes many contaminates, including paints and other organics. The resulting aluminum oxide, called dross, is skimmed off and processed to extract pure aluminum. This process is so efficient that most aluminum cans are recycled back into new aluminum cans. The turnaround time from end-use to new cans is 60–90 days.[31] The process is energy intensive, but the energy consumed is approximately 5% of that required to win pure aluminum metal from its ore, bauxite. This energy saving drives the recycling and efficiency not found in other materials. In 2005 Alcoa constructed a plant in Iceland to take advantage of geothermally generated electricity. Such strategies led to Alcoa being recognized by the World Economic Forum as one of the most sustainable corporations in the world.[32]

The highest aluminum recycling rate belongs to Brazil: 98.4%. Japan is also a standout at 77.1%, reported in 2015. Europe recycles 76.3%, the United States 49.8% residentially and 63.6% industrially.[33] Plastic competes with aluminum for beverage containers; for water, plastic is the container of choice. Given the financial benefit of recycling aluminum and the environmental impact of plastic in the environment, the alternatives are either to use aluminum or to develop a highly recyclable plastic. Since plastic recycling will never have aluminum's financial incentive, perhaps there needs to be a regulatory solution creating value. The creation of a plastic circular economy similar to that of aluminum will require another approach and the participation of all stakeholders.

Is a plastic circular economy possible?

The use of plastics is philosophically challenging, as they provide many benefits. In packaging, they keep food fresh during transport without adding significant weight. The savings in fuel cost, food safety, and reduction of spoilage constitute a game-changer. Low production cost from petroleum makes it more attractive than bio-based plastics. The two obvious complications are its dependence on nonrenewable petroleum and its environmental persistence—from decades to centuries. Because many plastics are lower in density than water, and are disposed of in waterways, they can be transported long distances and dispersed to an extent that the energy required to collect them is environmentally unsound. They need to be systematically collected after use, recycled, and redeployed to prevent them from entering the environment. The question is how to create value in light of the current economic model.

One way is by adopting the concept of design for recycling,[34,35] which specifies each step. Use renewably sourced materials, and consider the upstream and downstream environmental effects of this choice. Make the manufacturing process flexible; a newly available material is unlikely to be a drop-in replacement. Keep in mind efficient transportation of the product from factory to user. Work in cooperation with legislative bodies to incentivize the cleaning and collection of the product at its end of use. And enable the breakdown and recycling processes by rational choices of source materials and chemical design.

Reduction and reuse should be top priorities. Recycling is the last stop in the material chain for a circular economy. Globally significant gains have been made. Germany, Austria, and South Korea have recycling rates of 50%–55%. In Japan, up to 98% of metals are recycled.[36] The world must make products easier to recycle, scale up recycling infrastructure in developed and developing countries, impose stricter controls on shipments of waste around the world, and ensure that recycling workers have safety rights.

Certain materials have a limited life cycle, particularly those that are now mechanically recycled. Most plastics become brittle as the polymer chain lengths are shortened during the heating and recasting process, making this means of asset recovery less attractive. Transitioning away from such materials and methods to a circular economy will require innovation, investment, cooperation, and focus. The new economy will demand redesign at the manufacturing, retail, and customer systems levels. Materials such as polyester PUs will require innovative redesign to create the required characteristics at an acceptable price. The innovations are not just a STEM issue but also need to engage product design, marketing, sociology, and other disciplines. There must be a broad-spectrum examination to build the infrastructure, demand, and cultural changes necessary for success.

Global cooperation at a governmental level will be required to accomplish these goals on an international scale. Legislation that fosters circularity must be sought out and rewarded, and eventually required. Some countries are already starting to enact rules and guidelines to encourage circular economies. The challenge is to design an economy that is energy efficient and capable of restoring resources. It is to be hoped that innovation and a desire to protect the planet will drive and sustain this challenge.

The cost barrier

Only about 9% of plastic capable of being recycled is recycled.[37,38] The low cost of simply making plastics from petroleum renders the value of the material to be recycled low. The fixed costs associated with recycling—the effort and energy—further decrease its net value. Given the fluctuating price of petroleum, the cost of PET-based water bottles is difficult to determine, but it has been around 2–3 cents for the 16oz. (500mL) size, which is considerably less than that of a recycled replacement.

There are six common types of plastics: polystyrene (PS), polypropylene (PP), low-density polyethylene (LDPE), polyvinyl chloride or plasticized polyvinyl chloride (PVC), high-density polyethylene (HDPE), and polyethylene terephthalate (PET). Only PET, HDPE, and PVC products are recycled under curbside programs. PS, PP, and LDPE are not recycled because they are more difficult and expensive to process. Further complicating things, the lids and caps on bottles are not recycled and have to be separated out.

Recycling plastic is done in two steps.[39] First is collecting, sorting, and physical processing (shredding or grinding), and washing to decontaminate. The goal is to create clean particles of uniform size with a common chemistry. Second is melting it. The melted plastic can be either molded into the desired object or pelletized for use as the feedstock in another production application.

Sorting by hand is labor intensive and costly when hourly wages and benefits are considered against the economic value of the material. A mitigating step is to presort before collection, either residential or industrial. This places the burden on the entity seeking to recycle the materials. Presorting would require a shift in the current process in the United States, where mixed recyclables are placed in a single blue bin. Some jurisdictions in Europe use a multicomponent sorting scheme to increase the rate and financial efficiency. An alternative is to use an automated system. Development of this technology will vastly increase the speed and efficiency of recycling without requiring a cultural change from the mixed bins. However, for it to be effective, there must be standards and practices in the manufacturing process to make sorting compatible with the technology.

For example, the use of certain pigments to color the plastic may interfere with an automated system's ability to sort properly on the fly.

The material handling barrier

Cleaning is the second large barrier. With aluminum, melting burns away organic contaminants such as food residue and painted labeling. With plastics, the burning point temperature is too close to that required to combust residues. Cardboard pizza boxes become contaminated with oil from the pizza. That makes recycling too expensive, because the oil would have to be extracted, typically with a solvent. The need to remove residues or labels is time consuming and uses resources such as water. A solution to this is to require the waste generator to clean and remove labels from sorted materials. This small act done residentially dramatically increases the value of the recycling stream. Sorting and pretreatment for metal and glass is not required but is essential for plastics, the value of which is inherently low.

Mixed materials are used widely in food packaging. Different material types are combined to create the desired characteristics. Brewed coffee is sold in a disposable paper cup with a plastic lining. Potato chips are placed in bags of multiple layers, and the outer layer has a metalized surface upon which the printing is done. These layers keep the chips fresh by preventing air and moisture from entering or leaving the bag. Flexible packaging for food uses thin-film plastics for bags, wrappers, and labels. This type of plastic waste represents a significant portion of the total that enters the waste stream. Food safety and preservation are valuable traits and are not to be discontinued lightly. The reduction of spoilage and the positive impact on health are significant. Depending on the product, food packaging provides protection for those products that are negatively affected by moisture, air (oxygen), or both. Single-layer films are poor gas and moisture barriers, hence the use of multilayered films. Table 10.1 presents some characteristics and protective properties of film materials.

While paper, metal foil, cardboard, and plastic liners are all individually recyclable, multilayer materials are not now good candidates.[40] The film with a metallic layer in the potato chip bag and the coffee cup can be recycled only if the materials are separated. Worse than not recycling at all, separating these materials from the recyclable ones would require expenditure of additional effort and energy (think cost). To further complicate matters, different disposable coffee cup vendors are not even obliged to use the same type of plastic for the liner; so even if the cups were segregated and the liners removed, the plastic waste stream would still be mixed.

An understanding of the effects of morphology on the recyclability of plastic is critical. PET used in water bottles has a crystalline structure that is impermeable to water. The crystallinity prevents any water intrusion,

TABLE 10.1 Food packaging film material characteristics.

Type of polymer	Characteristics of value in food packaging
Polypropylene (PP)	Moisture barrier; denser and more heat resistant than PE
Polyethylene (PE)	Moisture barrier
Polyamide (nylon)	Aroma and O_2 barrier; stiff
Ethylene vinyl alcohol (EVOH)	Excellent O_2 and moisture barrier; transparent; not useable in high temperature processes
Ethylene vinyl acetate (EVA)	Good for sealing
Polylactic acid (PLA)	Biodegradable
Metallocene linear low-density polyethylene (mLLDPE)	Good optical and mechanical properties

which is great for storing water. However, it blocks microbial action. PET is polyester; hence the ester link should be subject to hydrolysis, either chemical or biological. This lack of water intrusion makes these bottles essentially impervious to microbial breakdown in the environment and difficult and more expensive to recycle. Instead of depolymerization, PET water bottles are often simply ground up, and the material is added to new PET to make more bottles, thereby reducing the amount of new material. This crystallinity makes the resulting bottles inferior in quality to virgin plastic bottles and hence limits the recycled value. Even if a clean waste stream of PET is available, there is another barrier to recycling based on contamination, a regulatory one.

The regulatory barrier

The current regulatory environment, especially that associated with food packaging, does not fully support recycled plastics.[41] The fear of contamination carries over to the new products because, unlike aluminum, the process does not remove all contaminates. Mixed sources also have the problem of other additives. For example, plasticizers are added to increase flexibility. Suppose PET from a product with this plasticizer is comingled with PET from water bottles intended for repurposing as food packaging. There is valid concern that these "legacy" chemicals from one application are not acceptable in other applications, especially those associated with food. Recycling would then require separate streams of plastics based on the source application and the new application. Combined with the issue of cost and quality is the logistical problem of separation and verification. The problem now grows to one that science often encounters with the public: the use of language.

Several studies have verified the safe use of recycled plastic. Scientists report findings based on measurements, and all measurements have some uncertainty. The problem now becomes, can any regulatory body assure anyone with 100% certainty that no ill will befall any user of recycled plastics? While the concern is valid, the reverse is rarely evaluated, that is, comparing the statistical potential for harm in food packaging from recycled plastics to the tangible harm these plastics have on the environment and public health if not recycled. So long as there is uncertainty around using recycled materials, many manufacturers will oppose establishment of a robust circular economy for plastics.

The other main barrier to recycling PET is the contamination issue, which has a ramification of a different type: legal.[42,43] In an article published September 19, 2017, the industry association Plastics Recyclers Europe reported that degradable plastics entering the traditional plastic waste stream negatively affect the functional characteristics of recycled plastics. A test demonstrated that substances used in degradable plastics (e.g., starch, PLA, and polybutylene adipate terephthalate) led to holes and specks in a film produced from the recycled plastic waste. Therefore they called for "separate streams not only for bio-waste but also degradable plastics in order to make sure that degradable plastics do not enter waste streams of conventional plastics," so we should "avoid putting in jeopardy the efforts of making plastics more circular."

n a press release in 2017, they pointed out the legal uncertainty involving recycled plastic for food packaging. The regulatory policy was defined in 2008 by the European Commission, document No. 282/2008. Since that time, several scientific studies have evaluated the safety of recycled plastic for food packaging. The European Food Safety Authority points to "more than 140" scientific reports on the subject. Europe is waiting for the European Commission to move the regulatory policy to include recycled plastic in food packaging. So long as there is no regulatory protection for these materials, their use cannot move forward, and the industry's ability to grow is stymied. The recyclers and the environmentalist both need these new markets opened for the industry to grow and a plastic circular economy to be realized.

Since plastic is challenging to recycle, the tendency is to recover (the last of the Four Rs) the energy content by incineration. When incinerated, some plastics such as PVC can produce chlorinated products (e.g., chlorinated furan), which are both toxic and persistent. While great for food protection, multilayer packaging is very difficult to recycle, as each layer must be separated and recycled separately. As a result, multilayer food packaging is neither recyclable nor compostable. However, work is ongoing. BASF has developed a packaging material consisting of six different layers. The innovation is that this combination of films is entirely compostable.[44]

Can you improve recycling by raising the value through a fee structure?

California Redemption Value (CRV) is a fee paid by beverage distributors in the state.[45] It is intended to make up the cost differential between a recyclable container's scrap value and the cost of recycling it (including a reasonable rate of return). The charge is passed to the retailer and on to the consumer. It is then up to the consumer to recoup the deposit by recycling the container. The value depends on the size of the container: five cents for containers of 24 fluid ounces (710 mL) or less, 10 cents for larger ones. To differentiate the financial impact of the fee compared to the product costs, the retailer produces a receipt to show that the charge for the bottled drink is, say, $1.00 plus 5 cents CRV. The scheme covers aluminum, plastic, glass, and bimetal containers.

From the program's implementation, the recycling rate of the approved containers increased to 82%, exceeding the statutory 80% goal. The total number of currently recycled containers has tripled. However, in 2019, over 250 of California's recycling redemption and processing centers downsized or closed, terminating 750 employees and liquidating assets to pay creditors. Their stated justification was the state's reduction of fees paid to the recycler combined with minimum wage increases; growing operational costs; and the declining value of recycled aluminum, glass, and PET.

Recycling diverts materials from entering landfills or being incinerated, reducing the demand for land and resources associated with traditional waste disposal. It provides green jobs in a community, enhancing the local economy. It engages the public in an act of environmentalism and raises community awareness. Probably most important globally, it reduces the energy used and CO_2 released into the atmosphere by manufacturing. However, beyond the low demand for a mixed plastic waste stream, there are other issues with implementation. Theft from recycle bins and just plain littering as people search through bins for higher value items result in either lost revenue or the expense of installing locks on bins or buying tamper-proof containers such as those made by Bigbelly. Another issue is the declining value of materials from recyclable containers and the low market value of materials accepted at local recycling centers. Visits of many small recycling operations to consumer locations such as grocery stores have declined in frequency, and recycling has moved toward centralized larger-scale operations, which enhances the financial incentive. This barrier is especially true for smaller municipalities. In California, inflation has reduced the incentive, as the CRV has not kept pace[45]; that cost the state "at least $308 million in 5-cent deposits on cans and bottles in 2018," according to the *Los Angeles Times*.[46] Critics of recycling point to the energy used and pollution created, making "reduce" and "reuse" more environmentally beneficial than "recycle." There is a fear that the recycling public overestimates the positive impact of recycling and offsets that with less sustainable behavior. To what degree should government regulation be implemented? Given the impact on the local and global environments and the finite supply of accessible oil, the idea of using recycling mandates directed at the manufacturer and the consumer is not out of the question. The educational part is to make people aware, and the regulatory part is to make them responsible for the costs when they do not engage in recycling or using recyclable goods. Manufacturers will make and sell goods that consumers will purchase. They respond to market demands. So, absent pressure from consumers or government, producers (brands and converters), will tend to pick the cheapest option, as that is often the consumer's choice.

The EPA reports that the recycling rate for plastic bottles in 2018 was about 28% in the United States, indicating that significantly more are discarded than are recycled.[47] Is it a lack of value, or is there a lack of demand for recycled plastics?

Chemical recycling

The traditional Four Rs: Reduce, Reuse, Recycle, Recover. There is a strong argument for reducing and reusing materials. It costs nothing and saves energy, landfill space, and impact on the environment. With plastics, recycling involves chemistry. The technologies are of four types: primary, secondary, tertiary, and quaternary. These constitute a spectrum of reuse, recovery, and depolymerization.

Primary

This is the processing of waste into a product with characteristics like those of the original product. It is closed-loop recycling, e.g., PET bottles collected, heated, and recast into new ones. The problem is that the heat treatment to melt the plastic breaks bonds in the polymer chain, lowering the overall molecular weight of the polymer and changing its physical characteristics so that bottles made from it are inferior to those made from petroleum feedstock. This mechanical recycling is also called downcycling, as there is a loss in product quality. To reduce the cost of sorting, new technologies such as laser sorting use emission spectroscopy to differentiate polymer types. This improves the ability to separate complex mixtures and can handle up to 860,000 spectra per second and scan each flake. It can sort different black plastics, a problem with previous automatic systems. Laser sorting is likely to increase the separation of waste electrical and electronic equipment (WEEE) and automotive plastics. It can also separate polymers by type or grade, and separate polyolefins such as PP from HDPE. Primary recycling, while providing some relief to landfills, rarely reduces production costs, and ultimately there is a limit to the number of times a mechanically recycled plastic can be incorporated into materials. Contamination removal is an added expense.

Secondary

This is also mechanical. The plastic is shredded or pulverized and the powder is added to plastics with compatible chemistry. Rebonded flexible foam, "rebond," is made with pieces of chopped flexible PU foam bound together. The principal product is carpet underlayment, about 90% of the market. Thermoset plastic can be ground into a powder and used as an inert filler to produce other composite materials; this is a less significant form of plastic recycling.

Both primary and secondary require physical sorting, grinding, washing, and extruding. Advances in automated sorting should increase the net value of these approaches, as the cost of hand sorting reduces the valorization of the waste stream. Primary and secondary are both downcycling: the new products are inferior.

Tertiary

This category, depolymerization, effectively undoes the synthesis reaction, converting the polymer back to its original units.[39,48,49] When produced safely and efficiently, the resulting monomers can be used to make the same or a new plastic virtually identical to that made from virgin feedstock. Tertiary recycling is a far more significant accomplishment in the grander scheme of things, as it means these plastics could be recycled indefinitely. This solution is energetically, materially, and environmentally sustainable. It is not downcycling but chemical recycling in the truest sense. Methanolysis, hydrolysis, glycolysis, and aminolysis take advantage of the chemical lability of bonds containing heteroatoms such as nitrogen and oxygen. In PUs, this means using the ester and urea lineages in the polymer.

Methanolysis is the degradation of polyesters to the methyl esters of the diacids and diols. Disadvantages include the high cost of separation and refining of the mixture of the reaction products. Also, if water perturbs the process, it poisons the catalyst and forms various azeotropes. Before, methanolysis and glycolysis were the methods applied on a commercial scale. The lack of usefulness of recovering dimethyl terephthalate rendered the methanolysis of PET obsolete.

Hydrolysis and glycolysis are the extensively used depolymerization methods for rigid and flexible PUs. The plastics are ground to create more surface area for reaction. The ground material is incrementally added to the reaction mixture of nucleophile and catalyst. This method works most efficiently for long-chain polymers, and the efficiency decreases with increasing crosslinking. Other glycols can act as nucleophiles.

Hydrolysis of PUs is the reaction of water and catalyst at high temperature and pressure. The initial reaction is for the urethane bonds to undergo decarboxylation to produce the amine and CO_2. Further reaction breaks the ester bonds, resulting in a mixture of diols and diacids or hydroxy acids, depending on the starting chemistry of the polymer. The resulting polyols, diacids, and amines from the isocyanates can be separated. The polyols and diacids can be directly used to recreate polyester polyols. The amines can serve as the feedstock for the production back into diisocyanates. These components can be reformulated to make a completely different product with very different physical properties, and this can be done repeatedly, a much better atom economy with no loss in product quality.

In glycolysis, glycol, a shortened name for ethylene glycol, a two-carbon diol, is combined with mixed industrial and postconsumer PUs. When heated, a transesterification occurs, breaking the ester bond in the polymer and forming a new ester bond between the diacid and the glycol, which results in the release of the original polyols. When separated, these polyols can be reformulated into new materials. The newly esterified diacids can be recovered and the glycol removed, recovering the diacid and the glycol to be used again. Glycolysis is the most industrially relevant chemical recycling method for PUs so far.

The solubility of the polyols is critical. If glycolysis is carried out in excess glycol, the resulting mixture contains two phases. The top layer or less dense phase is the starting polyols, and the bottom, denser layer consists of the excess glycerol and the other reaction products—glycols, carbamates, amines, and low-molecular-weight polyols. This system works well for high-molecular-weight polyols. Active research on the glycolysis of PUs has resulted in a variety of methodologies, with two-phase glycolysis having better reaction efficiency and higher purity than single-phase.

Typical catalysts for glycolysis are sodium hydroxide, potassium hydroxide, sodium acetate, potassium acetate, stannous octanoate, and amines. The choice depends on the reaction temperature or the glycol used. Keeping the temperature low reduces thermal decomposition and side reactions such as elimination.

Ammonolysis of PU foam using ammonia or ammonium hydroxide decomposes PUs. Ammonia is more nucleophilic than either water or glycol. This increased nucleophilicity combined with using excess ammonia requires lower temperatures and reaching reaction completion in less time. The reaction produces polyols, amines, and unsubstituted urea. Polyols produced by this process have characteristics favorable in the reformulation to create rigid foams. There are initial positive results in the use of supercritical ammonia. The cost of this technology at scale can be explored only if a viable economic model is built around the use of the resulting depolymerization products.

Quaternary

This method retrieves the energy content of waste plastics by incineration, an alternative to simple recovery. Pyrolysis and hydrogenation break down the polymer and convert it into base chemicals, just not the typical starting base ones of polyester PUs (diols, diacids, and diisocyanates). In pyrolysis, the PU is broken in an oxygen-free environment and releases gases and oily residues resulting from this "cracking" of the polymer chain. The reaction can also be run with a small amount of oxygen, resulting in a mixture of carbon monoxide and hydrogen, known as syngas. Syngas can be used to make methanol as a base chemical or fuel. The fuel can be used to generate electricity and recover the energy content of plastics. Hydrogenation, like pyrolysis, creates gases and oils from used PUs through heat and pressure. These products can be used to reassemble base chemicals to produce new plastics. Catalytic hydrogenation is an active area of research to identify a catalyst that can efficiently depolymerize PUs.

Enzymatic degradation

Background

There are three principal chemical linkages in polyester PUs that have direct analogs in biology: the ester bond, a peptide bond, and a carbamate bond. An ester is essential in various types of lipids. An ester bond is formed in the reaction of an acid and an alcohol. The carboxylic group in effect loses its hydroxyl group, which is protonated with the hydrogen from the alcohol, resulting in the loss of water and the linking of the acid to the alcohol through an oxygen bridge. In biology, lipids or triacyl glycerides are formed by joining a glycerol molecule to three fatty acids with ester links. There are other ester-like linkages that naturally occur in biology, such as the thioester bond. A thioester bond differs from an ester bond in that the bridging oxygen atom is replaced with a sulfur atom. The ester bond, when it undergoes hydrolysis, reincorporates the water molecule eliminated during synthesis and releases a tremendous amount of energy. The hydrolysis of the ester bonds yields 9 kcal/g of energy.

The peptide bond is the second most biologically abundant. It connects amino acids to form long chains called polypeptides. Like the ester reaction, the peptide bond is a condensation reaction between a carboxylic acid and an amine. The carbon of the carboxylic acid is connected to the nitrogen of the amine, forming an amide bond, and results in the loss of water. Peptide bonds also undergo hydrolysis; a water molecule is taken up, reversing the synthesis reaction, thereby separating the connected amino acids.

The carbamate bond is less common than the ester and peptide types. Several are known in biological systems and are present in important proteins such as hemoglobin, urease, and phosphotriesterase. A very important one is involved in the capture of CO_2 by plants. Carbamate bonds are biologically formed in a posttranslational modification known as carbamylation. Polymers whose units are joined by the divalent carbamate group —NH—(C=O)—O— constitute the PU family of plastics. Enzymatic degradation is described in detail in Chapter 4 and well reviewed in the literature.

Since Otto Bayer's process in 1937,[50] PUs are obtained mainly by the polyaddition between polyols and diisocyanates. The result is a urethane bond—the linkage of an ester and an amide—and a carbamate bond. The polyol of a PU can be composed of extended chains of esters, a polyester. PUs contain amorphous regions, known as soft segments, based principally on long polyester polyols, and organized hard segments, consisting of the isocyanate groups and the chain extender moieties based on short molecules. Material properties and behavior can be modulated by varying the chemical nature and content of the soft and hard segments.

The chemistry of biological systems and those of polyester PUs overlap. Chapter 4 describes the natural ability of microorganisms to biodegrade polyester PUs.[51,52] The process starts with an organism excreting an extracellular enzyme which attaches to PU. It catalyzes the hydrolysis of the carbamate and ester linkages into the smaller polymer units or even down to the monomer units that the polymer was formed from. These molecular fragments are transported into the cell and serve as a carbon source, which releases CO_2—the process of respiration. As long as the material is a polyester-based polymer, whether petroleum or bio-based, several microorganisms can consume it. Over 90 bacteria and fungi are known to degrade petroleum-based plastics.[53] (The surface morphology has a lot to do with biodegradation rate. The crystallinity of PET is slow or absent because the enzyme has difficulty attaching to the surface. Therefore, even though PET is a polyester, its biodegradation is slow. Biodegradation rate depends on the surface morphology and the molecular weight of the polymer. This is why an exact definition is difficult and typically done operationally by monitoring the loss of CO_2, mass loss of the substrate, reduction of the molecular weight, and loss of the mechanical characteristics.[54])

Fig. 10.1 illustrates an enzyme interacting with a polymer strand of PU. Enzymatic recycling is a promising strategy to depolymerize waste plastics into monomers.[55,56]

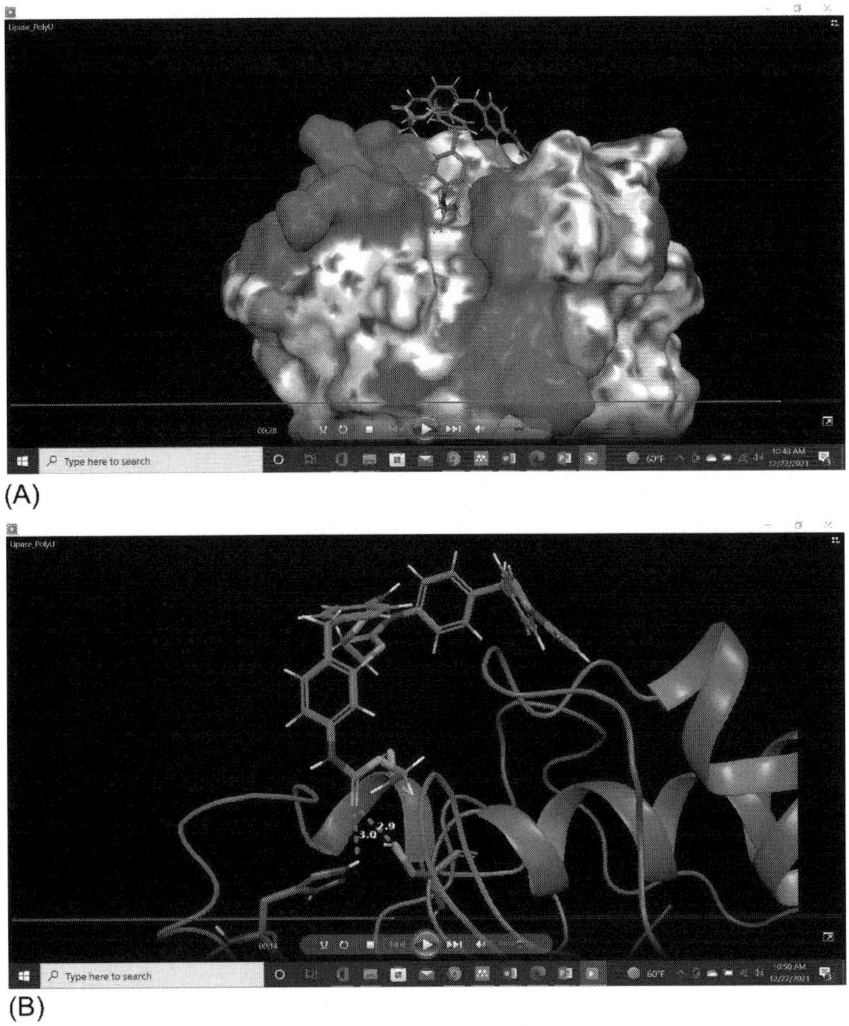

FIG. 10.1 An enzyme interacting with a polymer strand of polyurethane. (A) a space filling model. (B) thread and fold model of a lipase. In both images, a polyester polyurethane is shown exploring the active site.

Why enzymatic depolymerization

Enzymatic depolymerization of synthetic polymers is an appealing approach to reduce a large part of the environmental pollution from these materials and to address waste management issues in a green way. This soft process is considerably more eco-efficient than chemical ones, because it takes place at moderate temperatures without chemical catalysts and is therefore less polluting. Furthermore, controlled enzymatic depolymerization generates building blocks which can be valorized to develop new macromolecular architectures, a promising biotechnical recycling process. A prominent recent example is the recycling of polyesters that led to an efficient depolymerization process of PET-based packaging.[57] Continuing research on enzymatic depolymerization of PU is still required to be done at scale to determine the value proposition.[58,59]

As outlined at the start of the chapter, biodegradation by itself is not the solution. Nor is recycling. A 100% recyclable plastic product usually contains less than 20% recycled material! In addition, cleaning up plastics for melting and recasting is an expensive and water-intensive process, making result often more expensive than virgin plastic, and only 9% of potentially recyclable plastics are ever recycled. A recent study by the World Wildlife Fund reported that the top five major consumer goods brands combined used only 8% recycled plastic material, out of a total of 4.2 million metric tons of plastic sold.[60]

The goal is to create innovative materials that are designed for recycling and would avoid degradation of the environment if improperly discarded. Depolymerization has some promise, and appropriate waste streams may provide enough value to make it economically and environmentally possible. The drawback is the separation step: separating the desired monomers from the reaction mixture. This step is now inefficient in terms of mass recovery and cost. Biological or enzymatic depolymerization holds the greatest promise. The reaction conditions and media are much easier to contend with. The cost and efficiency of the enzymes might be market-ready, but the biotech industry has the greatest chance of finding a disruptive solution, the kind required to solve De Chant's four Earths dilemma. A specific need is new polyester PUs derived from renewable, sustainable feedstock and designed to biodegrade or be recycled at the end of life—in particular, bio-based polyester polyols made from renewable algae feedstocks that meet commercial specifications.[61,62] Enzymes that can efficiently depolymerize these bio-based PUs need to be mass produced. The search for the best enzymes and their microbial production will lean heavily on the growing biotech sector.

The Bioloop

The University of California San Diego-based Center for Renewable Materials calls this process the "Bioloop"[63] (Fig. 10.2). It differs from current mainstream plastic recycling. Rather than the mechanical downcycle, heat, and recast approach, the polymer is depolymerized back to starting raw materials. These monomers are recovered and used to create new PUs. This system follows the same mechanisms as biodegradation except that, rather than having the depolymerized monomers metabolized by microorganisms, they are recovered and used to produce new materials, and the material can be cycled many times without significant loss of quality.

Techno-economic and life cycle analyses elucidate the economic, environmental, and energy benefits of a circular carbon economy based on depolymerizing and recycling polyester-based plastics. The operational scale of recycling is an essential factor. The gap between bench and pilot scale in terms of cost and efficiency is enormous. Only at a significant scale can those analyses give an accurate picture. There must be a partnership between academic researchers, government agencies, and manufacturers.

The next frontier in plastics research is identifying and producing biobased products that enable a circular carbon economy. The plastic can be enzymatically or chemically degraded back into raw materials to allow new rounds of polymerization and product generation. There is no existing commercial process to achieve this, which makes widespread industrial investment too risky. But with support from governmental agencies, academia can demonstrate technical feasibility and economic viability, paving the way for industry to adopt this game-changing technology.

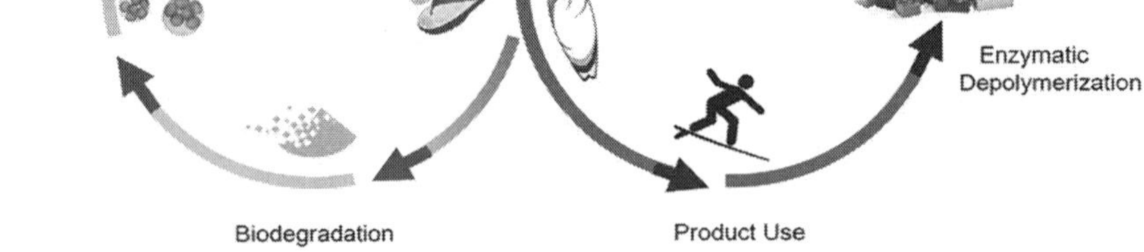

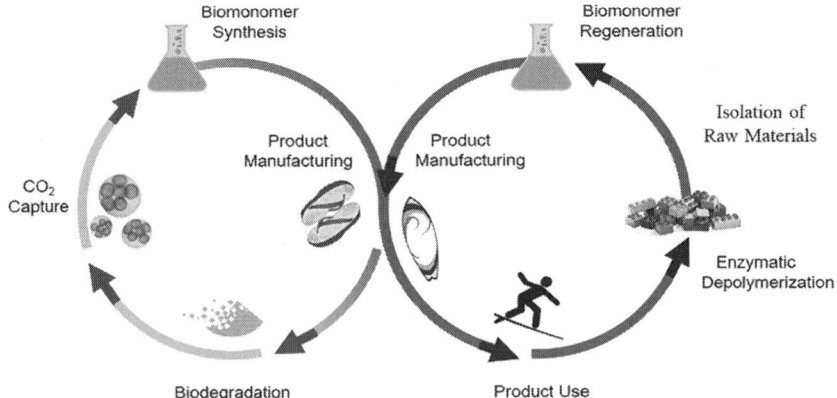

FIG. 10.2 The Bioloop cycles showing biodegradation and the depolymerization and reuse of polyester polyurethanes.

Conclusion

A plastic-free future? Not realistic. Plastic food packaging is unavoidable, given the health and other benefits. Packaging film prolongs shelf life, locks in protective gases such as nitrogen, prevents air and moisture from reaching the product, and provides a barrier to bacterial contamination.

What can we do? As consumers, we can provide our own reusable containers when buying, and the retailers can insist on packaging designed to be recycled. Reducing or eliminating single-use materials is possible, but it will involve us all. For materials that can be recycled, we the consumers, through small acts, can enhance the value of the waste stream by removing labels, rinsing containers, and sorting the plastic by type. We cannot recycle or landfill our way out of plastic pollution. Biodegradation, while a plus, is an inadequate alternative. We must insist that manufacturers, distributors, and retailers adopt materials that can be part of a proper circular economy. Those actors will sell what consumers will purchase. They are not likely to initiate renewable, sustainable, recyclable plastics widely, because their experience indicates that consumers might care about the environment but make most of their decisions with their wallets. As consumers, we have to consider the end-of-life aspect of our material choices and factor that into purchasing decisions.

The Waste & Resources Action Programme—WRAP—is a United Kingdom-based charity that regularly publishes on life cycle analysis. They compared several end-of-life scenarios and found that those using recycling resulted in better environmental outcomes than those using energy recovery (incineration) or burying in a landfill.[64] A systemic change toward a circular economy envisions the best economic and environmental outcomes. Compared with business as usual, the Bioloop has the potential to reduce the volume of waste plastics entering the oceans by 80%, generate savings of $200 billion per year, reduce greenhouse gas emissions by 25%, and create 700,000 net additional jobs by 2040.

Businesses are cognizant of the issue but will not invest without more assurance of consumer acceptance. What is required is innovation. And that will require governmental support financially and legislatively. The governmental input has to be thoughtful. While PLA for drinking straws is helpful, the legislation exempted fast food establishments, and the PLA straws are biodegraded only if they find their way to an industrial composter. The CRV program adds value to the waste stream but is effective only if priced to incentivize recycling and keep pace with inflation and recycling costs.

Advances in biochemistry, genetics, and machine learning have set the stage for a new version of the Green Revolution, one focused on energy, nutrition, and material science. A recent Department of Energy initiative,

BOTTLE (Bio-Optimized Technologies to Keep Thermoplastics out of Landfills and the Environment), focuses on "advancing plastic recycling technologies and the manufacturing of new plastics that are recyclable by design."[65] By combining metabolomics and genetic engineering, and guided by machine learning, the objective is to establish a biotech industry that sustainably produces the base chemicals, materials designed for recycling, and the enzymes to efficiently recover the source materials to create a renewable, sustainable, bio-based circular economy. This is the disruptive technology needed to allow higher standards of living for all without exhausting the natural resources of the Earth, thereby avoiding the conundrum of De Chant's analysis.

Close-up: A plastic circular economy

Bio-based surfboards, skis, flip flops, cosmetics, and films? Yes, it can be done and is being done.

The world of industrial chemicals and polyurethanes is undergoing three revolutions at the same time. There are historic shifts in circular feedstocks, applications, and consumer demand. At the heart of this is the global drive toward a circular economy, yet this sector has been remarkably bereft of the mandates, incentives, credits, and subsidies that are the hallmark of the drivers for battery-electric vehicles, renewable energy, and fuels.

And, it's happening all around the world.

In Colorado, Polaris Renewables has developed nonisocyanate polyurethane, which can be made from commercially available oils from linseed or soybeans or even algae or food wastes.

In Germany, Covestro's "DreamResource" project demonstrated that they can use up to 20% CO_2 instead of oil in rigid polyurethane foam. And the company launched its line of Baycusan eco series of bio-based polyurethane film formers, consisting of almost 60% renewables.

In Canada, researchers at Memorial University have described a process to turn fish heads, bones, skin, and guts into a biodegradable replacement for polyurethane.

In California, Algenesis Materials has launched its renewable isocyanates—and, a bio-based surfboard.

Also in California, Checkerspot has developed its Algal Core, a lightweight wood-polyurethane composite made from algal high density rigid foam. The WING technology has been expanding into bio-based wicking finishes, polyurethane for ski applications, and a sidewall that outperforms traditional ABS sidewalls at room temperature.

Consumer demand for new options is inspiring new sustainable brands to capture market share and older brands to adapt to the circular opportunity. It is WNDR Alpine, Volvo, Ford, Bridgestone, Patagonia, Cargill, BMW, Burger

King, Chanel, Crocs, Old Navy, and Adidas. The best is yet to come, and coming soon.

James Lane

Publisher and Editor, *The Daily Digest*, Key Biscayne, FL, United States

References

1. DeChant T. If the World's Population Lived Like.... https://persquaremile.com/2012/08/08/if-the-worlds-population-lived-like/.
2. Spaldin NA. Fundamental Materials Research and the Course of Human Civilization. Published online August 3; 2017. http://arxiv.org/abs/1708.01325. Accessed 12 December 2021.
3. Schollenberger CS, Stewart FD. Thermoplastic polyurethane hydrolysis stability. *J Elastomers Plast.* 1971;3(1):28–56.
4. Schollenberger CS, Stewart FD. Thermoplastic polyurethane hydrolysis stability. *Angew Makromol Chem Appl Macromol Chem Phys.* 1973;29:413–430.
5. Xie F. Degradation and stabilization of polyurethane elastomers. *Prog Polym Sci.* 2019;90:211–268.
6. Kalmykova Y, Patrício J, Rosado L, Berg PEO. Out with the old, out with the new—the effect of transitions in TVs and monitors technology on consumption and WEEE generation in Sweden 1996-2014. *Waste Manag.* 2015;46:511–522. https://doi.org/10.1016/j.wasman.2015.08.034.
7. Goecopure. *What is Biodegradation?* https://www.goecopure.com/what-is-biodegradation.aspx. Accessed 5 May 2022.
8. Vallero D, Blight G, Letcher TM, Vallero DA. Municipal landfill. In: *Waste: A Handbook for Management.* Elsevier Academic Press; 2019.
9. Chian ESK, DeWalle FB. Characterization of soluble organic matter in leachate. *Environ Sci Technol.* 1977;11(2). https://doi.org/10.1021/es60125a003.
10. Greene J. Biodegradation of compostable plastics in green yard-waste compost environment. *J Polym Environ.* 2007;15(4). https://doi.org/10.1007/s10924-007-0068-1.
11. Environmental Protection Agency U, Conservation M, Branch R, Conservation R, Division S, Office of Resource Conservation E. Advancing Sustainable Materials Management: 2014 Fact Sheet Assessing Trends in Material Generation, Recycling, Composting, Combustion with Energy Recovery and Landfilling in the United States; 2014.
12. Paiva Junior CZ, Peruchi RS, de Fim FC, de Soares WOS, da Silva LB. Performance of ethylene vinyl acetate waste (EVA-w) when incorporated into expanded EVA foam for footwear. *J Clean Prod.* 2021;317. https://doi.org/10.1016/j.jclepro.2021.128352.
13. Vural Gursel I, Moretti C, Hamelin L, et al. Comparative cradle-to-grave life cycle assessment of bio-based and petrochemical PET bottles. *Sci Total Environ.* 2021;793. https://doi.org/10.1016/j.scitotenv.2021.148642.
14. Domenek S, Ducruet V. Characteristics and applications of PLA. In: *Biodegradable and Biobased Polymers for Environmental and Biomedical Applications;* 2016. https://doi.org/10.1002/9781119117360.ch6.
15. Ong SY, Chee JY, Sudesh K. Degradation of polyhydroxyalkanoate (PHA): a review. *J Siberian Federal Univ Biol.* 2017;10(2):21–225. https://doi.org/10.17516/1997-1389-0024.
16. Andrady AL. The plastic in microplastics: a review. *Mar Pollut Bull.* 2017;119(1). https://doi.org/10.1016/j.marpolbul.2017.01.082.

17. Cole M, Lindeque P, Halsband C, Galloway TS. Microplastics as contaminants in the marine environment: a review. *Mar Pollut Bull.* 2011;62(12). https://doi.org/10.1016/j.marpolbul.2011.09.025.
18. Prata JC, da Costa JP, Lopes I, Duarte AC, Rocha-Santos T. Effects of microplastics on microalgae populations: a critical review. *Sci Total Environ.* 2019;665. https://doi.org/10.1016/j.scitotenv.2019.02.132.
19. Scott G. 'Green' polymers. *Polym Degrad Stab.* 2000;68(1). https://doi.org/10.1016/S0141-3910(99)00182-2.
20. Crainer S. Squaring the circle. *Bus Strateg Rev.* 2013;24(4). https://doi.org/10.1111/j.1467-8616.2013.00988.x.
21. Kirchherr J, Reike D, Hekkert M. Conceptualizing the circular economy: an analysis of 114 definitions. *Resour Conserv Recycl.* 2017;127. https://doi.org/10.1016/j.resconrec.2017.09.005.
22. Prieto-Sandoval V, Jaca C, Ormazabal M. Towards a consensus on the circular economy. *J Clean Prod.* 2018;179. https://doi.org/10.1016/j.jclepro.2017.12.224.
23. Velenturf APM, Purnell P. Principles for a sustainable circular economy. *Sustain Prod Consum.* 2021;27. https://doi.org/10.1016/j.spc.2021.02.018.
24. Geissdoerfer M, Savaget P, Bocken NMP, Hultink EJ. The circular economy – a new sustainability paradigm? *J Clean Prod.* 2017;143. https://doi.org/10.1016/j.jclepro.2016.12.048.
25. Ostle G. The changing face of Cuba: more than just old cars and cigars. *Int Pap Board Ind.* 2016;59(3):20–27.
26. Daly H. Economics, Ecology, Ethics—Essays Toward a Steady-State Economy. Published online; 1980.
27. Peterson M. Cradle to cradle: remaking the way we make things. *J Macromark.* 2004;24(1). https://doi.org/10.1177/0276146704264148.
28. Suárez-Eiroa B, Fernández E, Méndez-Martínez G, Soto-Oñate D. Operational principles of circular economy for sustainable development: linking theory and practice. *J Clean Prod.* 2019;214. https://doi.org/10.1016/j.jclepro.2018.12.271.
29. Aluminum. Production D. Aluminum 1. 2020;1(703):2019–2020.
30. Tyabji N, Nelson W. Despite a 90% Increase in Primary Aluminum Production. Columbia University Earth Institute; 2012. July.
31. Capuzzi S, Timelli G. Preparation and melting of scrap in aluminum recycling: a review. *Metals (Basel).* 2018;8(4). https://doi.org/10.3390/met8040249.
32. Yu M, Gudjonsdottir MS, Valdimarsson P, Saevarsdottir G. Waste heat recovery from aluminum production. In: 2018. Minerals, Metals and Materials Series; https://doi.org/10.1007/978-3-319-72362-4_14. vol. Part F6.
33. Aluminiumtoday n.d. International Aluminium Institute publishes global recycling data. Accessed 21 December 2021. https://aluminiumtoday.com/news/international-aluminium-institute-publishes-global-recycling-data.
34. Leal JM, Pompidou S, Charbuillet C, Perry N. Design for and from recycling: a circular ecodesign approach to improve the circular economy. *Sustainability (Switzerland).* 2020;12(23). https://doi.org/10.3390/su12239861.
35. Ragaert K, Huysveld S, Vyncke G, et al. Design from recycling: a complex mixed plastic waste case study. *Resour Conserv Recycl.* 2020;155. https://doi.org/10.1016/j.resconrec.2019.104646.
36. EPA. (n.d.) National Overview: Facts and Figures on Materials, Wastes and Recycling. https://www.epa.gov/facts-and-figures-about-materials-waste-and-recycling/national-overview-facts-and-figures-materials.
37. Geyer R, Jambeck JR, Law KL. Production, use, and fate of all plastics ever made. *Sci Adv.* 2017;3(7):e1700782. http://advances.sciencemag.org/.

38. Gillies R, Jones P, Papineschi J. Recycling—WHO Really Leads the world? (issue 2). Eunomia; 2017:1–39. Published online http://www.eunomia.co.uk/reports-tools/recycling-who-really-leads-the-world-issue-2/.
39. Thiounn T, Smith RC. Advances and approaches for chemical recycling of plastic waste. *J Polym Sci.* 2020;58(10):1347–1364. https://doi.org/10.1002/pol.20190261.
40. Burgess M, Holmes H, Sharmina M, Shaver MP. The future of UK plastics recycling: one bin to rule them all. *Resour Conserv Recycl.* 2021;164. https://doi.org/10.1016/j.resconrec.2020.105191.
41. Vollmer I, Jenks MJF, Roelands MCP, et al. Beyond mechanical recycling: giving new life to plastic waste. *Angew Chem Int Ed.* 2020;59(36). https://doi.org/10.1002/anie.201915651.
42. de Römph TJ, van Calster G. REACH in a circular economy: the obstacles for plastics recyclers and regulators. *Rev Eur Comparative Int Environ Law.* 2018;27(3). https://doi.org/10.1111/reel.12265.
43. Yu Yangyao LL. Legal regulation of recycling of food contact plastic packaging materials in China: a case study on polyethylene terephthalate (PET) beverage bottles. *Food Sci.* 2019;40(19).
44. BASF. (n.d.) Multilayer Packaging: Innovative and Sustainable. https://www.basf.com/global/en/who-we-are/sustainability/whats-new/sustainability-news/2019/multi layer-packaging.html.
45. Berck P, Blundell M, Englander G, et al. Recycling policies, behavior and convenience: survey evidence from the calrecycle program. *Appl Econ Perspect Policy.* 2021;43(2). https://doi.org/10.1002/aepp.13117.
46. Consumer, n.d. Half a Nickel: How California Consumers Get Ripped Off On Every Bottle Deposit They Pay. https://www.consumerwatchdog.org/report/half-nickel-how-california-consumers-get-ripped-every-bottle-deposit-they-pay, Accessed: 2022-11-17.
47. EPA. (n.d.) Plastics: Material-Specific Data. https://www.epa.gov/facts-and-figures-about-materials-waste-and-recycling/plastics-material-specific-data.
48. Ragaert K, Delva L, van Geem K. Mechanical and chemical recycling of solid plastic waste. *Waste Manag.* 2017;69:24–58. https://doi.org/10.1016/j.wasman.2017.07.044.
49. Rahimi AR, Garciá JM. Chemical recycling of waste plastics for new materials production. *Nat Rev Chem.* 2017;1:1–11. https://doi.org/10.1038/s41570-017-0046.
50. Sharmin E, Zafar F. Polyurethane: An Introduction. Published online; 2012:3–16.
51. Tokiwa Y. CBP, UCU, & AS. Biodegradability of plastics. *Int J Mol Sci.* 2009;10(9):3722–3742.
52. Mohanan N, Montazer Z, Sharma PK, Levin DB. Microbial and enzymatic degradation of synthetic plastics. *Front Microbiol.* 2020;11:2837.
53. Jumaah OS. Screening of plastic degrading bacteria from dumped soil area. *IOSR J Environ Sci Toxicol Food Technol.* 2017;11:93–98.
54. Harrison JP, Boardman C, Ocallaghan K, Delort AM, Song J. Biodegradability standards for carrier bags and plastic films in aquatic environments: a critical review. *R Soc Open Sci.* 2018;5(5). https://doi.org/10.1098/rsos.171792.
55. Grima S, Bellon-Maurel V, Feuilloley P, Silvestre F. Aerobic biodegradation of polymers in solid-state conditions: a review of environmental and physicochemical parameter settings in laboratory simulations. *J Polym Environ.* 2000;8(4):183–195. https://doi.org/10.1023/A:1015297727244.
56. Montazer Z, Najafi MBH, Levin DB. Challenges with verifying microbial degradation of polyethylene. *Polymers (Basel).* 2020;12(1). https://doi.org/10.3390/polym12010123.
57. Gamerith C, Zartl B, Pellis A, et al. Enzymatic recovery of polyester building blocks from polymer blends. *Process Biochem.* 2017;59:58–64. https://doi.org/10.1016/j.procbio.2017.01.004.

58. Singh A, Rorrer NA, Nicholson SR, et al. Techno-economic, life-cycle, and socioeconomic impact analysis of enzymatic recycling of poly(ethylene terephthalate). *Joule.* 2021;5 (9):2479–2503. https://doi.org/10.1016/j.joule.2021.06.015.
59. Knott B.C., Erickson E., Allen M.D., et al. Characterization and engineering of a two-enzyme system for plastics depolymerization Proc Natl Acad Sci U S A doi:https://doi.org/10.1073/pnas.2006753117/-/DCSupplemental.
60. Overview M. ReSource Footprint Tracker; 2020. October.
61. Phung Hai TA, Neelakantan N, Tessman M, et al. Flexible polyurethanes, renewable fuels, and flavorings from a microalgae oil waste stream. *Green Chem.* 2020;22 (10):3088–3094. https://doi.org/10.1039/d0gc00852d.
62. Gunawan NR, Tessman M, Schreiman AC, et al. Rapid biodegradation of renewable polyurethane foams with identification of associated microorganisms and decomposition products. *Bioresour Technol Rep.* 2020;11:100513. https://doi.org/10.1016/j.biteb.2020.100513.
63. Bonanomi L. (n.d.) Re-Thinking the Recycling of Plastics Through Materials Regeneration. https://renewablematerials.ucsd.edu/research/bio-loop.html.
64. WRAP - The Climate Crisis: Act Now (n.d.) Accessed 4 May 2022. https://wrap.org.uk/.
65. BOTTLE (n.d.) Bio-Optimized Technologies to keep Thermoplastics out of Landfills and the Environment (BOTTLE) | About BOTTLE. Accessed 4 May 2022. https://www.bottle.org/about.html.

CHAPTER

11

The bioplastics market: History, commercialization trends, and the new eco-consumer

Thomas Cooke[a] *and Robert S Pomeroy*[b]

[a]Algenesis Materials, Cardiff, CA, United States, [b]University of California San Diego, La Jolla, CA, United States

The history of bioplastics

The first plastics *were* bioplastics. This truth is often in conflict with our association of plastics with petroleum. Manufactured bioplastics date back to the 19th century with the invention of Parkesine, a plastic derived from cellulose. Throughout the century and into the early 20th, inventors at chemical companies achieved important scientific milestones in bioplastics using various biological feedstocks. There was a wide range of applications for these nascent technologies—everything from billiard balls, films, coatings, and packaging to automotive parts.[1]

In response to the Industrial Revolution, the modern environmental rights movement also began in the 19th century, driven by pollution and public health concerns. The leaders of this movement were scientists, philosophers, and writers who understood humanity's dependence on the natural world. They advocated protecting the environment, conserving natural resources, and regulating pollution through government action. The movement created a clash between opposing political philosophies, which is still being fought today. The environmental movement seeded early demand for green materials like bioplastics and encouraged clean-tech innovation throughout the 20th century.[2]

The plastics market has historically been linked to the petroleum oil market. This relationship is the competition between petroleum oil and

Rethinking Polyester Polyurethanes
https://doi.org/10.1016/B978-0-323-99982-3.00003-1

bio-based oil feedstocks as precursors to plastic materials. The commercial relationship is clear. As petroleum oil prices have fluctuated, so has the competitiveness of bioplastics as an eco-friendly replacement.[3]

The oil and energy crisis of 1973 spiked crude oil prices and was an early driver for the further development of bioplastics. By the 1980s, the overproduction and oversupply of oil made it less urgent to find alternatives to petroleum-based plastics. Marlborough Biopolymers, one of the first specific bioplastics companies, was founded in 1983. Another notable entrant into the early market was Novamont. They established themselves in 1990 and are still viewed as an industry leader.

The 1990s and 2000s were when the larger chemical companies displayed a renewed interest in the bioplastics market due to pressure from environmental groups and consumers. The discovery of the plastic garbage patches in the five ocean gyres generated outrage in the mid-1990s and continues to be a significant story in the global news media. Established petrochemical firms such as BASF, Cargill, and Dow increased their investment in bioplastics research. They fielded new commercial products into the space during this period. The rise of social media in the mid-2000s played a pivotal role in spreading information on the environmental crisis. These new viral communication channels allowed consumers to apply increased pressure on governments and corporations with minimal effort. We saw a new wave of innovation in bioplastics and renewable materials in response.

Newer biotech companies such as Corbion and Avantium have successfully developed partnerships with major food and beverage companies. Large brands such as Coca-Cola, Pepsi, and Nestlé have felt immense pressure from consumers to do something about single-use nonbiodegradable plastic packaging. Single-use packaging is one of the most visible parts of the global plastic problem. Its volume and connection with marine pollution tend to take center stage. An example of recent innovations in packaging is the PlantBottle by Coca-Cola, developed with Avantium. The primary challenge with packaging as an entry point for bioplastics is cost and market pricing. Single-use plastic packaging products are incredibly cheap and tend to be viewed as commodities. Any price increase is viewed as a margin drag.

The fashion industry (apparel, footwear, and accessories) is another large sector under social pressure to change. Fashion is ranked between the 4th and 10th most polluting industries worldwide. Therefore it has been under pressure from consumers to become more sustainable and adopt more eco-friendly materials. It is an attractive entry point for bioplastics because of its high profitability and loyal consumer base, among other factors. However, it creates tremendous pollution. It uses lots of plastic, particularly in footwear and accessories, and is adept at creating emotive marketing stories.

Consumers' willingness to pay a green premium is a hotly debated topic that impacts the commercial viability of bioplastics. Millennials and Gen Z generations form a large and growing new eco-consumer group. Young people are highly concerned about the environment and understand their power to affect change. Their impact on the market is predicted to be significant. Green price premiums will be explored later in the chapter.

Current bioplastics market size and share

The bioplastics market has successfully weathered the challenges posed by the COVID-19 pandemic and is poised for continued growth.[4,5] The crash in global oil prices in 2020 threatened to suck the life out of the resurging bioplastics industry by creating a vast gulf in prices between bioplastics and fossil-based plastics. Fortunately, the oil price normalized quickly, and the positive market forces listed in Table 11.1 are still firmly in play. The global market is predicted to grow by 36% over the next 5 years. Global production capacity is forecasted to increase from 2.1 million metric tons in 2020 to 2.8 million in 2025. The dollar value of the industry was $9.17 billion in 2020 and is expected to grow to $20.19 billion in 2025.

Despite their clear environmental advantages, bioplastics still have only about a 0.5% to 1% global market share. The primary reason for this low penetration is the higher cost. The current average is $2/kg to $6/kg, compared with $1/kg to $2/kg for traditional plastics. The higher R&D

TABLE 11.1 The primary market forces at work in the bioplastics industry.

Positive forces	Negative forces
Climate change	Cost
Volatility of oil pricing and supply	Historical performance reputation
Sustainability	Greenwashing
Biodegradability	Lack of collection infrastructure
Corporate environmental policies	Lower petroleum oil pricing
Government involvement	Food supply considerations
Recent performance evaluation	Lack of industrial composting
Market demand and supply	Genetic modification
	Consumer ambivalence

and production costs combined with smaller production volumes are the drivers behind the cost differential.

Bioplastics and biopolymers are used in various industries, such as packaging, consumer goods, automotive and transportation, textiles, and agriculture and horticulture. There are many types of bioplastics across resin types, and each has its unique properties and end-uses. Biodegradables make up about 60% of the bioplastics market, while nondegradables make up about 40%. The biodegradable bioplastics segment is forecasted to have the highest compound annual growth rate over the next 5 years (Fig. 11.1). Innovative biopolymers, such as bio-based, drive the current market growth (polypropylene and especially polyhydroxyalkanoates). Bio-polyols and bio-based polyurethane (PU), which are the focus of this book, make up a significant class of next-generation biopolymers primed for growth. The global market for bio polyols was $4.52 billion in 2020 and is expected to grow by approximately 10% a year over the forecast period 2021 to 2027.

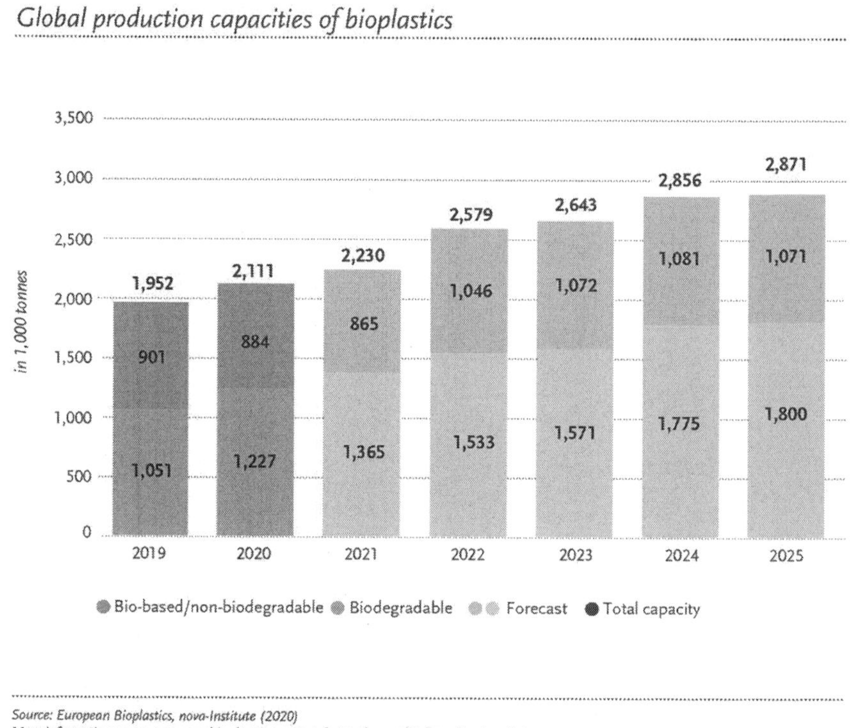

FIG. 11.1 Current and projected global production capacities of bioplastics. *Courtesy: European Bioplastics.*

The highest volume application for bioplastics is still packaging, which held about 47% market share as of 2020. The consumer goods category is gaining share at an accelerated rate (albeit from a lower starting point), based upon large corporations' sustainability goals related to renewable materials.[6–8]

The reputation of bioplastics

The historical reputation has been affected by the negative market forces mentioned in Table 11.1. Three of those have had an outsized impact:

- Bioplastics were not initially able to match the physical material properties of fossil fuel plastics and, over time, became viewed as inferior from a performance perspective.
- Some bioplastic manufacturers, unfortunately, engaged in greenwashing campaigns. The resulting public confusion over the benefits of the technology in some cases drove public outrage about being deceived with false advertising.
- Bioplastics cost significantly more than fossil fuel plastics. They have failed to gain any substantial share of the plastic market. Critics argue that current bioplastics are not a real solution to plastic pollution.

The "brand reputation" has been damaged over the years. The pressure to solve the environmental challenges associated with fossil fuel plastics forced early bioplastic companies to launch products that could not compete with incumbent technologies on material performance or price. They believed that the environmental benefit would offset these challenges but failed to understand actual consumer behavior. Market demand, greed, and a lack of integrity lead to greenwashing. Some in the industry chose to market their new technologies as completely biodegradable when they degraded only under industrial composting conditions. The message caused confusion and anger among consumers and led to the early products being somewhat tainted. Greenwashing will be explored in more detail later in the chapter.

The costing challenge is the most difficult of all. The cost advantage that fossil fuels have over renewable technologies is due to a hidden carbon subsidy. Elon Musk gave a passionate speech in Paris in 2015 calling for a carbon tax. His remarks hinged on a simple observation: pumping carbon into the atmosphere equates to an enormous global fossil fuel subsidy. The CO_2 contributes to climate change and prevents humanity from transitioning out of fossil fuel and into renewable energy. He cited an International Monetary Fund study that calculated the environmental damage caused by each ton of carbon emitted. This study estimated a hidden carbon subsidy of $5.3 trillion per year. That figure has been updated to $7.6 trillion as of 2020. If the government successfully implemented a carbon tax, renewable energy and materials would immediately be cost-competitive with their fossil fuel counterparts.

The issue of a carbon tax is politically and economically complicated. Pundits argue that a carbon tax would lead to inflation that would have a disproportionately negative impact on developing countries struggling to raise living standards. There have been other successful tactics to help level the playing field on cost. The electric vehicle industry has successfully implemented carbon tax credits. The Biden administration proposed the Clean Electricity Payment Program legislation, which takes this idea a step further for clean energy, using a carrot-and-stick approach. It may be the best shot at a progressive climate law this decade. The bioplastics industry needs to push for this kind of legislation.

The good news is that progress continues to make bioplastics more competitive than petroleum-based plastics, and that legislation was debated ahead of COP26 in Glasgow. Production capacity growth leads to cost reductions that make bioplastics more economically competitive. The market demand for these technologies far outstrips the available supply, so supply solutions are rapidly developed. There is growing public demand that governments respond to the climate crisis with progressive climate policies, aggressive goals, and new legislation to lower emissions and address pollution. Finally, the bioplastics industry is making new technology breakthroughs driven by funding through government grants and R&D at top universities. There is a new wave of public, corporate, and venture capital investment into next-generation technologies that can help bioplastics live up to its environmental promises.

The commercialization of new bioplastic technologies

The bioplastics market is full of new and exciting technologies at various stages of market penetration. One example is polyethylene furanoate, PEF, a new polymer expected to enter the market in 2023. It is comparable to polyethylene terephthalate, PET, but is fully bio-based and features superior barrier properties, making it an ideal material for beverage bottles.

Algenesis Corporation was founded by faculty from the University of California San Diego in 2016. The company strives to develop fully biodegradable bio-based PU made from algae and other plant-based feedstocks with support from the US Department of Energy, the National Science Foundation, and the university itself. Bio polyol formulations were developed, leading to high bio-content PU foams that were 100% biodegradable. The work started in rigid foams and proceeded toward proof of concept. Since two founders were surfers, naturally they built an algae surfboard prototype. News of this sustainable innovation went viral, leading to major footwear companies inquiring about flexible foams. Those companies collaborated with Algenesis to push the performance metrics to acceptable levels for the industry. The foam material was branded Soleic, which means sun oil. By 2019 Soleic biodegradable footwear foams matched industry performance specifications, and commercial discussions commenced.

In 2020 the first commercial deals were being inked for a market launch. Then the pandemic hit, which impacted everything. Algenesis was forced to remotely commercialize the nascent material technology during the pandemic, which proved highly challenging. The company moved forward and signed an essential partnership with a chemical company in Asia to provide contract manufacturing and commercialization support services. The Soleic biodegradable PU foam was poised for a market launch with two different brands at the time of this writing (spring 2022).

The global bio polyol market is expected to grow from about $2 billion in 2020 to $6 billion in 2025. The global market for PU was valued at $70.67 billion in 2020 and is expected to grow to 5 billion in 2025.[9] Footwear is an excellent application for flexible biodegradable PU foam. Footwear has been an attractive entry point for large chemical companies that launch novel foam technologies. A good example is the expanded thermoplastic urethane foams developed by BASF in collaboration with Adidas, branded as Boost foam.

Many improvements to Soleic biodegradable PU are in the pipeline. In addition to the bio polyol, Algenesis has invented several novel bio-isocyanates, leading to 100% bio-content foams. There is also work being done on feedstocks. The initial Soleic launch used plant feedstocks, but the aspirational goal is to switch to algae-based feedstocks. Price and supply constraints are the current challenge there. Companies like Global Algae Innovations, a partner to Algenesis, are receiving new grant awards to scale algae feedstock production. Bio-refineries are a supporting business to biopolymers that must be funded to unlock the benefits of algae feedstocks. Increased supply will also lower the price. The advantage of algae is that it is the world's most photosynthetic-efficient plant. It grows on nonarable land, in nonpotable water, and does not compete with food.

Algenesis collaborated with Business Insider on a 7-min Soleic technology explainer video posted to YouTube in February 2021. It has been viewed 2.7 million times and has over 1900 public comments. There is clear demand for these new eco-innovations, particularly from the Millennial and Gen Z generations, who are very concerned about climate change and pollution.[10,11]

The evolution of the eco-consumer

The rise of conscious consumption[12] has its roots in various progressive social movements that started after World War II:

- fair trade movement of the 1950s.
- civil rights and social justice movement of the 1960s.
- environmental movement of the 1970s.
- climate movement of the 2000s.

Today an entire economy has developed around sustainable products. Some data:

- The sustainable products market grew 5.6 times faster than conventional products between 2013 and 2018.
- The sustainable products market is expected to grow to over $150 billion in 2021.
- The resale market grew 25 times faster than the retail market in 2019.
- 72% of respondents to a survey indicated they were buying more sustainable products and that they expected to increase this demand in the future.
- 65% of shoppers are more likely to buy eco-friendly products and 47% reported they would be willing to pay more.

Gen Z and Millennial consumers are driving a transformational shift in consumer preferences and spending power. These two groups are projected to grow to 70% of the global population by 2028 versus 60% today. Sustainability and corporate values play a significant role in purchase intent for the younger generations. They are not as much of a driver for older generations. A Cowen Research study from 2019[13] reported that 72% of Gen Z and Millennial consumers view social impact and sustainability as being "very important" versus "somewhat important," which was well above older demographics, which averaged 53%.

Social media plays a significant role in shaping the consumption preferences of younger generations. Gen Z and Millennials use platforms such as Instagram for brand engagement at a much higher rate than older consumers. Moreover, they do this to discover new brands and products that match their preference for sustainable and social values. Their preferred sales channel is also digital, favoring e-commerce websites and Amazon over physical stores.

The COVID-19 pandemic created seismic shifts in retail and e-commerce in 2020 that may forever change the way people shop. Twenty-one major North American retailers went bankrupt in 2020, while e-commerce sales increased by 32.4% and went from 15.8% of total retail sales to 19.6%. This change represents the equivalent of 5 years' worth of growth in 1 year, according to new data from IBM's US Retail Index.[14,15]

Social media also sends viral messages about climate change and pollution to these younger generations. Young people are dealing with eco-anxiety as they contemplate their future on this planet. They feel that the older generations have been irresponsible and ignored the severity of the problem and the scientific consensus on fighting climate change.

In 2018 at age 15, Greta Thunberg started her school strikes for climate, which led to a global movement under the name Fridays for Future. This cultural tipping point galvanized the younger generations' cry for government and corporate change. Her sudden rise to fame made her a leader in the activist community and a target for critics.

Thanks to social media, it has never been easier to publish messages to a large audience, so we have seen increased activist movements. Many groups are explicitly attacking the problem of global plastic pollution. These groups are calling for new bioplastic materials, better waste management policies, and related beach and ocean cleanup.

These trends bring far more awareness to brands and products that are messaging sustainability and social impact. Sustainable materials are seeing increased demand as a result. Material trends have shifted from recycle stories to renewable plant-based material narratives. There has also been a big focus on lowering the carbon footprint of products through better material inputs and better manufacturing processes. Recently, biodegradable materials have returned to the forefront of the conversation as public concern over solid waste continues to increase and new technologies like Soleic come to market.

Barriers to commercialization

Cost

Higher cost remains the number one reason why green materials aren't adopted at a higher percentage of the marketplace. The latest wave of materials still costs more to produce and therefore carries a higher price tag when sold to corporate customers. The opportunity is for early adopters to aggressively market these new bio-based technologies and begin the scale-up process. Lower costs will result from the economies of scale in production.

Currently the debate is whether or not consumers are willing to pay more for green products—the "green premium." The jury is still out on this. However, recent Consumer Insights data indicates that younger consumers are willing to pay more than older ones. A 2019 Cowen report suggests that Gen Z and Millennials are willing to pay up to 15% more at retail. The exact percentage depends on which Consumer Insights reports you read.

When designing green material technologies, it's essential to factor in the economics during the R&D process. The goal is a balance between achieving a novel green material that meets physical specifications while optimizing for cost, supply availability, production specifications at the factory, and scale-up potential. More work needs to be done on the supply chain for bioplastics to gain economies of scale and reduce production costs. When optimizing for production, factories always seek efficiencies in speed, quality, and cost. Experiences at Algenesis are informative here. Early Soleic biodegradable PU formulations were problematic for the factory. These materials had higher viscosities than the polyether chemistry. The formulations had shorter cream times and longer demold times than traditional petroleum PUs. These challenges only increased the net cost of parts made from the material, because production ran slower and with a higher rejection rate. The factories had very little interest in working with the material, because the production specifications were not aligned with their expectations. Substantial effort and innovative formulation chemistry were necessary to overcome these challenges.

Feedstock

Another hurdle specific to bioplastics is the feedstock dilemma. The ideal feedstock to make bioplastics is photosynthetic algae. Algae is the most efficient photosynthetic plant on Earth; it can grow on nonarable land and in almost any kind of water. Photosynthetic algal plastics have the lowest carbon footprint and don't compete with food. The marketing departments of many top companies who want these materials have fallen in love with this story and won't accept any substitutes. The challenge with algae feedstocks is the high price and limited supply. These challenges must be overcome. In the meantime, innovative companies have shifted to plant-based feedstocks that are lower in cost and don't have the supply constraints associated with algae. The best of these are non-food-based, but many come from sources like sugar cane and corn and compete with food. Large corporate customers refuse to buy these for fear of being called out in the media for taking away food to make bioplastics.[16]

Resistance to change

This is a hurdle facing any new bio-based material technology company. It shows up in many forms. If you are in the green materials business long enough, you will hear every excuse from a potential customer. Established brands purchasing green materials are often caught in a tug of war between doing the right thing for the planet and pleasing shareholders. The shift from a shareholder economy to a stakeholder economy is relatively new, and old habits are hard to break. The first step to success is finding customers who share your values.

In our experience, there are several specific areas where resistance to change negatively affects the market adoption of new bio-based materials. The first is resistance to compromise on physical performance properties. This opposition is especially true in footwear, where decades of running shoe marketing convinced consumers that higher performance was always better. Brands such as Nike and Adidas designed sports performance products that were highly valued for their lighter weight, greater energy return, and lesser compression. Most shoes see only casual use and will never be pushed to their performance athletic limits. However, regardless of that, most shoe companies place a high value on lofty performance metrics and refuse to compromise for the planet.

As a society, we have developed an unfortunate addiction to perfection. This drive for the perfect material is what unwittingly led into the plastics crisis in the first place. Formulation chemists listened to decades of customer complaints about undesirable physical properties in the materials. Hydrolysis and UV resistance are good examples. We don't want our colors to fade or change tones in sunlight. We want the materials to be perfectly waterproof and highly durable. Talented chemists found solutions for all these problems. However, this resulted in materials that didn't degrade in nature for 400 to 500 years. An example from Algenesis: One customer's team asked for biodegradable foams that did not hydrolyze or turn yellow when exposed to UV light. This request baffled the scientists and resulted in some laughs. However, the unfortunate fact is that most companies do not understand material science at all. Hindsight is 20/20, and it's clear now that chemical companies overdesigned petroleum materials compared to the average life cycle of the product.

Funding

Funding models for new start-ups is another area where resistance to change hurts a company's chances of a successful market entry. Venture capital proved to be a poor funding model for risks versus rewards associated with early renewable energy and material companies. A recent shift toward private and angel funding has helped improve the situation. Government grants are also an excellent source of nondilutive financing for scientific founders looking to develop new bio-based material technologies.

Greenwashing

Another challenge that new eco-consumers face is separating fact from fiction in advertising, particularly when evaluating environmental material claims. Greenwashing is a huge problem, and it shows up in various forms. TerraChoices's *The Sins of Greenwashing: Home and Family Edition*[17] lists the seven sins of greenwashing as follows:

(1) *Hidden Trade-off* claims that a product is "green" based on an unreasonably narrow set of attributes without attention to other important environmental issues.
(2) *No Proof*: a claim that cannot be substantiated by easily accessible information or a reliable third-party certification.
(3) *Vagueness*: a claim so poorly defined or broad that its real meaning is likely to be misunderstood by the consumer.
(4) *Worshipping False Labels*: a claim that, through words or images, gives the impression of a third-party endorsement where none exists.
(5) *Irrelevance*: a claim that may be truthful but which is unimportant or unhelpful to consumers seeking environmentally preferable products.
(6) *Lesser of Two Evils*: a claim that may be true within the product category, but that risks are distracting consumers from the more significant environmental impact of the category as a whole.
(7) *Fibbing*: a simply false claim. This is the rarest sin.

There is a need for more watchdog groups to help protect consumers and the planet from misleading and false advertising. In the United States, the Federal Trade Commission provides voluntary guidelines for environmental marketing claims but only recently began enforcing their Green Guide. For the bioplastics industry and any green material technology company, the best watchdog groups would be those versed in the material science of the market sector. This need for educational expertise makes the top scientific universities ideally suited to forming antigreenwashing coalitions in their countries.

Unfortunately, most companies do not have rigorous enough standards to ensure that greenwashing does not occur. The best companies have PhD scientists who clearly understand what is beneficial to the planet regarding green material technology and sustainable production methods. Most companies want to do the right thing. However, they lack the scientific education or staff to determine the best sustainable strategies. The worst companies lack integrity and are willing to make patently false claims in advertising to appear relevant.

There has been a recent trend in the industry toward "science-backed" sustainability strategies. These corporate policies require scientists to verify sustainability claims. They include robust legal criteria for sustainability messaging coming from their marketing departments. Some of the ideas in these programs can help, starting with obvious things like making specific claims versus general or generic claims. These strategies also demand proof of benefits and certifications from third-party scientific labs to verify the claim. This validation is an essential tactic moving forward to combat greenwashing. Corporations and consumers must educate themselves to make a real impact.

The future of the bioplastics market

The global bioplastics market is forecasted to reach $16.8 billion by 2030, at which time the total plastics market is expected to be $850.1 billion. If these values hold, bioplastics market share will go only from about 1% to 2%.[14,15,18,19] For the markets in which Algenesis competes, bio polyol would be $9.6 billion, while global PU would be $88.8 billion. This suggests that bio-based PU could achieve a market share of almost 11% of the total PU market. From an environmental perspective, this is not very inspiring. So more aggressive work needs to be done to accelerate the transition to bio-based, renewable materials. In short, all the barriers to bioplastic commercialization must be prioritized and overcome one by one. The most challenging obstacles are higher cost, meeting performance specifications, and overcoming greenwashing.

The electric vehicle (EV) industry is an interesting analogy to bioplastics. EV total sales growth and market share growth are much stronger than those of bioplastics. The EV industry has benefited from innovators such as Elon Musk at Tesla producing innovative, high-quality cars. It is a classic example of the green bundle concept in marketing, combining top-notch performance, sexy design, and substantial environmental benefits. The bioplastics industry needs a killer application that can generate viral public relations and advance awareness of new technologies. The area where EVs are way ahead of bioplastics is fighting the war on cost by pushing carbon tax credit legislation. This is an important tactic that the bioplastics industry needs to adapt quickly. Deloitte[20] forecasts that EVs will make up 32% of all new car sales globally by 2030. Admittedly, tax credits are easier to implement with expensive cars versus commodity items. Even if governments only eliminated import duties on products built from renewable materials, that would be a good start and would help level the playing field on costs.

The legislative battle in this war against fossil fuel plastics is important. The petroleum industry will not go quietly into the night. Big Oil[21] is looking to grow plastics as they see the demand for fossil fuel waning in the energy sector. This is a war that needs to be fought and won, with consumers demanding that governments and corporations abandon nondegradable fossil fuel plastics.[17]

There are reasons for hope. The fossil fuel plastics sector may not grow as forecasted given several key trends in society that act as negative market forces:

(1) Rising carbon emissions are not "cool" in the age of the Paris Climate Agreement.

The annual greenhouse gas emissions from global oil refining and that from the production of plastics are similar, about 5%. What is deceiving is that there are so many more tons of oil refined that plastic produced, meaning that plastics production is a very carbon-intensive process.[22–24] Suppose plastic demand continues to grow as forecasted. A report by SYSTEMIQ[25] calculates that it would use 19% of the entire remaining global carbon budget. For one sector to double its carbon footprint while the rest of the world plans to phase out emissions makes no sense. Policymakers cannot allow this to happen.

(2) Plastic produces external costs that are almost equal to its total market value.

Like the energy industry, the plastics industry imposes costs on society that it doesn't have to pay, which are referred to as "externalities." Plastic emits carbon, generating air pollution. After use, it must be collected and sorted, yet much of it winds up in the oceans. The latest research suggests that plastics have a total unpriced externality cost of between $800 and $1400 per ton when the costs of expelled CO_2, collection and sorting, and cleanup in the ocean are considered. The unpaid external costs are being paid today, overwhelmingly by poor people living in developing countries. It is a human rights abuse by wealthy plastics companies. Legislators need to find ways to end these subsidies, ideally with a carrot-and-stick approach to carbon taxes and renewable materials tax credits.

(3) The plastics industry is extraordinarily wasteful.

About 36% of all plastics are single use, the worst form of waste. Approximately 40% of this waste is mismanaged, with 5% ending up in the ocean, 22% in open burning, and 14% in terrestrial leakage. Recycling rates are pathetic, with only about 9% of total plastics being recycled and less than 5% of that is substituting for virgin plastic. Finally, there are not many guidelines or regulations on plastic products, so just about anything goes. Rather than improve the underlying products, the industry has engaged in propaganda and misdirection. In short, the fossil plastics industry is a bloated behemoth that is ripe for disruption.

(4) The public is waking up to the enormous costs of plastic.

The public and lawmakers are becoming more concerned about climate change and plastic pollution, especially in the ocean. It is hard to imagine

TABLE 11.2 Findings of a poll on plastics conducted by IPSO.

Statement	Response
Manufacturers should be obliged to help with the recycling and reuse of packaging that they produce	80% Either Strongly or Tended to Agree with the statement
Single-use plastics should be banned as soon as possible	75% Either Strongly or Tended to Agree with the statement
I feel better about brands which make changes to achieve better environmental outcomes	77% Either Strongly or Tended to Agree with the statement

fossil fuel plastics investors believing that the sector might somehow be immune to these negative forces. An IPSO (Independent Press Standards Organization) poll in 2019 shows clearly that the public wants to reduce plastics and favors a ban on single-use plastics (Table 11.2).

In Europe, regulators cracked down, imposing an 800 euro per ton tax on unrecycled plastic waste. The majority of forecasted demand growth for plastics is expected to come from China and other emerging markets. There are signs that those nations have stepped up to curtail plastic use. China recently banned a range of single-use plastics, and many other countries are expected to follow suit. New York banned plastic bags last year, which was a big step to reducing single use in the state.

So, what must be done to ensure that we reduce fossil fuel plastic use as quickly as possible and accelerate the transition to renewable bio-based plastics? Fig. 11.2 shows projected reduction of the plastics growth curve if the five specified interventions are implemented.[25] The three most scalable and cost-effective solutions are reducing demand through elimination or reuse, substituting other products like compostables or paper, and better recycling. These ideas constitute the System Change Scenario (SCS). Under SCS, total demand for global plastics plateaus in 2020 and peaks in 2030. It achieves a net reduction of 200 million metric tons of plastic production per year by 2040. A percentage of this net reduction, perhaps 50 to 100 million metric tons of the annual output by 2040, could be awarded to next-generation bioplastics. SCS is also cheaper for the industry than business as usual. Investment in new technologies goes up, while virgin production and conversation are sharply reduced.

To achieve SCS and maximize the future market size for bioplastics, three groups need to step up and take swift, aggressive action.[21–24]

- Consumers: The general public typically underestimates the power they wield to make environmentally conscious purchasing decisions and pressure corporations and governments to change their policies.
- Corporations: Companies of all sizes, but in particular the Fortune 500, need to implement scientifically backed sustainability strategies that include switching to renewable, bio-based materials.

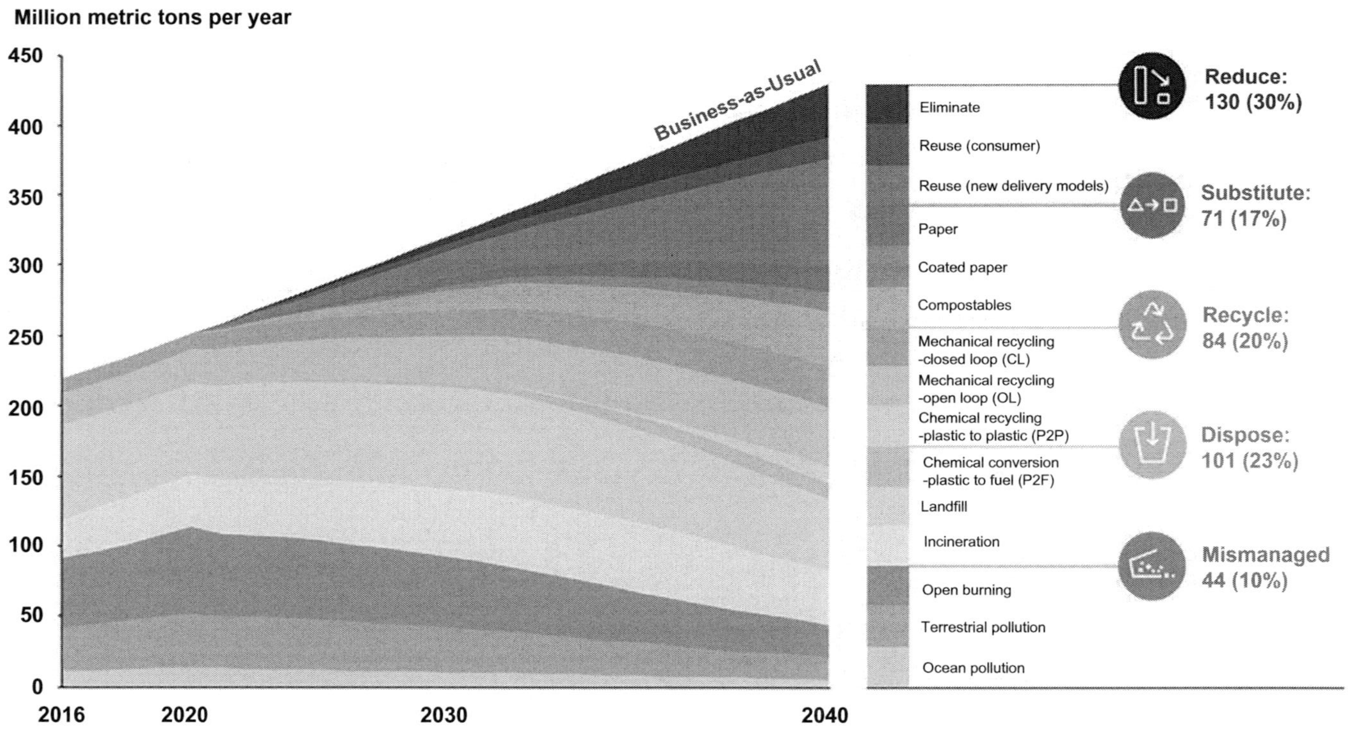

FIG. 11.2 Projected reduction of the plastics growth curve by implementation of five strategies. *Courtesy SYSTEMIQ.*

- Governments: The world's top governments need to end the lip service and summon the political will to pass the necessary legislation to accelerate the transition from the fossil fuel era into the renewables era. This change is just as crucial for the materials sector as for the energy sector.

Conclusion

Humanity must exit the fossil fuel era, for both energy and materials, as soon as possible. If we are successful in this, we can avoid the worst possible outcomes of climate change and pollution. Regardless of our actions, there will be damage to contend with. The Paris Climate Accords and the recent COP26 conference in Glasgow clearly state the goals and targets that must be achieved. Environmental activists, especially the youth climate movement, have correctly stated that world leaders are not doing enough. While there have been positive measures taken to protect the Earth and an increasing focus on sustainability, what we hear from our leaders is mostly rhetoric and is essentially failing to implement the legislative policies we desperately need. To have a chance at achieving net-zero emissions by mid-century and to keep 1.5 degrees in reach, every industry has to change, and fast.

The green materials industry needs more collective effort toward influencing government legislation. Laws are required to incentivize the adoption of bio-based materials and punish the continued use of fossil fuel materials. These efforts will help level the playing field on cost and market pricing. As mentioned earlier, the bioplastics industry needs a green tax credit similar to what was employed in the electric vehicle industry. Bioplastics companies need to form a coalition and start lobbying in Washington, DC.[26]

A much more comprehensive and aggressive strategy must be applied to supplant petroleum plastic for the bioplastics industry. We know what must be done to bend the plastic curve downward. We must do better as a society than 2% to 11% market share for bioplastics by 2030; otherwise, we'll never reach net zero by 2050 or limit warming to 1.5 degrees. For the sake of the planet, the bioplastics industry and all industries looking to exit fossil fuel materials need to think about *how* they will achieve their goals to disrupt and gain share. Here are some final thoughts and ideas for the future of the bioplastics industry:

- Establish a global bioplastics industry coalition to lobby governments worldwide for the legislative changes the industry needs across the board.

- Focus the initial wave of legislation on fighting the war on cost. These new laws should include carbon taxes on fossil fuel materials and tax credits for renewable bio-based materials.
- Encourage entrepreneurship through increased funding from government grants and angel/impact investing groups. This will ensure that the next generation of bioplastics is even more competitive.
- Establish a coalition to combat greenwashing and aggressively prosecute offenders who don't meet new science standards.
- Continue to push for supply solutions to photosynthetic algae-based feedstocks. This is the future of the industry.

We cannot sit back and wait for other organizations to do the work. We have to band together and act now. We are fighting for a safer and more secure world.

> The best way to predict the future is to create it. ***Peter Drucker***.

Close-up: Challenges of commercialization of new bio-based products (or any new technology)

Investors (including both angel investors and VCs) are always looking for the next new thing, which usually embodies some new technology. Investors seek to find that elusive "white space," where something new meets an unmet market need. VCs invest other people's money and so have a needed timeline for an exit. Therefore they are unlikely to want to invest in a company where the science is so novel that it is hard to distinguish between research and product engineering. On the other hand, angels invest their own money, so are more likely to invest in a breakthrough technology that still needs to be proven.

Bringing a breakthrough technology to market is both difficult and expensive. The difficulty is often underestimated by academics, who think of a project as "done" when the peer-reviewed paper is published. Turning that technology into a cost-effective product and then convincing customers to change the existing way they do something is a daunting task. Most scientists would assume that it is relatively easy to change buying behavior, because there is a commercial benefit to make the change. Unfortunately, that is not the way most buyers (both businesses and consumers) behave. But, having a plan that incorporates the new technology to convince the buyers to buy your new thing is critical to commercial success.

This is an area where the technologists can partner effectively with their angel investors, particularly if those angels have skills that the founding team lacks and can use.

Start-ups need to be careful of who invests—it's more than just money. Investors become owners of the company and have certain rights as owners. Setting clear expectations with investors on timing and deliverables, and then communicating regularly, is a critical element of a successful start-up. Having a lead investor, who can serve as a board member, aids in this process.

There is a clear market trend toward environmentally friendly products. Consumers want to buy products that help the environment. Clearly, bio-based products can fit that bill. The narrative (often coupled with a marketing plan) will help the company deliver a science-based story that gives customers a reason to change behavior and buy your product.

Daniel Rosen

CEO and President, Dan Rosen & Associates, Kirkland, WA, United States

References

1. Bioplastics News. The History of Bioplastics; 2022. https://bioplasticsnews.com/2018/07/05/history-of-bioplastics/. Accessed 4 February 2022.
2. Environmentalism. History of the Environmental Movement. Britannica; 2022. https://www.britannica.com/topic/environmentalism/History-of-the-environmental-movement. Accessed 4 February 2022.
3. Bioplastics Magazine. What the Oil Crash Means for Bioplastics; 2022. https://www.bioplasticsmagazine.com/en/news/meldungen/20200722-What-the-oil-crash-means-for-bioplastics.php. Accessed 4 February 2022.
4. Ecommerce Trends Amid Coronavirus Pandemic in Charts; 2021. https://www.digitalcommerce360.com/2021/02/19/ecommerce-during-coronavirus-pandemic-in-charts/. Accessed 4 February 2022.
5. TechCrunch. COVID-19 Pandemic Accelerated Shift to E-Commerce by 5 years, New Report Says; 2022. https://techcrunch.com/2020/08/24/covid-19-pandemic-accelerated-shift-to-e-commerce-by-5-years-new-report-says/. Accessed 4 February 2022.
6. Markets and Markets. Bioplastics & Biopolymers Market Global Forecast to 2026; 2022. https://www.marketsandmarkets.com/Market-Reports/biopolymers-bioplastics-market-88795240.html. Accessed 4 February 2022.
7. European-Bioplastics e.V. Market Update 2020: Bioplastics Continue to Become Mainstream as the Global Bioplastics Market is Set to Grow by 36 Percent Over the Next 5 Years; 2022. https://www.european-bioplastics.org/market-update-2020-bioplastics-continue-to-become-mainstream-as-the-global-bioplastics-market-is-set-to-grow-by-36-percent-over-the-next-5-years/. Accessed 4 February 2022.
8. Bioplastics Market Size, Share, Growth Report, 2021-2028; 2022. https://www.grandviewresearch.com/industry-analysis/bioplastics-industry. Accessed 4 February 2022.
9. MarketWatch. Global Microcellular Polyurethane Foam Market Size, Share Remuneration to Surge At 4.3% CAGR Through 2025—Industry Report; 2022. https://www.marketwatch.com/press-release/global-microcellular-polyurethane-foam-market-size-share-remuneration-to-surge-at-43-cagr-through-2025- - -industry-report-2022-03-23?mod=search_headline. Accessed 23 March 2022.
10. Polyurethane Market Size, Share & Trends Report, 2021-2028; 2022. https://www.grandviewresearch.com/industry-analysis/polyurethane-pu-market. Accessed 4 February 2022.

11. IMF F&D. The True Cost of Reducing Greenhouse Gas Emissions; 2019. https://www.imf.org/external/pubs/ft/fandd/2019/12/the-true-cost-of-reducing-greenhouse-gas-emissions-gillingham.htm. Accessed 15 February 2022.
12. Bloomberg. CEPP: Biden's Green Power Plan Helps Solar, Wind and Bankers; 2022. https://www.bloomberg.com/opinion/articles/2021-09-10/cepp-biden-s-green-power-plan-helps-solar-wind-and-bankers. Accessed 15 February 2022.
13. Cowen. Gen Z/Millennials: Sustainability Supports Durability; 2022. https://www.cowen.com/insights/gen-z-and-millennials-are-driving-force-in-scaling-digital-and-sustainability/. Accessed 23 March 2022.
14. Green and Bio Polyols Market Size, Share, Growth, Forecast till 2030; 2022. https://www.decisionforesight.com/reports/green-and-bio-polyols-market. Accessed 4 February 2022.
15. MENAFN.COM. Global Polyurethane Market Worth US$ 88.76 Billion by 2030; 2022. https://menafn.com/1103037291/Global-Polyurethane-Market-worth-US-8876-billion-by-2030-with-a-CAGR-of-250. Accessed 4 February 2022.
16. Sardon H, Mecerreyes D, Basterretxea A, Avérous L, Jehanno C. From Lab to Market: Current Strategies for the Production of Biobased Polyols. Published online; 2021. https://doi.org/10.1021/acssuschemeng.1c02361.
17. The SinS of GreenwaShinG Home and Family EdiTion. www.ulenvironment.com.
18. Bioplastics Market is Projected to Reach $16.8 Billion by; 2021. https://www.globenewswire.com/en/news-release/2021/09/06/2291947/0/en/Bioplastics-Market-is-Projected-to-Reach-16-8-Billion-by-2030-AMR.html. Accessed 4 February 2022.
19. News Desk24. Plastic Market Report 2021-2030 Breakdown by Manufacturer, Products, Applications, Region; 2022. https://www.mynewsdesk.com/se/news-desk24/pressreleases/plastic-market-report-2021-2030-breakdown-by-manufacturer-products-applications-region-3091314. Accessed 4 February 2022.
20. Electric Vehicle Trends | Deloitte Insights; 2022. https://www2.deloitte.com/us/en/insights/focus/future-of-mobility/electric-vehicle-trends-2030.html. Accessed 4 February 2022.
21. Big Oil's Hopes are Pinned on Plastics. It Won't End Well.—Vox; 2022. https://www.vox.com/energy-and-environment/21419505/oil-gas-price-plastics-peak-climate-change. Accessed 4 February 2022.
22. Cabernard L, Pfister S, Oberschelp C, Hellweg S. Growing environmental footprint of plastics driven by coal combustion. *Nat Sustain.* 2022;5(2):139–148. https://doi.org/10.1038/S41893-021-00807-2.
23. Futurity. Plastic is Even Worse for the Environment than We Thought; 2022. https://www.futurity.org/plastic-carbon-emissions-2668492/. Accessed 23 March 2022.
24. Stanford News. Measuring Crude Oil's Carbon Footprint; 2022. https://news.stanford.edu/2018/08/30/measuring-crude-oils-carbon-footprint/. Accessed 23 March 2022.
25. SYSTEMIQ. Breaking The Plastic Wave: Stop Plastic Pollution; 2022. https://www.systemiq.earth/breakingtheplasticwave/. Accessed 15 February 2022.
26. Acemoglu D, Akcigit U, Hanley D, Kerr W. Transition to clean technology. *J Polit Econ.* 2016;124(1):52–104. https://doi.org/10.1086/684511.

CHAPTER

12

The future of biobased polymers from algae

Stephen P. Mayfield and Michael D. Burkart

University of California San Diego, La Jolla, CA, United States

Introduction

Just 200 years ago, most of the world's population was living in poverty,[1] unsure of where their next meal was coming from and unconcerned about the sustainability of their lifestyle. The industrial revolution was already underway, but it had yet to impact much of the world, as it was not yet supercharged with the petroleum that would soon power internal combustion engines and bring the changes, good and bad, that the petroleum revolution would unleash. As the industrial revolution, and then the agricultural (green) revolution, kicked into high gear in the 20th century, the world's population began to climb. Over the last 50 years, the world's consumption—of energy, food, water, minerals, or any product—has soared, increasing almost 30-fold. We used fossil fuels to drive both the industrial and agricultural revolutions to propel world populations, and with it, consumption, to levels that are unsustainable and potentially on a path to catastrophe. One example is the availability and use of plastics. Plastics, produced almost exclusively from petroleum, are ubiquitous, inexpensive, functional, and persistent. They highlight the good and the bad that the petroleum revolution has had on our lives and the planet. In the 21st century, we must move to a more sustainable world, with forethought and design and a full accounting of the life cycle costs of all products, not merely their production costs. We need to do this immediately, while there is still time to mitigate the most egregious damage.

Algae: The beginning

Algae made our lives possible. Algae are not just the green slime growing in the birdbath or the seaweed along the beach; they are the

Rethinking Polyester Polyurethanes
https://doi.org/10.1016/B978-0-323-99982-3.00012-2

primary producers in all aquatic systems and are therefore the critical organisms on which all aquatic life depends. Algae are also the reason that animals, including humans, can thrive on this planet. They are the primary enablers of the economy, as they were the primary producers of fossil petroleum and natural gas. Four billion years ago, when the earth was new, the atmosphere consisted primarily of carbon dioxide (CO_2), and oxygen was in limited supply, with most of it trapped in CO_2. There were ancient prokaryotic chemotrophs, but no animals could survive in that environment, as there was no reduced carbon suitable as a food source and no oxygen for respiration. Into that world came cyanobacteria, and with them the process of photosynthesis. It took some time for algae to work their magic, but utilizing sunlight as their energy source, they began to turn CO_2 into algae biomass and secrete oxygen into the atmosphere. As the algae multiplied, the seas filled with algae biomass in the form of lipids, proteins, and carbohydrates, and the atmosphere began to fill with oxygen. Over the next billion years, algae turned a once unforgiving planet into a world filled with organisms of every imaginable form. About 800 million years ago, the first animals evolved, as there was now sufficient oxygen and food in the form of reduced carbon to make their lives possible. About 400 million years ago, vascular plants evolved from algae, allowing land-based herbivores to thrive, which eventually led to mammals appearing about 200 million years ago, and finally to humans just a few million years ago. The lineage of life on earth as we know it is very clear, and it all started with algae, and they remain an essential part of the environment today.

Petroleum is ancient algae

Given that algae fixed all of the CO_2 in our early atmosphere into biomass, that leads to one important question: Where did all that carbon go? Much of it eventually formed the enormous deposits that we call fossil fuels, specifically petroleum and natural gas. You likely know that coal is fossil plant matter, but not many know that petroleum and natural gas (methane) are primarily fossil algae accumulated over hundreds of millions or even billions of years. As algae "fixed" CO_2 from the atmosphere into biomass, they settled to the bottom of shallow seas and were covered with silt and sand. Over millions of years under high pressure and heat inside these geologic formations, this biomass was transformed into crude oil, with the nitrogen, phosphate, metals, and oxygen precipitating or volatilizing, leaving primarily hydrocarbons behind. This petroleum sat underground in vast reserves for hundreds of millions of years, sequestering that carbon from exposure to the atmosphere and holding enormous amounts of stored energy in these liquid and gaseous hydrocarbon deposits.

Our petroleum addiction

Over the last 50 years, the world's population has doubled, and the global material footprint for goods has grown nearly 500%.[2] To say that it is unsustainable does not begin to describe what this overconsumption has done to the planet. If we look just at plastic waste, the numbers are staggering. Plastic scarcely existed outside of research labs before 1950, and yet today we produce close to 500 million metric tons every year, with the vast majority never being recycled. An estimated 2600 pounds of plastic has accumulated for every person alive today, and it continues to accumulate at a rate of 300 pounds per person per year. To be sure, plastics have an essential role in modern society, but we have become addicted to their convenience and low cost. Most people do not consider the consequences of using these materials for everything from single-use packaging to shoes. We need to rethink how we make, use, and dispose of these materials, and we must come up with solutions that retain the essential benefits while reducing the environmental pollution. The question is, how do we do this in a way that doesn't cost too much or set back our standard of living? Indeed, one option requires that all plastics be recycled—not recyclable, but completely recycled! This requirement will likely require significant government involvement, both carrots and sticks—economic incentives to recycle and significant penalties for material not recycled. Another option is to make plastics biodegradable, much like paper or wood, which can either be recycled or naturally biodegrade when left in the environment. Plastic production choices can also have enormous environmental impacts. We must choose to utilize those processes that require less energy, water, and land to make plastics, and make sure they are easy to fully recycle or, ideally, both recyclable and biodegradable. There are no simple solutions, but the consequences of not changing how we make, use, and dispose of plastics are too great to ignore.

The origins of plastics

Since plastics are derived from petroleum and petroleum comes from ancient algae oil, the obvious next question is, can we make plastics from cultivated algae oil or even plant oils? The answer is, of course, yes! The development of plant-based renewable plastics is already well underway. Major considerations in this technology are what plastics should be made, and what features should they embody. One highly desirable feature is biodegradability, so that plastic products that end up as terrestrial or ocean garbage quickly biodegrade and do not become persistent pollution. Biodegradability would also avoid the production of microplastics, found to concentrate in the food chain and eventually end up in us. It is quite possible to make plastics from algae and plant-produced metabolites. We have made polyurethane (PU) from algae that is fully biodegradable and meets quality specifications for commercial products, and these plastics can degrade in the natural environment in less than a year. Now that we have biodegradable PUs, the challenge will be to implement them at scale and reasonable cost, using sustainable, nonfossil starting materials. Because PUs are used in everything from high-end fashion to commodity building materials, there is an opportunity to start with expensive, low-volume products, such as shoes, and build less valuable commodity products over time. In this way, consumer habits can adapt to these more environmentally responsible options as availability increases and cost decreases.

First-generation renewable plastics

The first target for sustainable development was redirecting algae and plant oils that had already been optimized for biofuels toward use as plastic precursors. Because algae oils typically contain at least 50% unsaturated fatty acids, we recognized that these unsaturations could serve as useful handles for chemical modification. We first turned our attention to PU, which is made from two components, polyols and diisocyanates. The polyols provide two or more alcohols that become chemically crosslinked by diisocyanates to give the polymer matrix structure. Vegetable oils had already been used to produce polyols for PUs through simple transformations, and we realized that these methods could be applied to algae oil. Here, the unsaturations in the oil were epoxidized, and subsequent ring opening with water or hydroxylated nucleophiles provided a simple approach to polyol synthesis, making polyols that were suitable for rigid foams in products like insulation, automobile dashboards, and surfboard cores. Our first foray into algae PU was to develop a rigid surfboard core. That program made headlines in magazines around the world and sparked an interest in other PU-based products.

Flexible PU foams were our next target. These were much more difficult to implement due to the requirement that they demonstrate advanced material properties dictated by existing products, such as shoe footbeds or outsoles. PU formulations for major athletic shoe manufacturers are quite advanced, providing performance enhancements for athletic activities. We quickly realized that we needed to improve the ability to tailor our PUs through molecular control of the polyols, and to do this we needed to develop new chemistry with our algae oil precursors. Our recent advancements include producing drop-in polyester polyols and diisocyanate crosslinkers using chemical transformation from isolated unsaturated fatty acids. This required significant investment into downstream chemical transformations, but we can now prepare PUs that approach 100% renewability from fatty acids isolated from wild-type algae biomass. These are first-generation monomers, since they require no genetic manipulation to prepare novel monomers or their precursors.

Second-generation renewable plastics

The next step will require genetic modification of algae to achieve novel plastic precursors. These will come from pathways that may not naturally exist in algae but may be found in other organisms, such as bacteria or fungi, and translated into an algae host. We have spent the last 30 years developing all of the genetic tools needed to engineer the algae genome and are now at a critical stage to realize these advancements. For instance, we are working on modifying algae fatty acid biosynthesis to develop new monomers for PU applications. These ongoing programs are expected to further improve PU diversity with superior material properties.

Algae still the base of the global carbon cycle

After the first algae revolution, the one that made earth habitable for animals and deposited all the world's crude oil underground, algae kept on growing. Today they make up the most diverse set of organisms on the planet and the foundation of the global food web. Take, for instance, omega-3 fatty acids. For decades, medical science has known that these polyunsaturated oils are essential for human health, including neural development and cardiovascular function. These have traditionally been sourced from fish oils. However, fish do not make omega-3s. These essential metabolites are produced by marine algae and translated up the food chain to eventually arrive in seafood. Based on this knowledge, algae aquaculture has been shown to provide the most sustainable means to produce omega-3s for medicines and nutritional supplements. This situation is emblematic of many opportunities for algae. Sourcing our molecular and energy needs from organisms at the base of the ecosystem offers the most efficient and ecological option.

Algae are the most productive photosynthetic organisms by far; some produce biomass 20 times faster than the most efficient terrestrial crops. This efficiency offers an opportunity to make various products, including food, feed, fuel, and other commodities, in a very efficient manner, sequestering CO_2 in the process. There are many ways to measure the efficiency of algae production: one good example is oil. The highest yielding oil plant is palm, which produces, on average, 600 gal per acre per year. In comparison, some algae can have 70% oil by weight and produce over 3000 gal per acre per year. Algae can be grown on nonarable land, using nonpotable water, and strains can be grown in any climate; there is no need to cut down rain forests to grow algae. In addition to the oil produced, this same algae would also make about 10 tons of protein per acre per year, which could be used as a coproduct for animal feed, significantly reducing the overall carbon footprint of the oil production. Bioplastics made in this way would also reduce overall energy and water use. Taken together, is there another process that can make bioplastics as sustainably as algae production?

Fuel, food, and bioplastics

Algae have already been shown capable of agricultural biomass production of important commodities. Between 2005 and 2015, because of the increasing cost of petroleum and the threat of climate change, there was great interest in the use of algae to produce renewable fuels. Research efforts in academic labs and small cleantech companies demonstrated the potential to do so at scale. This required a combination of new and old technologies, some borrowed from other areas, including methods initially developed for aquaculture, water treatment, and petroleum refining. Algae cultivation provided protein and carbohydrates suitable for animal feed and oils suitable for renewable fuel production.

Importantly, these large-scale proof-of-concept studies successfully produced renewable gasoline from algae cultivation for approximately $8 per gallon, just over $2 per kilogram, at a time when gasoline cost over $5 a gallon in California. Although algae gas was not price-competitive with petroleum, it became clear that economies of scale and continued incremental improvements in the technology could result in renewable fuels that would someday compete. These important demonstrations showed that algae biomass could be grown in real time to replace fossil fuels and provided tremendous hope to those paying attention to climate change. Unfortunately, enthusiasm was dashed by the collapse of crude oil prices starting in 2014. Saudi Arabia and Russia decided to increase production, which, together with the increase from fracking in the United States, caused an oversupply. Oil dropped as low as $30 a barrel and remained there until 2021. This, along with denial of climate change and the role of fossil fuels in it, essentially killed investment in and support for renewable fuel. Companies that had shown such great success with algae biofuels were forced to focus on other research areas or go out of business. The sector has yet to recover.

Although this was a setback for the renewable fuels industry, and hence for climate remediation, it also resulted in the realization that if algae can come close to competing with the cheapest commodity on the planet (petroleum fuel), then surely there must be opportunities in other products with higher value. Indeed, various petroleum-derived products have a significantly higher value than fuel, including petrochemicals. The petrochemical market is cumulatively as valuable as that for fuels, even though petrochemicals use less than 20% of crude oil output. Although algae fuels had not been a commercial success, the door was opened, and petrochemical replacements were an obvious target market.

Biodegradation

Plastic development and engineering have transformed the world. Plastic is a component of most consumer products, and incorporation of plastics into new applications continues to grow. While this has brought us greater convenience and lower prices, it has come at a cost to the environment. Most plastics cannot biodegrade, and only a tiny portion are recycled. For decades, materials scientists and engineers aimed to make products that would last forever, as it was never conceived that they would have an end-of-life impact on the environment. Now we recognize that end-of-life consequences must be considered in design, manufacturing, distribution, and recycling.

All life forms biodegrade; it is the ultimate recycling of all natural materials *via* the carbon cycle. Using this as inspiration, we have chosen to design biodegradability into every new plastic material we develop. This is accomplished by making sure that the underlying chemical bonds of each polymer are breakable by enzymes from microorganisms, which limits the types of bonds that can be used. To date, we have focused mainly on developing PUs, specifically polyester PUs, as both ester and urethane (carbamate) bonds are common in biology and are therefore readily biodegradable. PUs are also a very large category of plastics that make up hard foams, soft foams, thermoplastics, films, and adhesives that can be made fully biodegradable. The gold standard for quantifying biodegradation of material uses respirometry, tracking microbial CO_2 production in a controlled environment with an instrument called a respirometer. We have used this methodology to validate the biodegradability of our PU materials in standard home compost and soil conditions, with an aim for complete biodegradation within 3 to 6 months. We have ongoing studies to track every atom of our polymers as microorganisms biodegrade them, and believe that this approach will become a requirement across the plastic development sector.

Recycling vs. monomer reuse

Recycling involves collection, washing, chipping, and then remelting and recasting of thermoplastics. Because of the degradation of plastic in the environment, and the contaminants that are not washed off, there is a practical limit on the amount of recycled material that can be used—notably the many items that already have about 20% recycled content. Depolymerization and molecular recycling of PUs work fundamentally differently, so the practical limit of reused monomers is theoretically 100%. The biodegradation process involves the secretion of PU depolymerizing enzymes by microorganisms, followed by the microorganisms consuming the released polymer degradation products, which are the same molecules originally used to build the PU monomers. With this in mind, we have begun to develop a process for the enzymatic depolymerization of PU products, with the subsequent recovery and reuse of the released monomers. The released monomers are purified so that only intact ones are recovered. This could prepare new PU products that contain 100% recovered material and still meet commercial specifications, demonstrating a viable circular carbon economy. We call this process Bio-Loop. It follows the same mechanisms as biodegradation, except that the depolymerized monomers are not metabolized by microorganisms back into CO_2. Instead, they are captured, purified, and reused to make new materials. Damaged monomers are separated from intact ones, and contaminating materials are removed. Hence, we can produce new monomers from 100% recycled materials without significant loss of quality, unlike traditional recycling.

Technologies on the horizon that can help us realize a more sustainable future

Although the increases in world population, consumption, and plastic pollution over the last 50 years are alarming, they are small compared to the growth in scientific advances. The growth in computing power has been exponential. The growth in biotechnology has been essentially infinite, because that discipline did not exist 50 years ago. The combination of just these two areas offers real hope for developing advanced, sustainable materials and renewable energy solutions that can have a profound positive impact on the environment in the coming years. If we consider just advanced metabolic cell engineering coupled with artificial intelligence for the design of new polymers, we have the potential to supercharge the rate at which new sustainable biodegradable materials can be developed. Molecules made from living organisms can be orders of magnitude more complex than those from traditional chemistry, which offers the potential for completely new advanced high performance materials. To achieve this, we can turn to AI to help predict material function, as it would be impractical to attempt to create and screen the enormous number of possible materials that could be made using biological substrates.

The opportunity in biotechnology lies particularly in DNA sequencing and synthesis. As we uncover new metabolic pathways for natural products from diverse organisms, we can also elucidate the enzymes that catalyze these processes. We can already engineer new metabolic pathways from one organism into another, and these tools are improving rapidly. Algae offer an ideal photosynthetic host for the production of new monomers and other fine chemicals, and we believe that they will play a major role in biomanufacturing. In their production, we will not only sequester CO_2 from the atmosphere but also take into account the life cycle of the products and plan for their end-of-life. It may be possible to replace petroleum altogether, and these solutions must provide better outcomes than the current fossil oil dependency.

How you can make an impact

We need to change the way we live and quickly move toward a more sustainable lifestyle. Everyone must be part of this. We need to think about conserving water and energy, reducing air and water pollution, and finding ways to decrease our carbon footprint. One relatively easy change is to move away from single-use plastics. As a consumer, you have the power to influence how products are made and what they are made of. You can choose to purchase products that contain less plastic. When new products that contain biodegradable or truly recyclable plastics are available, you can support them, knowing that they will likely be more expensive than the polluting alternative. These are ways that you can use your consumer choices to make a difference.

There are other things you can do if you are willing to speak up. You can directly tell the companies making your favorite products that you want them to remove nonbiodegradable plastics. You can support the creation of local, state, and national laws requiring the manufacturers of plastic products to make them fully recyclable, take responsibility for their collection and recycling, or make them biodegradable in an ordinary home compost. After all, local municipalities bear the burden of recycling or landfilling all the wasteful plastic that producers are adding to their products.

Conclusion

We have reached a critical point in human history where we have pushed the population, lifestyle, and consumption to levels that are simply no longer sustainable. We now have to make hard choices or face the brutal reality of creating an inhospitable world. Although we have become accustomed to plastics' convenience and low cost, those come with a significant environmental cost. This cost is one area where we can change the way we make, use, and dispose of these materials, which will still allow us to retain the essential benefits but decrease most of the environmental costs. Of course, this will not come free, and it will require us to develop new and likely more expensive alternatives. We will need to change the way we deal with these materials in manufacturing and recycling. The development of biofuels from algae has already demonstrated proof of concept for these goals. Although there will undoubtedly be trade-offs, we are convinced that real opportunities exist to make these changes without severely affecting our current way of life.

Close-up: The future of the bio-based polymer industry

The field of synthetic polymer chemistry is over 100 years old. Despite that age, nearly all synthetic polymers are sourced from fossil-based raw materials. No surprise here. The petrochemical industry has made fossil feedstocks cheap and readily available. Polymer scientists in turn developed a wide range of novel materials to address various customer needs. However, myriad studies confirm that today's linear plastic economy is far from sustainable. So, what can we do? First, we can agree that there is no single solution to the issue of plastic waste. Instead, it will take multiple solutions working in concert. Examples include bio-based raw materials, biodegradable polymers, government penalties and incentives, changing consumer behavior, improved mechanical recycling, new chemical recycling, product redesign, and more. Additionally, all these technologies must somehow fit into a circular economy.

For industry, bio-based polymers are an exciting solution that attacks a core polymer sustainability issue—replacing fossil fuels. Numerous bio-based feedstocks are available and include sugars, cellulose, and fatty acids sourced from both edible and nonedible plant biomass. The biomass can be raised specifically for this purpose or be a waste by-product of other industries. Synthetic microbiology provides an additional means of converting bio-based feedstocks into value-added monomers and polymers. This tremendous potential is fueling many bio-based polymer commercialization efforts, but success will be contingent on several factors. The polymer must perform as well as or better than incumbent materials. The raw material source must be available in sufficient quantity and consistent quality. This, combined with an efficient production process, will help achieve the lowest possible cost. There must be a market need for the polymer. Finally, partnerships between industry, academia, governments, and start-ups will be critical to minimizing risk and maximizing results. It will be hard work getting there, but the future is bright for bio-based polymers.

Joshua Speros
BASF Venture Capital, Boston, MA, United States

References

1. The Short History of Global Living Conditions and Why It Matters That We Know It—Our World in Data. https://ourworldindata.org/a-history-of-global-living-conditions-in-5-charts. Accessed 27 February 2022.
2. Wiedmann T, Lenzen M, Keyßer LT, Steinberger JK. Scientists' warning on affluence. *Nat Commun.* 2020;11(1):1–10. https://doi.org/10.1038/s41467-020-16941-y.

Index

Note: Page numbers followed by *f* indicate figures and *t* indicate tables.

C

T

U

V

W

CPI Antony Rowe
Eastbourne, UK
November 10, 2023